AF598411

WITHDRAWN

Advanced Computing in Electron Microscopy

Second Edition

Earl J. Kirkland

Advanced Computing in Electron Microscopy

Second Edition

Earl J. Kirkland
School of Applied and Engineering Physics
Cornell University
212 Clark Hall
Ithaca, NY 14853
USA
ejk14@cornell.edu

ISBN 978-1-4419-6532-5 e-ISBN 978-1-4419-6533-2
DOI 10.1007/978-1-4419-6533-2
Springer New York Dordrecht Heidelberg London

Library of Congress Control Number: 2010931437

© Springer Science+Business Media, LLC 2010
All rights reserved. This work may not be translated or copied in whole or in part without the written permission of the publisher (Springer Science+Business Media, LLC, 233 Spring Street, New York, NY 10013, USA), except for brief excerpts in connection with reviews or scholarly analysis. Use in connection with any form of information storage and retrieval, electronic adaptation, computer software, or by similar or dissimilar methodology now known or hereafter developed is forbidden.
The use in this publication of trade names, trademarks, service marks, and similar terms, even if they are not identified as such, is not to be taken as an expression of opinion as to whether or not they are subject to proprietary rights.

Printed on acid-free paper

Springer is part of Springer Science+Business Media (www.springer.com)

Preface

Preface to Second Edition

Several new topics have been added, some small errors have been corrected and some new references have been added in this edition. New topics include aberration corrected instruments, scanning confocal mode of operations, Bloch wave eigenvalue methods and parallel computing techniques. The first edition included a CD with computer programs, which is not included in this edition. Instead the associated programs will be available on an associated web site (currently `people.ccmr.cornell.edu/~kirkland`, but may move as time goes on).

I wish to thank Mick Thomas for preparing the specimen used to record the image in Fig. 5.26 and to thank Stephen P. Meisburger for suggesting an interesting biological specimen to use in Fig. 7.24.

Again, I apologize in advance for leaving out some undoubtedly outstanding references. I also apologize for the as yet undiscovered errors that remain in the text.

Earl J. Kirkland, December 2009

Preface to First Edition

Image simulation has become a common tool in HREM (High Resolution Electron Microscopy) in recent years. However, the literature on the subject is scattered among many different journals and conference proceedings that have occurred in the last two or three decades. It is difficult for beginners to get started in this field. The principle method of image simulation has come to be known as simply *the multislice method.* This book attempts to bring the diverse information on image simulation together into one place and to provide a background on how to use the multislice method to simulate high resolution images in both conventional and scanning transmission electron microscopy. The main goals of image simulation include understanding the microscope and interpreting high resolution information in the recorded micrographs. This book contains sections on the theory of image formation and simulation as well as a more practical introduction on how to use the multislice method on real specimens. Also included with this book is a CD-ROM with working programs to perform image simulation. The source code as well as the executable code for IBM-PC and Apple Macintosh computers is included. Although the programs may not have a very elegant user interface by today's standards (simple command line dialog), the source code should be very portable to a variety of different computers. It has been compiled and run on Mac's, PC's and several different types of UNIX computers.

This book is intended to be at the level of first year graduate students or advanced undergraduates in physics or engineering with an interest in electron microscopy. It assumes a familiarity with quantum mechanics, Fourier transforms and diffraction, some simple optics and basic computer skills (although not necessarily programming skills) at the advanced undergraduate level. Prior experience with electron microscopy is also helpful. The material covered should be useful to students learning the material for the first time as well as to experienced researchers in the field. The programs provided on the CD can be used as a black-box without understanding the underlying programs (with a primary goal of understanding the transmission electron microscope image) or the source code can be used to understand how to write your own version of the simulation programs.

Although an effort was made to include references to most of the appropriate publications on this subject, there are undoubtedly some that were omitted. I apologize in advance for leaving out some undoubtedly outstanding references. I also apologize for the as yet undiscovered errors that remain in the text.

I wish to acknowledge the support of various funding agencies (principly DOE, NSF and NIH) that have supported my research efforts over the past several decades. My research experience has substantially contributed to my understanding of the material covered in this book.

I also wish to thank Dr. David A. Muller and Dr. Richard R. Vanfleet for providing many helpful suggestions and help in proof reading the manuscript and to thank Dr. M. A. O'Keefe for providing helpful comments on electron microscopy and image simulation.

Earl J. Kirkland March, 1998

MATLAB(R) is a registered trademark of The Mathworks, Inc.

The Matlab and other programs listed in this book are supplied for instructional purposes, AS-IS WITHOUT ANY WARRANTY, WITHOUT EVEN THE IMPLIED WARRANTY OF MERCHANTABILITY or FITNESS FOR A PARTICULAR PURPOSE to the extent permitted by law. Effort has been made to insure the programs are correct, but neither the author or the publisher shall be held responsible or liable for any damage resulting from the use or failure to use these programs.

Contents

Chapter 1
Introduction

Abstract This chapter has a brief summary of various ways that a computer and computation can be used in electron microscopy. There is also a short summary of the organization of this book and a list of symbols.

1.1 Computing in Electron Microscopy

Electron microscopy continues to push the limits of resolution. At high-resolution, image artifacts due to instrumental or specimen limitations can greatly complicate image interpretation. The computer is finding an every increasing role in interpreting high resolution transmission electron micrographs as well as extracting additional information from the recorded images. Computer technology has been progressing at a very rapid pace over the past several decades. The rate of improvement in computing is certainly much faster than the rate of improvement of the electron microscope. A very powerful computer is now much less than 1% of the cost of a respectable electron microscope even though this level of computer hardware used to cost much more than a high-performance electron microscope. It is very worthwhile to try to exploit the computer in electron microscopy in any way possible to extract more information about the specimen or to reduce the cost or effort required to obtain this information. Various applications of computing to electron microscopy may be arranged in the following categories.

image simulation: Numerically, calculate electron microscope images from first principles and a detailed description of the specimen and the instrument. Usually involving various nonlinear imaging modes and dynamical scattering in thick specimens.

image processing: The inverse of image simulation. Try to extract additional information from the experimentally recorded electron micrographs by applying numerical computation to the digitized micrographs.

E.J. Kirkland, *Advanced Computing in Electron Microscopy*,
DOI 10.1007/978-1-4419-6533-2_1, © Springer Science+Business Media, LLC 2010

instrument design: computer aided design (CAD) in electron optics. Numerical calculation of electron optical properties (i.e., aberration, etc.) of magnetic and electrostatic lens and deflectors in the electron microscope to optimize the performance of the instrument.

on-line control: Directly control the operation of the microscope and record images and spectra directly from the instrument. The computer is directly wired into the electron microscope electronics.

data archiving: Save the recorded data. Manage the large volume of data generated when recording a series of images.

Image simulation of electron micrographs has a long history and is the principle topic of this book. There are two general types of image simulation. One group of methods involves Bloch wave eigenstates and a matrix formulation in reciprocal space (Bethe [24], Howie and Whelan [163]) and the other group involves mathematically slicing the specimen along the beam direction (the multislice method). The multislice method (Cowley and Moodie [63], Lynch and O'Keefe [231], Goodman and Moodie [127], Ishizuka and Uyeda [179], Van Dyck [357]) is usually more flexible for a computer simulation of crystalline specimens with defects or interfaces as well as completely amorphous materials. Bloch wave solution are more amenable to analytical calculations with pencil and paper for small unit cells and can provide valuable insight into the scattering process.

Attempts to analytically derive the theory of image formation in the electron microscope for specimen with large unit cells (and defects and interfaces) quickly arrive at equations that do not have a closed form analytical solution or are too difficult to easily interpret. The only recourse is a numerical solution. Image simulation numerically computes the electron micrograph from first principles. Starting from a basic quantum mechanical description of the interaction between the imaging electrons in the microscope and the atoms in the specimen the wave function of the imaging electrons may be calculated at any position in the microscope. If the optical properties of the lenses in the microscope are known, then the two dimensional intensity distribution in the final electron micrograph can be calculated with a relatively high precision. Image simulation can provide several sources of additional information about the specimen. First, it can reveal which features of the image are due to artifacts produced by aberrations in the electron microscope and which image features are due to the specimen itself (and possibly relate features in the image to unsuspected properties of the specimen). Image simulation is an aid in interpreting the image recorded in the electron microscope. Second, it is relatively simple to change instrumental parameters in the simulation that would be difficult if not impossible to change in practice. For example it is easy to change the beam energy or spherical aberration to an arbitrary value to see what happens. It is much easier to use image simulation to determine what type of instrument is required to investigate a particular specimen than it would be to build each type of electron microscope and see what happens. Image simulation can be used as both an aid in image interpretation and a means of exploring new types of imaging in the microscope.

Image processing is the inverse of image simulation. Starting from recorded experimental images the computer can process the micrographs to improve their interpretability or to try to recover additional information in the micrographs. Image processing includes image enhancement such as simple contrast stretching or noise cleaning as well as image restoration or image reconstruction. Image restoration attempts to deconvolve the transfer function of the instrument from a single image to improve the apparent resolution of the recorded image. Image reconstruction attempts to combine several images (such as a defocus series) into one image with more information. In bright field phase contrast microscopy a series of images taken at different defocus values (a defocus series) together contain more information than any single image in the series. The computer can be used to reconstruct all of the information in the defocus series into a single image with more information (usually this means higher resolution) than any single isolated image in the series (see for example Kirkland [199, 202], Kirkland et al. [210] and Coene et al. [52]). Image reconstruction is inherently more difficult than image simulation because it must invert a complicated nonlinear process.

Computer aided instrument design is a broad and rich field all of its own. This subject has been recently reviewed by Hawkes and Kasper [151, 152] and will not be considered here.

On-line control of the electron microscope received considerable attention in the literature in the last decade (see for example Smith [327], Skarnulis [321], Kirkland [203]) but is now becoming mainly the province of the commercial instrument manufacturers. The original equipment manufacturers are perhaps in a better position to interface directly with the inner working of the instrument. Many new electron microscopes now come equipped with a computer to record the data (spectra and images) and possibly control the instrument. This can take the form of automatic alignment and focusing or simply a replacement for a traditional collection of knobs and switches. Aberration corrected instruments have become so complicated that computer control is required and direct manual control is not practical. The related topic of telemicroscopy or remote access (see for example Fan et al. [97], Zaluzec [386] or O'Keefe et al. [273]) involve accessing a microscope (or other instrument) over a network (the world wide web) from a computer in a location far away from the instrument. Many software packages now exist to remotely control a computer in a general sense and can easily be used to access computer controlled instruments without writing new specialized software.

Data archiving is now common place in many fields. Storing electron micrographs is in fact similar to storing any other type of image data. Digital storage has the advantage that the data will not degrade with time (as a photograph might) and the data can be readily transmitted electronically to any location. Also, digital storage can take up less space than a traditional collection of film or plates. Electron microscopy can readily take advantage of every day advances in data storage.

1.2 Organization of this Book

The level of discussion in this book is approximately at a beginning graduate student or advanced undergraduate student interested in the theory of image formation in the transmission electron microscope. It is assumed that the reader has some familiarity with quantum mechanics, Fourier transforms, and elementary optics. This book is not an introduction to computer programming. The reader is also assumed to understand the basic principles of computer programming. Numerical computer programming is discussed in a high level abstract manner.

Chapter 2 begins with an overview of the electron microscope instrument (scanning and conventional, STEM, and CTEM), the fundamental physics of electron dynamics, and the optical aberrations of electron lenses. Chapter 3 contains a theory of image formation for very thin specimens ignoring the geometrical thickness of the specimen. The transfer function is presented and investigated for several imaging modes. Chapter 4 is somewhat of a detour. It discusses numerical sampling and the fast Fourier transform (FFT) that will be needed in later chapters. Chapter 4 can be skipped if the reader is familiar with these topics. Chapter 5 starts the discussion of calculation methods, beginning with the electron atom interaction and very thin specimens. Chapter 6 is a long theoretical discussion of methods of calculating the propagation of the electron through thick (usually less than a few thousand Angstroms however) specimens. The Bloch wave eigenvalue and multislice methods are presented and discussed. Chapter 7 gives several applications of the multislice method. Some simple examples are first worked through in an educationally approach so the reader can learn how to use the method, then several more complicated examples are given to illustrate some interesting features of the electron microscope image. Chapter 8 documents how to use the programs used in this book and available to be downloaded from an associated web site.

Table 1.1 show some frequently used symbols. Some symbols may be used for more than one thing, but the usage should be clear from the context.

Table 1.1 Some symbols and their descriptions

Symbol	Description
a,b,c	Unit cell size of the specimen in x,y,z directions
a_0	Bohr radius (0.529 Ang.)
c	Speed of light
e	Charge on the electron
m_0	Rest mass of the electron
m	Total mass of the electron
V	Accelerating voltage
h	Planck's constant ($\hbar = h/(2\pi)$)
λ	Electron wavelength
χ	Phase error due to aberration of a focused electron wave
α	Electron scattering half angle
β	Condenser illumination half angle
x,y	Position in the image plane
z	Position along the optic axis
k	2D spatial frequency in the Fourier transform of the image plane
$K = k(C_s\lambda^3)^{1/4}$	dimensionless spatial frequency
Δf	Defocus
$C_S = C_{S3}$	Third order spherical aberration
C_{S5}	Fifth order spherical aberration
σ	Electron interaction parameter
$\frac{\partial\sigma}{\partial\Omega}$	Partial cross section for scattering

Chapter 2
The Transmission Electron Microscope

Abstract This chapter gives a short description of the physical instrumentation of the transmission electron microscope (fixed beam and scanning modes). It starts with the fundamental physics of electron dynamics for energies in the range 100–1000 keV. Some types of magnetic lenses and aberration correctors used to focus the electrons in the microscope are discussed. Various approximation used in modeling the microscope are introduced. Optical aberrations are defined, and general methods of aberration correction are described briefly.

2.1 Introduction

The modern transmission electron microscope has evolved over most of the twentieth century into a rather complex instrument. The twenty first century is bringing forth a sequence of commercial aberration correction devices further complicating the instrument (as well as adding to their expense) in a never ending quest for higher resolution. The Conventional Transmission Electron Microscope (CTEM) was first invented in the early 1930s by Knoll and Ruska [214, 307] as an extension of earlier work to perfect the oscilloscope. Early microscopes had a resolution that was no better than a light microscope but there was considerable speculation at the time that atomic resolution should be possible. These speculations have been realized in current commercial instruments. For his work on the CTEM, Ruska shared the 1986 Nobel prize in physics with Binnig and Rohrer [28] for their invention of the scanning tunneling microscope (STM). The Scanning Transmission Electron Microscope (STEM) was invented shortly after the CTEM in the late 1930s by von Ardenne [360]. The utility of the STEM was greatly increased in the late 1960s by Crewe et al. [69] with the addition of a cold field emission gun (FEG) source with a small source size and high brightness. Both CTEM and STEM form an image from the electrons that are transmitted through a thin specimen and usually require a relatively high electron energy (100–1,000 keV). The beginning

E.J. Kirkland, *Advanced Computing in Electron Microscopy*,
DOI 10.1007/978-1-4419-6533-2_2, © Springer Science+Business Media, LLC 2010

chapters of Heidenreich [154] or Hall [141] and the two contributed volumes edited by Hawkes [149] and Mulvey [258] have a more complete discussion of the history of the transmission electron microscope.

The scanning electron microscope (SEM) is related to the STEM in that both scan a focused electron beam across the specimen. However, the SEM usually uses an electron beam of lower energy (approximately 1–30 keV vs. 100 keV to 1 MeV for CTEM and STEM) and forms an image signal from the secondary or back scattered electrons from a bulk specimen (the specimen is not necessarily thin). This results in a lower achievable resolution but has the advantage of being able to view the surface of bulk specimen without thinning. The SEM mode was first investigated around the same time as the CTEM and STEM were invented, and was refined by Oatley's group at Cambridge University The history of the SEM has been reviewed by Oatley et al. [265] and McMullan [237]. The SEM itself has been reviewed by many authors (for example Goldstein et al. [122], and Reimer [294]) and computer calculation of SEM images has been summarized by Joy [190]. Although the SEM is equally important it will not be discussed further in this book.

A schematic cross section of a modern high resolution aberration corrected CTEM instrument is shown in Fig. 2.1. The specimen is loaded in the gap of the objective lens just above the objective aperture (not shown in the drawing) mechanism. Combined CTEM/STEM's and dedicated CTEM's are commercially available, however, few dedicated STEM's are commercially available (Nion Corp. is now reviving the art of dedicated STEM's and adding an aberration corrector). The instrument shown in Fig. 2.1 has a field emission gun or FEG (labeled Schottky electron source) instead of a thermionic source for improved brightness (and coherence). The condenser lenses transfer the beam onto the specimen. In a STEM scan coils would also be placed above the specimen. The aberration corrector corrects for the aberrations in the objective lens, and the projectors provide further magnification and transfer the image onto the detector (usually a CCD) at the bottom of the figure (detector not shown). In a CTEM the whole image is formed (in parallel) at one time whereas in the STEM a focused probe is scanned across the specimen in a raster fashion and the image is sequentially built up, one pixel (or image point) at a time. The CTEM is similar to a conventional light optical microscope and the STEM is similar to a scanning confocal light optical microscope (Wilson and Sheppard [383]).

The electron source can be a simple thermionic, LaB_6, or field emission (hot or cold) source. The field emission source has a high brightness and small source size that makes it useful for both conventional and scanning electron microscopes. The small source size of the field emission source is essential for high resolution STEM because the probe is essentially an image of the source (the smaller the source the smaller the probe and better the resolution). The increased coherence of the FEG can also significantly increase the information limit in the CTEM (Otten and Coene [277]).

The general shape of a simple magnetic electron lens is shown (on the left) in Fig. 2.2. A large DC current flows through a coil of wire which produces a magnetic field. The field follows the magnetic material (typically an iron alloy) until it

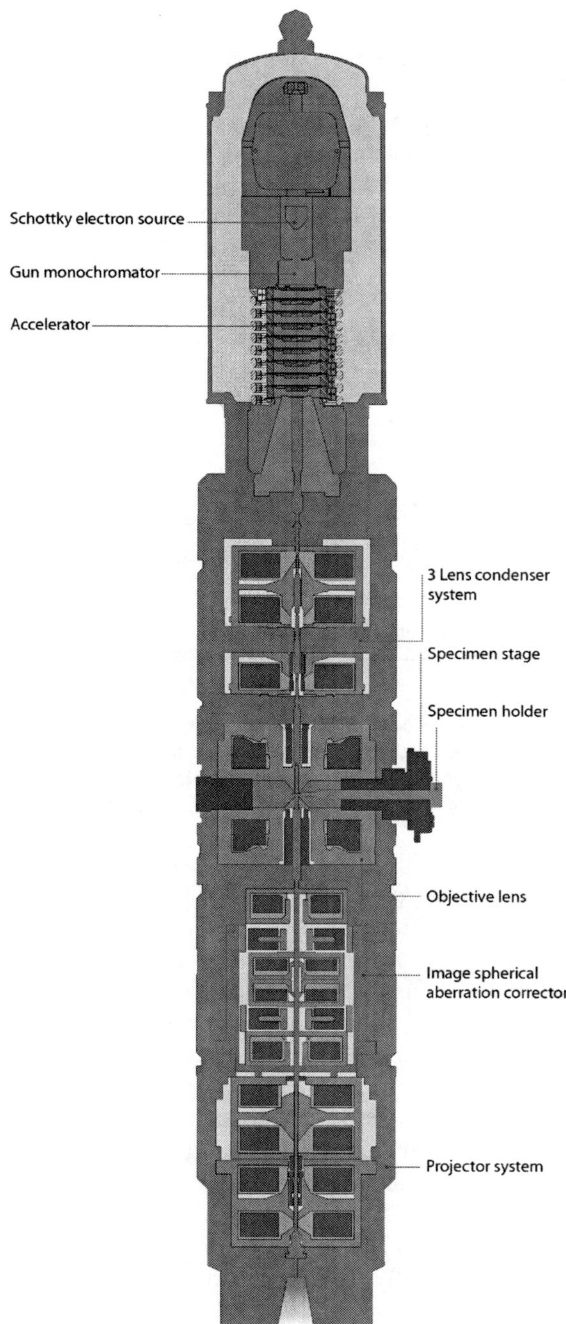

Fig. 2.1 Cross section of an FEI Titan series high resolution CTEM (*courtesy of FEI*). There will be a CCD camera or other detector at the bottom of the instrument which is not shown in this figure. There is an aberration corrector after the specimen to correct aberration of the objective lens

gets to the gap where it extends out toward the electron beam near the center of the lens. The specimen is usually placed in or near the gap of the objective lens. Lens designers expend a great amount of effort to shape the magnetic pole faces in the gap to shape the magnetic field and produce optimum focusing of the electron beam. The magnetic field forms a lens much like a glass lens for visible light. An analogous glass lens is shown on the right side of Fig. 2.2. Refer to the books on electron optics listed at the end of this chapter for more details on how magnetic lenses work.

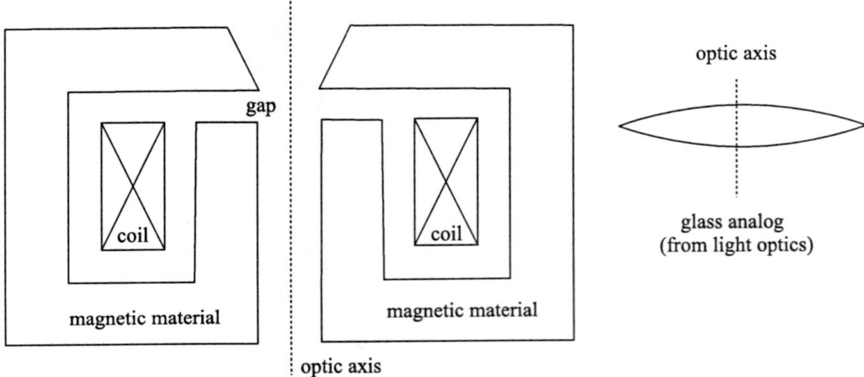

Fig. 2.2 Cross section of the general shape of a rotationally symmetric (about the optic axis), round lens (*on left*). The lens is symmetrical for rotation about the optic axis. The fringe fields in the gap focus the electron beam in a manner similar to that of a glass lens for visible light (*on right*). The electron beam travels up or down in this illustration (the optic axis, or z direction)

There are typically two condenser lenses that gather the electrons emitted from the source and transfer them to the specimen. With the exception of the electrostatic lens in the electron gun (i.e., electron source), lenses in the electron microscope are usually rotationally symmetrical magnetic lenses (Fig. 2.2). The condenser lenses can illuminate the specimen with a wide collimated parallel beam (in CTEM) or present the objective lens with a parallel narrow beam (in STEM). The objective lens images the specimen in CTEM mode or forms a small probe of atomic dimensions on the specimen in STEM mode. In the CTEM mode the objective forms a virtual image which is further magnified by several projector lenses. STEM and CTEM modes require the objective lens to be on opposite sides of the specimen. A combined CTEM/STEM will usually sacrifice the performance in one mode because the specimen and objective lens are difficult to move. Most instruments currently available were initially CTEM's that have been converted to a combined STEM/CTEM instrument so the STEM mode may not use the best portion of the objective lens field and hence has a reduced performance. STEM mode may be operated with or without postspecimen lenses. Operation without post specimen lenses can have advantages if there is significant inelastic scattering in the specimen (because of the inherent chromatic aberration of magnetic lens). STEM mode also can be particularly useful for atomic resolution spectroscopies (X-ray, electron energy loss, etc.) from a localized volume of the specimen. The CTEM can also yield near atomic resolution spectroscopy using imaging energy loss filters (for example see Reimer [296]). Herrmann [156, 157] and Smith [322, 323, 325] have reviewed the instrumental aspects of high resolution electron microscope operation.

2.2 Modeling the Electron Microscope

If considered as a whole the TEM is a rather complicated instrument. However much of it can be ignored when considering the specific features of a high-resolution image. The vacuum system is essential and the TEM only works if there is a vacuum, but once it is established (with considerable effort in some cases!), it has no further effect. In a similar vein the illumination system (i.e., condenser lens) once aligned can be reduced to the properties of the illuminating rays (i.e., coherence and angular distribution). In CTEM mode the projector lenses magnify the virtual image formed by the objective lens. Because any defects in the objective lens are greatly magnified and the angles into the projector lenses are greatly reduced the projector lenses have little effect on the final image resolution so can also be ignored (in some instances the projectors may be responsible for small distortion in the image as opposed to a reduction in resolution). The simplest model to adequately describe the high resolution imaging performance of a CTEM is shown in Fig. 2.3 and a similar model for a STEM is shown in Fig. 2.4. In each model the electrons will be assumed to be moving in the positive z direction (down in Figs. 2.3 and 2.4) and the image plane is assumed to be an x, y plane. The symbol α will denote the angle between the scattered electrons and the optic axis (and also the angle into the objective lens) in CTEM mode and the angle between the specimen perpendicular and the focused ray from the objective lens in STEM mode. The optic axis is typically perpendicular to the plane of the specimen.

The STEM may have two (or more) different types of detectors. The bright field (BF) detector (on the optic axis) detects the electrons that have passed through the specimen without significant deviation. The annular dark field (ADF) detector should detect the electrons that have been scattered to high angles (typically greater than three or four times the objective aperture angle).

The STEM has a set of scan coils positioned before the objective lens. These are usually arranged so that the illumination electron trajectories rock about the principle plane of the objective lens (so that the beam does not move across the objective aperture during scanning). Alternately a virtual objective aperture (VOA) may be place prior to the scan coils. At first glance it might seem as if the scan would cause the beam to move off the BF detector, however the scan angles are so small that this does not happen. The objective lens usually gives a large demagnification, so that the specimen is essentially at the focal plane of the objective lens. The focal length is typically of order 2–3 mm. Therefore, for a 1,000Å scan field the change in angle due to the scan is approximately 500Å/2 mm $= 2.5\times10^{-5}$. At 100 keV

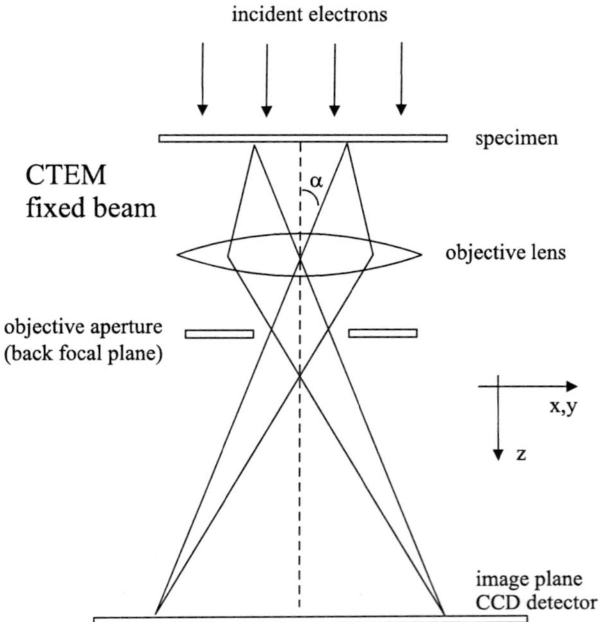

Fig. 2.3 Simplified model (not to scale) of a high resolution Conventional Transmission Microscope (CTEM). The condenser lenses (*above the top of the drawing*) and projector lenses (*below the bottom of the drawing*) will be ignored

typical angles on the BF detector are 2–3 mrad and about 40–200 mrad on the ADF detector. Therefore, scanning the beam should not significantly affect the angles onto the detector.

2.3 Relativistic Electrons

The electron has a relatively small mass so that even a 100 keV electron is traveling at approximately one half the speed of light. This means that quantities such as velocity, wavelength, etc. should be calculated relativistically. The total energy E_T of a charged particle with charge e and rest mass m_0 accelerated through a potential V is given by:

$$E_T^2 = (m_0c^2 + eV)^2 = p^2c^2 + m_0^2c^4 = m^2c^4, \tag{2.1}$$

where c is the speed of light in vacuum, $p = mv$ the particle's momentum, v its velocity, and m is the mass of the particle. From this expression it follows that the ratio of the electron's mass to its rest mass is:

$$\frac{m}{m_0} = \gamma = \frac{1}{\sqrt{1 - v^2/c^2}} = 1 + \frac{eV}{m_0c^2} \tag{2.2}$$

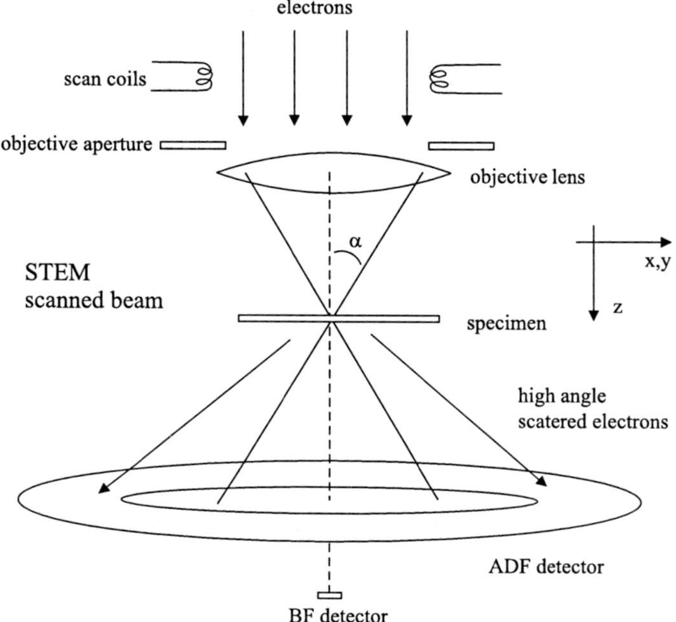

Fig. 2.4 Simplified model (not to scale) of a Scanning Transmission Electron Microscope (STEM). The condenser lenses (*above the top of the drawing*) are ignored. BF = bright field, ADF = annular dark field

and that the velocity of the electron v relative to the velocity of light is:

$$\frac{v}{c} = \left[1 - \left(\frac{m_0c^2}{m_0c^2 + eV}\right)^2\right]^{1/2} = \frac{[eV(eV + 2m_0c^2)]^{1/2}}{m_0c^2 + eV}. \tag{2.3}$$

For comparison the nonrelativistic velocity is $v/c = \sqrt{2eV/(m_0c^2)}$. The velocity is shown in Fig. 2.5 as a function of the kinetic energy eV of the electron.

The de-Broglie wavelength λ of the electron is:

$$\lambda = h/p, \tag{2.4}$$

where h is Planck's constant. Substituting this in the earlier expression for the total energy (2.1) yields:

$$\begin{aligned} (m_0c^2 + eV)^2 &= \left(\frac{hc}{\lambda}\right)^2 + m_0^2c^4 \\ \lambda &= \frac{hc}{\sqrt{(m_0c^2 + eV)^2 - m_0^2c^4}} \\ \lambda &= \frac{hc}{\sqrt{eV(2m_0c^2 + eV)}}. \end{aligned} \tag{2.5}$$

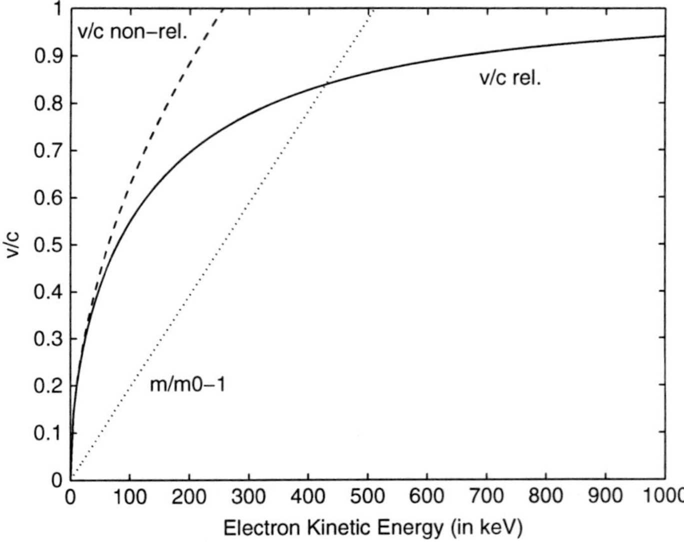

Fig. 2.5 Velocity v of the relativistic electron as a function of its kinetic energy eV. c is the speed of light in vacuum. For comparison the ratio of the electron mass to its rest mass (m/m_0) and the nonrelativistic velocity are also shown

The relevant constants have values of $m_0c^2 = 511$ keV, $hc = 12.398$ keV-Angstroms. For comparison the nonrelativistic result is $\lambda = hc/\sqrt{2m_oc^2eV}$. The electron wavelength is plotted as a function of kinetic energy in Fig. 2.6.

The Schrödinger wave equation of quantum mechanics is not relativistically correct. The electron is relativistic at the beam energies used in the electron microscope meaning that the Schrödinger equation should not be used directly The relativistic Dirac equation would be the correct wave equation for relativistic electrons, however it is significantly more difficult to work with mathematically (by hand or in the computer). It is now traditional to simply use the Schrödinger equation with a relativistically correct electron wavelength and mass. This approach has been compared to more accurate calculation using the Dirac equation by Fujiwara [119], Ferwerda [101,102], and Jagannathan et al. [181,182] and is usually accurate enough in the typical energy ranges used in the electron microscope. Op de Beck [75] recently found additional corrections to the use of the Schrödinger equation with relativistic mass and wavelength. The standard definition of the relativistic wavelength (2.5) can lead to an additional error of order 10% at 100 keV and 20% at 400 keV in the electron wavelength. This error mostly cancels if the optical parameters (defocus, spherical aberration, etc.) are consistently determined using the same electron wavelength (which is usually the case) because only their product enters. The remaining error produces an apparent change in the crystal thickness (or equivalently the beam energy) of a few percent in image simulations. Rother et al. [306] have done a detailed comparison of relativistic and nonrelativistic electron scattering and

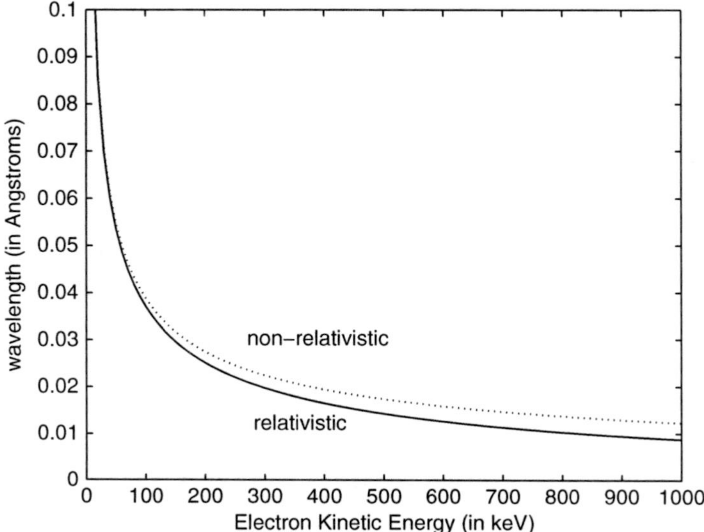

Fig. 2.6 Electron wavelength λ as a function of its kinetic energy. For comparison the nonrelativistic result is also shown

also found good agreement with the possible exception of very high angle scattering which might be a problem for ADF-STEM but probably not BF-CTEM. The form of the Schrödinger with a relativistic electron wavelength and mass as above will be used here because it is significantly easier to work with, however some caution must be taken in a strict quantitative interpretation.

2.4 Reciprocity

The reciprocity theorem of scattering theory as adapted for electron microscopy (see Pogany and Turner [286]) states that the electron intensities and ray paths in the microscope (including a specimen) remain the same if their direction is reversed and the source and detector are interchanged (i.e., the electrons trajectories and elastic scattering processes have time reversal symmetry). Cowley [58], Zeitler and Thomson [390], and Engel [92] have discussed the fact that this implies that the BF-CTEM and the BF-STEM should produce the same image. Reciprocity applies to all orders of elastic interaction of the electrons and the specimen so it is not necessary to restrict its implications to thin specimens. Figure 2.7 compares the ray paths in the CTEM with those in the STEM.

In the CTEM (on left in Fig. 2.7) the electrons start from a point source at the top and travel down. If the source is in the far field (either directly or by virtue of

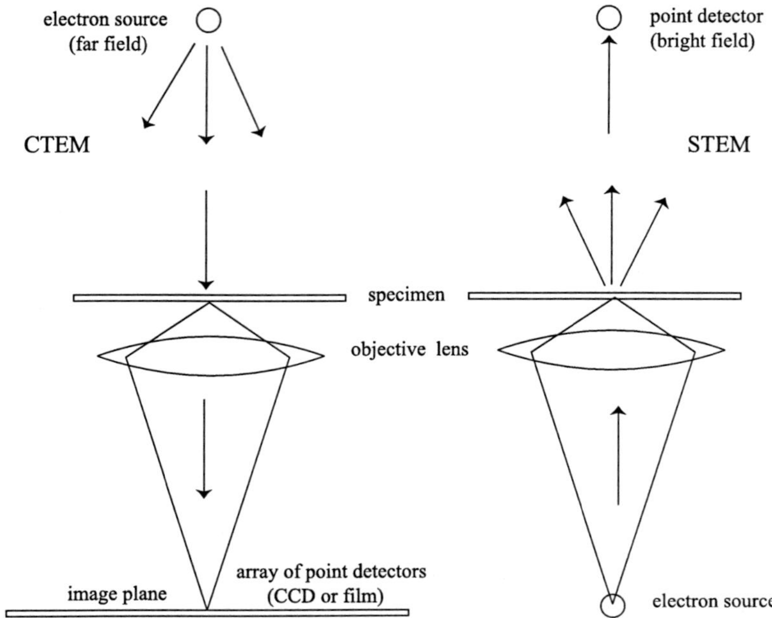

Fig. 2.7 Reciprocity applied to CTEM and STEM. The electrons are traveling downward in the CTEM (*left*) and upward in the STEM (*right*). The *arrows* indicate the direction of electron travel. Note that the geometry of the ray paths appear identical except for direction meaning that BF-CTEM is equivalent to BF-STEM

the condenser lens) then the specimen is illuminated by a nearly parallel beam. The objective lens then images each point on the specimen onto the image plane (film or other electron detectors such as a CCD). It is important to realize that the film or a CCD is an array of point detectors (one point detector for each image point in the specimen). In the STEM (on right in Fig. 2.7) the electrons start from a point source at the bottom and travel up (VG actually built their STEM's upside down but this orientation is only for this discussion and not necessary in practice). The objective lens forms a small probe on the specimen. The probe must be scanned to produce an image. A bright field detector is a small point detector on axis far away from the specimen (again in the far field). In each case the electron paths for only one image point are shown for simplicity. There is an obvious similarity between these two diagrams. If the direction of the electrons in the STEM is reversed and the source and detector interchanged then the BF-STEM and the CTEM are identical. Both have a point source and a point detector on opposite sides of the specimen (the ADF-STEM is distinctly different however). This means that the image in the BF-STEM is equivalent to the BF image in the CTEM. It is even possible to measure the aberrations in a STEM using methods developed for the CTEM (Wong et al. [384]). It is not strictly necessary to have an exact point source (CTEM) or detector

(STEM) for the equivalence of BF-STEM and BF-CTEM to hold. An increase in the illumination angles in the CTEM is equivalent to an increase in the size of the BF detector in the STEM.

2.5 Confocal Mode

Confocal mode is a combinations of both CTEM and STEM as in Fig. 2.8. The confocal optical microscope has demonstrated improved resolution and other advantages for many years. This mode has recently been adapted to electron microscopy (Zaluzec [116, 387]). This mode is somewhat more difficult to implement in electron microscopy due to the added instrumentation. The top of the instrument (Fig. 2.8) is a STEM column (Fig. 2.4). The electron beam is focused to a small spot on the specimen. The bottom portion of the instrument after the specimen is basically a CTEM (Fig. 2.3) and images the probe transmitted through the specimen onto a small detector. There also has to be some way to scan the probe across the specimen in a raster and synchronously descan the image onto the detector. This can be done with two sets of scan coils carefully aligned or physically moving the specimen back and forth with a piezoelectric actuator (Takeguchi1 [340]) leaving the electron beam and lenses in a fixed position. Neither mechanism is shown in Fig. 2.8 for simplicity (may be difficult to implement in an actual instrument) but can be ignored in the calculations that will come later.

2.6 Aberrations

Most objective lenses in use today are rotationally (cylindrically) symmetric as in Fig. 2.2. Lens designers expend a great deal of effort in shaping the pole faces to get the best possible lens (minimum possible aberrations). Computer aided design of magnetic lenses is a sophisticated field by itself (see for example the recent books be Hawkes and Kasper [151–153]). It is also possible to form a lens by superimposing multipole elements such as quadrupole, hexapole, octopoles, etc. Few if any microscopes currently use this approach for their primary focusing method because of its much greater complexity, however several forms of aberration correction employ multipole elements.

The symmetry and shape of the magnetic fields in the objective lens are determined from Maxwell's equations which prevent the magnetic lens from performing exactly like the familiar ideal lens in classical light optics. There is however a close analogy between electron optics and light optics. An equivalent index of refraction and ray paths for electrons can be defined and magnetic lenses do approximately focus an image. The wavelength of the high-energy electrons is much smaller than the dimensions of the lenses so it is appropriate to think about the geometric rays in the lens just like those in a light optical glass lens. Furthermore, the symmetries

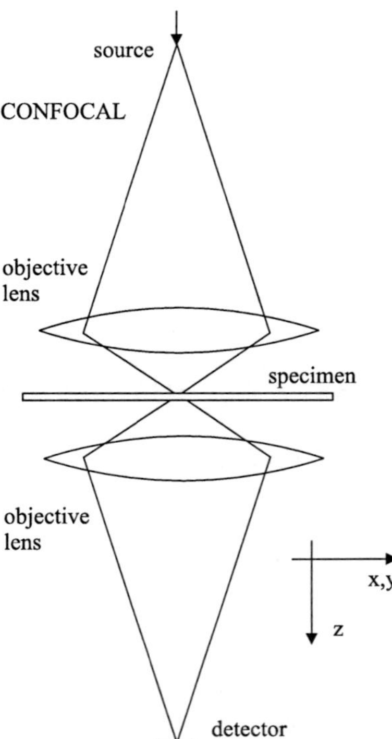

Fig. 2.8 Confocal mode is a combination of CTEM (*bottom*) and STEM (*top*). A STEM focuses a small probe onto the specimen (*top*) and a CTEM (*bottom*) images the probe transmitted through the specimen onto a small detector (*bottom*)

of Maxwell's equations are such that most of the primary light optical aberrations (see for example Born and Wolf [34]) also exist for magnetic lenses. These aberrations determine the deviation from an ideal lens and can be used to characterize the artifacts in the image.

In a high-resolution image only a small portion of the specimen near the optic axis is imaged. Therefore, only the electron trajectories (or rays) near the optic axis need be considered. This approach is called the paraxial ray approximation. If the electron microscope is well aligned then off-axis aberration are also negligible and the remaining lowest order effect is the third-order spherical aberration C_s. Physically the magnetic field further away from the axis is stronger than is required so that electrons traveling at larger angles (α) are deflected more than is required to focus them (as in Fig. 2.9). C_s produces a position error in the electron trajectory or ray that is proportional to the third power of this angle and a phase error in the electron wave function that is proportional to the fourth power of the angle.

Ideally a lens forms a spherical wave converging on or emerging from a single point as on the right side of Fig. 2.9. The aberrations cause the wave to deviate from a spherical surface with an error δ, and the phase error is $\chi = (2\pi/\lambda)\delta$. The error δ can be represented in a variety of basis functions, the most obvious being a power series in positional deviations (x, y) and angular deviations (α_x, α_y) from the optic axis. In high-resolution microscopy the specimen should always be very near the optic axis so positional deviations can be ignored leaving only the angular deviations

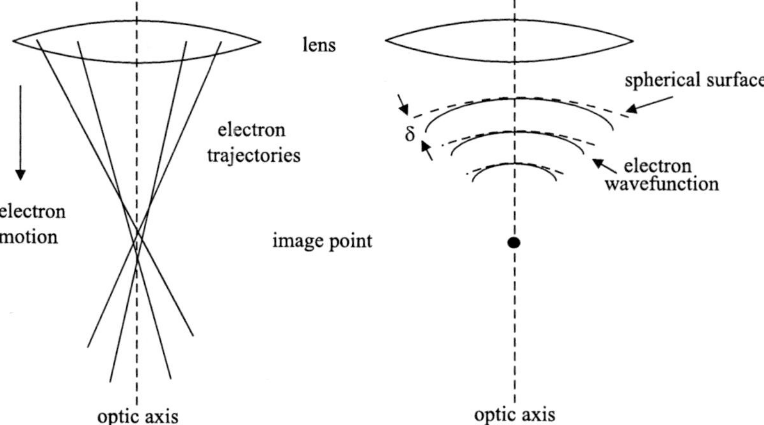

Fig. 2.9 The effect of spherical aberration C_s on the electron trajectories (*left*) and electron wave function (*right*). On the right the equivalent wave function without spherical aberration is shown as a *dashed line*

(larger for higher resolution). If the lens is perfectly symmetric then the deviation will not depend on the sign of α_x or α_y or which direction is chosen. Therefore, the deviation δ can only depend on even powers of $\alpha = \sqrt{\alpha_x^2 + \alpha_y^2}$, leaving:

$$\chi = \frac{2\pi}{\lambda}\delta = \frac{2\pi}{\lambda}\left(\frac{1}{2}C_1\alpha^2 + \frac{1}{4}C_3\alpha^4 + \frac{1}{6}C_5\alpha^6 + \cdots\right), \tag{2.6}$$

where the C_n coefficients have units of length. The deviations of the geometric rays are proportional to the derivative of χ and the leading numerical factors of each term are chosen to disappear after differentiation. $\Delta f = -C_1$ is defocus and $C_3 = C_{s3} = C_s$ is third-order spherical aberration. There are an infinite number of higher order terms. $C_5 = C_{s5}$ is the fifth-order spherical aberration. The subscript refers to the order of the geometric aberration (one less than the order of the wave aberration). If C_s is given without a numerical subscript it will be assumed to be the primary third-order spherical aberration.

Defocusing the lens also produces a deviation of the electron trajectory that departs from the ideal. Defocus can be changed by moving the specimen or changing the strength of the lens (proportional to the current through the lens coil). The amount of defocus, Δf is also defined as the deviation of the defocused image plane from the ideal Gaussian image plane. Defocus produces a phase error in the electron wave function but is proportional to the second power of the angle α.

Scherzer [311] found that a static, rotationally symmetric magnetic field with no sources on the axis will always produces a spherical aberration greater than zero because the expression for C_s can be written as the sum of quadratic terms. This is sometimes referred to as "Scherzer's Theorem." An electron microscope using this type of lens will have its instrumental resolution determined mainly by spherical aberration. Scherzer also noted that the defocus term can be used to partially offset the effect of C_s over some limited range of angles. For most microscopes, the net phase error is due to just spherical aberration and defocus as:

$$\chi(\alpha) = \frac{2\pi}{\lambda}\left(\frac{1}{4}C_s\alpha^4 - \frac{1}{2}\Delta f\alpha^2\right). \tag{2.7}$$

Defocus Δf varies with the strength of the objective lens, however C_s is essentially constant for a given specimen holder and beam energy. Equation (2.7) can be defined with the Δf term as positive or negative. Both sign conventions are used in the literature. With this definition, a positive Δf represents an underfocus (weaker lens current) of the objective lens. An accurate value of C_s is essential for image simulation or processing. Several methods for measuring C_s have been described by Budinger and Glasear [40], Krivanek [219], and others.

The semiangle into the objective lens α (also the angle between the incident and scattered rays in CTEM) is related to the spatial frequency k in the image plane by multiplication by the electron wavelength λ as:

$$\alpha = \lambda k. \tag{2.8}$$

The units of k are such that $1/k$ corresponds directly to a spacing d on the specimen (without multiplication or division by 2π). The literature on this subject can be very confusing because various authors add different factors of 2 and π. The simplest choice of $k = 1/d$ is used here (consistent with the optics literature but not with the traditional physics choice of $k = 2\pi/d$).

The aberration function $\chi(\alpha)$ rewritten as a function of k is:

$$\begin{aligned}\chi(k) &= \frac{2\pi}{\lambda}\left(\frac{1}{4}C_s\lambda^4k^4 - \frac{1}{2}\Delta f\lambda^2k^2\right)\\ &= \pi\lambda k^2(0.5C_s\lambda^2k^2 - \Delta f).\end{aligned} \tag{2.9}$$

Deviations from rotational symmetry are inevitable because of small machining errors in the magnetic lenses and small mis-alignments between lenses. The lowest order effect is the additional aberration of astigmatism and possibly coma which causes the defocus to vary with azimuthal angle ϕ.

$$\begin{aligned}\chi(k) = {} & \frac{\pi}{2}C_s\lambda^3k^4 - \pi\Delta f\lambda k^2 + \pi f_{a2}\lambda k^2\sin[2(\phi-\phi_{a2})]\\ & + \frac{2\pi}{3}f_{a3}\lambda^2k^3\sin[3(\phi-\phi_{a3})] + \frac{2\pi}{3}f_{c3}\lambda^2k^3\sin[\phi-\phi_{c3}],\end{aligned} \tag{2.10}$$

where f_{a2} is twofold astigmatism, f_{a3} is threefold astigmatism and f_{c3} is coma (Zemlin et al. [391]). ϕ_{a2}, ϕ_{a3}, and ϕ_{c3} are the azimuthal orientations of these aberrations. In principle astigmatism can be corrected but small amounts may remain in the image in practice. Astigmatism can result from small physical machining errors in the objective lens pole pieces or misalignments of the microscope column. Threefold astigmatism has recently been found to have a significant effect as resolution approaches or exceeds 2Å (Ishizuka [175], Krivanek [220]). Zemlin et al. [391] described a coma-free alignment procedure using the so-called Zemlim tableau.

It is possible to write the aberration function in dimensionless form by scaling with appropriate powers of λ and C_s:

$$K = k(C_s\lambda^3)^{1/4} \tag{2.11}$$

$$D = \Delta f/\sqrt{C_s\lambda} \tag{2.12}$$

$$D_{a2} = \Delta f_{a2}/\sqrt{C_s\lambda} \tag{2.13}$$

$$D_{a3} = \Delta f_{a3}/(C_s^3\lambda)^{1/4} \tag{2.14}$$

$$D_{c3} = \Delta f_{c3}/(C_s^3\lambda)^{1/4} \tag{2.15}$$

$$\begin{aligned}\chi(K) = {} & \pi(0.5K^4 - DK^2 + D_{a2}K^2\sin[2(\phi-\phi_{a2})] \\ & + \frac{2}{3}D_{a3}K^3\sin[3(\phi-\phi_{a3})] + \frac{2}{3}D_{c3}K^3\sin[\phi-\phi_{c3}]).\end{aligned} \tag{2.16}$$

The aberration function $\chi(K)$ is plotted in Fig. 2.10 for various amounts of defocus. $\chi(K)$ is the error (or deviation) of the wavefront from the ideal spherical wavefront.

The effect of the objective lens aberration function is to modulate different spatial frequencies (or different angles) by a complex function:

$$H_O(K) = \exp[-i\chi(K)] = \cos[\chi(K)] - i\sin[\chi(K)]. \tag{2.17}$$

This function is plotted in Fig. 2.11.

The nonsymmetrical aberrations, such as astigmatism etc. are not easily visualized in a simple graph. Some of these are illustrated in image form in Figs. 2.12 and 2.13 using computational methods that will be described later. These are images of a self-luminous point which is the probe function in STEM neglecting any remaining source size coming from the tip itself.

2.7 Aberration Correction

The most common electron lens is a rotationally symmetric lens similar to that shown in Fig. 2.2. The fringe field in the gap focus the electron beam. Lens designers carefully shape the pole faces (of the magnetic material) near the gap to minimize

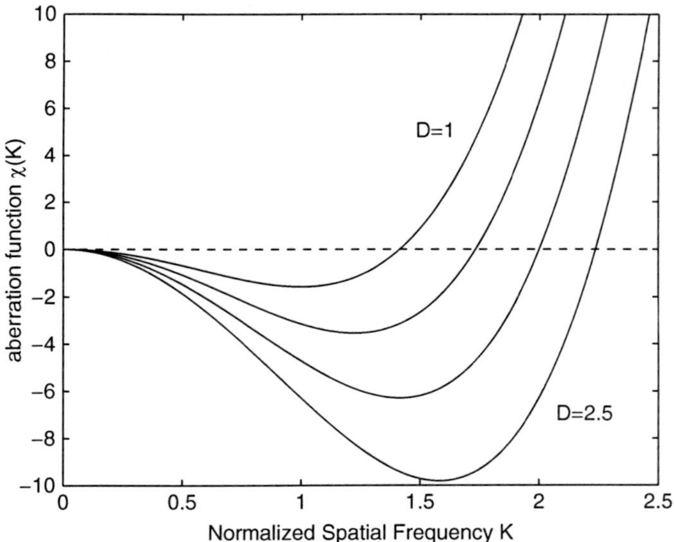

Fig. 2.10 The aberration function $\chi(K)$ as a function of the normalized spatial frequency K for different values of the normalized defocus D = 1.0, 1.5, 2.0, 2.5. Astigmatism and coma are assumed to be zero

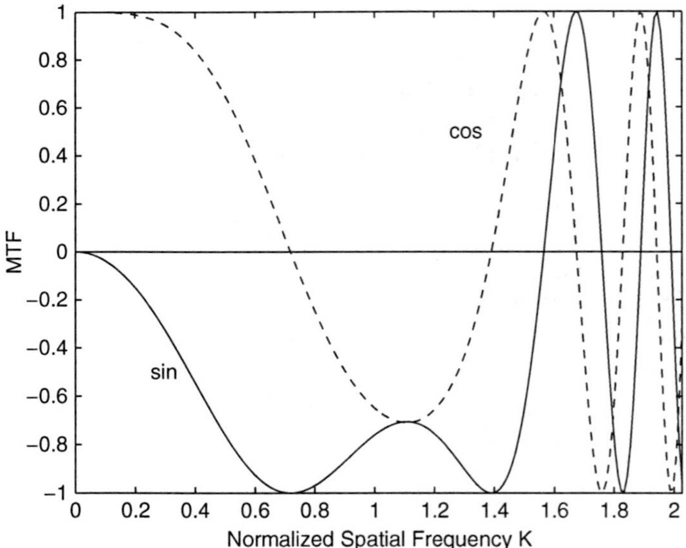

Fig. 2.11 The complex modulation function of the objective lens $H_O(K) = \cos[\chi(K)] - \mathrm{i}\sin[\chi(K)]$ as a function of the normalized spatial frequency K for the special case of $\Delta f = \sqrt{1.5C_s\lambda}$. Astigmatism and coma are assume to be zero

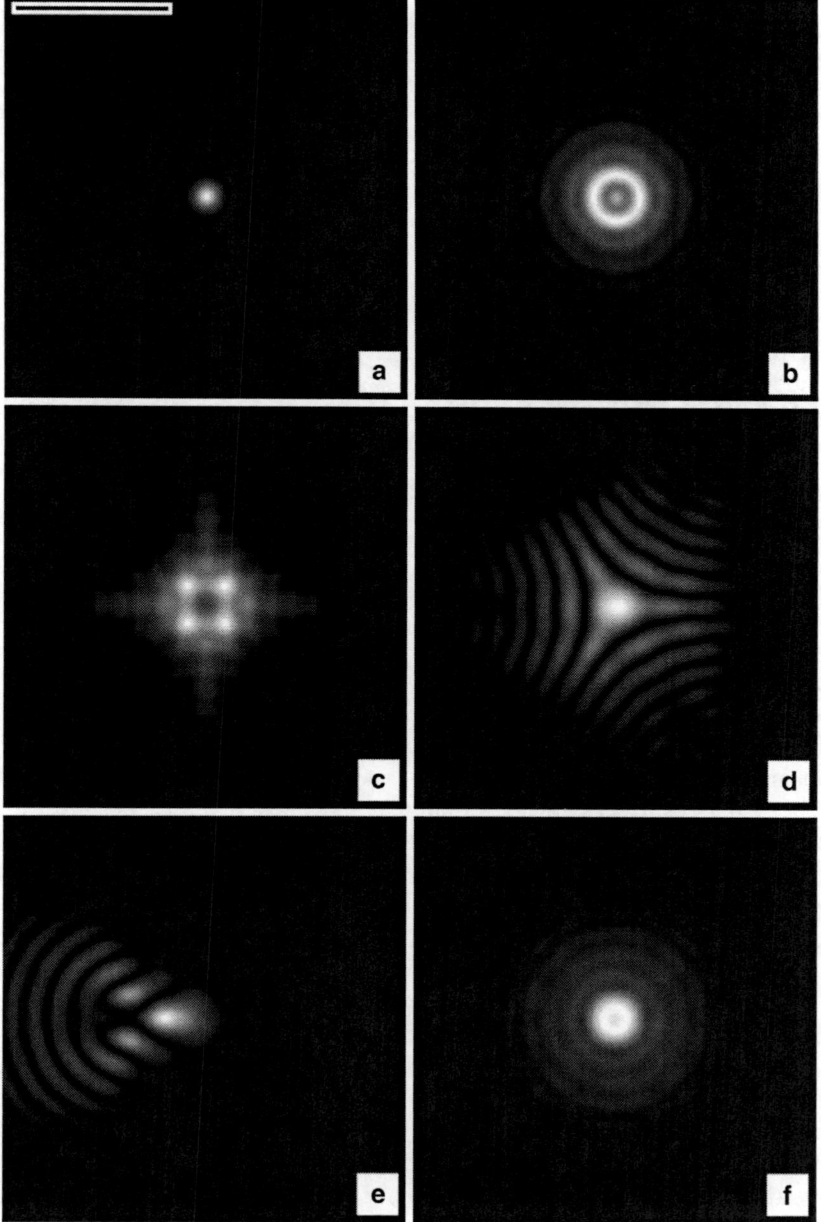

Fig. 2.12 Various single aberrations. All images are of a 200 keV probe (or self luminous point) with a 20 mrad apert. (a) no aberrations (scale bar 5 Å), (b) defocus of 100 Å, (c) twofold astigmatism of 100 Å, (d) threefold astigmatism of 10,000 Å, (e) coma of (g) 25,000 Å, and (f) spherical aberation of 0.057 mm

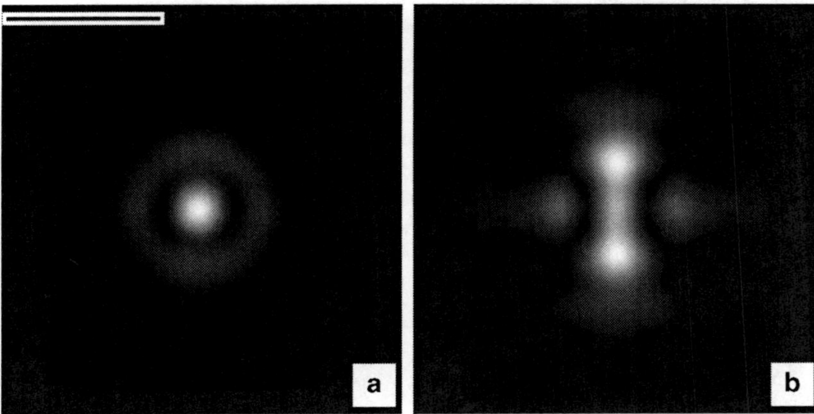

Fig. 2.13 Probe shape with combinations of aberrations. Both are for 200 keV, spherical aberration of 1.3 mm, an aperture of 10 mrad and (a) defocus of 800Å, and twofold astigmatism of 50Å, (b) defocus of 800Å, and twofold astigmatism of 200Å. The scale bar in (a) is 5 Å

the optical aberrations of the lens. Scherzer showed theoretically that a static round lens always has a positive spherical aberration. Theoretically, spherical aberration can be arbitrarily small (but still positive), but in practice has reached a practical limit determined by the minimum size of the gap that can be practically utilized (the specimen usually must be inserted in the gap) and the maximum field strength that can be obtained by magnetic materials. The current trend is to utilize nonrotationally symmetric multipole lenses to produce negative spherical aberration to balance other positive spherical aberration in the system in a process similar to the design of light optical systems (except for multipole elements). Other approaches such as radio frequency microwave cavities (for example Oldfield [274]) and foil lenses (with charges on the axis, for example [144]) have been tried but have not been aggressively pursued of late. Currently nonrotationally symmetric lenses have been developed commercially to correct for spherical aberrations.

Aberration correctors typically involve a rather sophisticated combination of multipole focussing elements. The general shape of a quadrupole element is shown in Fig. 2.14. The electron beam (optic axis) is into or out of the plane of the paper. A single quadrupole converges in one direction and diverges in the other direction (both are perpendicular to the optic axes) so more than one is needed to form an image. It is this ability to both converge and diverge that permits a negative spherical aberration unlike a round lens. A set of three or four quadrupoles can be made to function as a conventional lens.

Multipole elements can be formed with any even number of poles. A two pole element is just a deflector. Aberration correctors typically involve some combination of quadrupoles, hexapoles, and octopoles. A general form of a corrected microscope column is shown in Fig. 2.15. The corrector itself may involve on order of 100 elements. Attempts were made many decades ago to build correctors but until recently did not significantly improve the images probably due to the

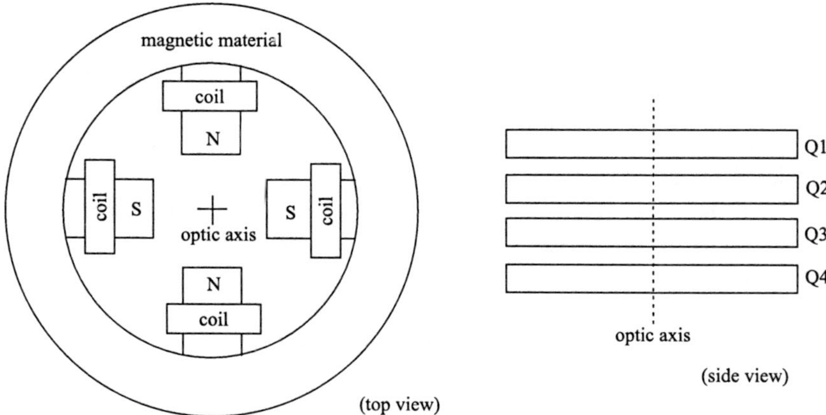

Fig. 2.14 General shape of a quadrupole lens. In the *top view (on left)* the electron travels into the plane of the page in the center of the lens. The beam converges in one direction and diverges in the other. A combination of quadrupole elements (*side view on right*) forms a lens. The magnetic field alternates direction in each element

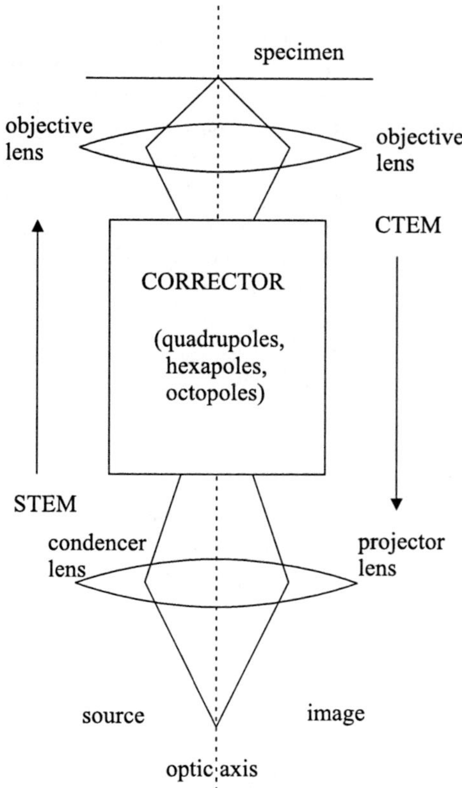

Fig. 2.15 General form of an aberration corrector. The electron beam is traveling in opposite directions in the STEM (*left*) and CTEM (*right*). Usually one or more traditional round lenses (positive C_s) are used in conjunction with a correctors (negative C_s)

difficulty of aligning such a complicated system manually. For example, Koops [215] demonstrated multipole correction of chromatic aberration with combined electric and magnetic dipoles. Rose [304], Hawkes [150], and Septier [316] have given a review of the history of the development of aberration correctors starting with Scherzer. Smith [324] has also reviewed current progress in aberration corrected microscopes. The recent success of correctors is largely due to advances in computer technology. Fast computer hardware and algorithms for automating the alignment of these complicated electron optical system is essential for utilizing this technology.

Multipole elements necessarily include many more nonrotationally symmetric aberrations. This complicates the imaging somewhat.

2.8 More Aberrations

A set of multipole elements (quadrupole, hexapole, octopole, etc.) may be used to correct for the unavoidable third-order (and possibly higher) spherical aberration of a rotationally symmetric round lens. However, in the process, these multipoles by definition manifest a series of new rotational aberrations. Optical aberrations are like many other annoying things in life. Getting rid of one aberration causes another to pop up to take its place. These new aberrations must also be corrected. Some of these new aberrations will be described next.

There are a variety of ways to express the deviation δ of the wavefront from an ideal spherical wave (Fig. 2.9). For image points on the optic axis due to small (nonzero) angles with respect to the optic axis δ could be expanded in a double power series in angles $\alpha_x^n \alpha_y^m$, where α_x and α_y are components of α in the x and y directions, respectively. For high resolution small lateral deviations (x and y) from the optic axis can be ignored. For example there are four third order terms with $n+m=3$ of α_x^3, $\alpha_x^2\alpha_y$, $\alpha_x\alpha_y^2$, and α_y^3. Any set of four linear combinations of these terms (that are not linearly dependent) would also work. The usual practice is to group these terms into a basis set of functions, each with a particular rotational symmetry (not apparent in this set of functions). A compact (slightly devious) mathematical notation using a complex angle $\omega = \alpha_x + \mathrm{i}\alpha_y$ having a specific rotational dependence can generate a sequence of terms separated into separate rotational orders. The imaginary component is a temporary mathematical convenience, and has no physical significance. Only the real part will be used in the end. A particular power is:

$$\omega^n = (\alpha_x + \mathrm{i}\alpha_y)^n = \alpha^n \exp(\mathrm{i}n\phi), \tag{2.18}$$

where ϕ is the azimuthal angle and α is the polar angle. This expression has a rotational order of n. Now expand $\chi = (2\pi/\lambda)\delta$ in a double power series of ω and its complex conjugate ω^* with complex coefficients C_{nm}.

$$\chi(\omega) = \frac{2\pi}{\lambda}\text{Real}\left[\sum_{n=1}\sum_{s=0}^{n+1}\frac{1}{n+1}C_{nm}(\omega^*)^{n+1-s}\omega^s\right] \tag{2.19}$$

$$= \frac{2\pi}{\lambda}\text{Real}\left[\sum_{n=1}\sum_{s=0}^{n}\frac{1}{n+1}C_{nm}\alpha^{n+1}\exp[\mathrm{i}(2s-n-1)\phi]\right], \tag{2.20}$$

where $m = 2s - n - 1$ is the rotational order of each term. Positive and negative m with the same magnitude yield the same thing. Several terms in this sequence may be repeated but only one of each form will be kept. The $n = 0$ term is just a deflection and need not be considered here. Terms with $m \neq 0$ will have two terms with the same nm because C_{nm} is complex. Expanding to fifth-order yields:

$$\begin{aligned}\chi(\omega) = \frac{2\pi}{\lambda}\text{Real}\Big[&\frac{1}{2}C_{10}\omega^*\omega + \frac{1}{2}C_{12}\omega^{*2} + \frac{1}{3}C_{21}\omega^{*2}\omega + \frac{1}{3}C_{23}\omega^{*3}\\ &+ \frac{1}{4}C_{30}\omega^{*2}\omega^2 + \frac{1}{4}C_{32}\omega^{*3}\omega + \frac{1}{4}C_{34}\omega^{*4}\\ &+ \frac{1}{5}C_{41}\omega^{*3}\omega^2 + \frac{1}{5}C_{43}\omega^{*4}\omega + \frac{1}{5}C_{45}\omega^{*5}\\ &+ \frac{1}{6}C_{50}\omega^{*3}\omega^3 + \frac{1}{6}C_{52}\omega^{*4}\omega^2 + \frac{1}{6}C_{54}\omega^{*5}\omega + \frac{1}{6}C_{56}\omega^{*6} + \cdots\Big],\end{aligned} \tag{2.21}$$

C_{10} is defocus, C_{12} the twofold astigmatism, C_{21} the coma, C_{23} the threefold astigmatism, C_{30} is third-order spherical aberration, etc. In terms of polar angle α and azimuthal angle ϕ this series can be written as:

$$\chi(\alpha,\phi) = \frac{2\pi}{\lambda}\sum_{mn}\frac{1}{n+1}C_{nm}\alpha^{n+1}\cos[m(\phi-\phi_{nm})], \tag{2.22}$$

where this C_{nm} is real and ϕ_{nm} is the real rotational angle of each aberration. n and m take the indicated values in (2.21). Writing out these terms in a Cartesian like form to a little over third order yields:

$$\begin{aligned}\chi(\alpha_x,\alpha_y) = \frac{2\pi}{\lambda}\Big[&\frac{1}{2}C_{10}\alpha^2 + \frac{1}{2}C_{12a}(\alpha_x^2-\alpha_y^2) + C_{12b}\alpha_x\alpha_y\\ &+ \frac{1}{3}C_{21a}\alpha_x\alpha^2 + \frac{1}{3}C_{21b}\alpha_y\alpha^2 + \frac{1}{3}C_{23a}\alpha_x(\alpha_x^2-3\alpha_y^2)\\ &+ \frac{1}{3}C_{23b}\alpha_y(\alpha_y^2-3\alpha_x^2) + \frac{1}{4}C_{30}\alpha^4 + \frac{1}{4}C_{32a}(\alpha_x^4-\alpha_y^4)\\ &+ \frac{1}{2}C_{32b}\alpha_x\alpha_y\alpha^2 + \frac{1}{4}C_{34a}(\alpha_x^4-6\alpha_x^2\alpha_y^2+\alpha_y^4)\end{aligned}$$

$$+ C_{34b}(\alpha_x\alpha_y^3 - \alpha_x^3\alpha_y) + C_{41a}(\alpha_x^4\alpha_y + 2\alpha_x^2\alpha_y^3 + \alpha_y^5)$$
$$+ C_{41b}(\alpha_x\alpha_y^4 + 2\alpha_x^3\alpha_y^2 + \alpha_x^5) + \cdots \Big], \qquad (2.23)$$

where the $C_{nma,b}$ coefficients are real valued. The sign of some of the $m \neq 0$ terms can be used either positive or negative (if used consistently) which can be a little confusing. This Cartesian like representation has some computational advantages because it does not require evaluating transcendental functions (like sin and cos) which may take a significant number of CPU cycles. There are many ways to represent the aberrations. Every author seems to have a favorite form. Krivanek et al. [221], Haider [139] et al., and Urban et al. [349] have listed the aberrations to fifth order or higher and Sawada et al. [308] have compared the various notations and provided a translation table of a sort. The notation here is similar to that of Krivanek [221]. The Zernike polynomials (for example Sect. 9.2 of Born and Wolf [34]) are orthogonal over a unit circle and Sheppard [320] has given a general discussion of orthogonal aberration functions.

2.9 Further Reading

Some Books on Electron Microscopy

1. P. R. Buseck, J. M. Cowley and L. Eyring, edit., *High-Resolution Transmission Electron Microscopy*, Oxford Univ. Press, 1988 [43]
2. J. J. Bozzola and L. D. Russell, *Electron Microscopy, Princ. and Tech. for Biologists, 2nd edit.*, Jones and Bartlett Pub., 1999 [36]
3. D. K. Bowen and C. R. Hall, *Microscopy of Materials*, MacMillan Press, 1975 [35]
4. M. De Graf, *Intro. to Conventional Transmission Electron Microscopy*, Cambridge Univ. Press, 2003 [129]
5. J. W. Edington, *Practical Electron Microscopy in Materials Science*, Van Nostrand Reinhold, 1976 [87]
6. B. Fultz and J. Howe, *Transmission Electron Microscopy and Diffractometry of Materials*, Springer, 2001 [120]
7. C. E. Hall, *Introduction to Electron Microscopy*, 2nd edition, McGraw-Hill, 1966 [141]
8. P. W. Hawkes, *Electron Optics and the Electron Microscope*, Taylor and Francis, 1972 [148]
9. R. D. Heidenreich, *Fundamentals of Transmission Electron Microscopy*, Wiley, 1964 [154]
10. P. Hirsch, A. Howie, R. B. Nicholson, D. W. Pashley, M. J. Whelan, *Electron Microscopy of Thin Crystals, second edition*, Krieger, 1977 [159]
11. S. Horiuchi, *Fundamentals of High Resolution Transmission Electron Microscopy*, North-Holland, 1994 [161]

12. R. Keyse et al., *Intro. to Scanning Transmission Electron Microscopy*, Springer, 1998 [193]
13. G. A. Meek, *Practical Electron Microscopy for Biologists, second edition*, Wiley, 1976 [240]
14. L. Reimer, *Transmission Electron Microscopy, third edition*, Springer-Verlag, 1993 [295]
15. L. Reimer, *Scanning Electron Microscopy*, Springer-Verlag, 1985 [294]
16. J. C. H. Spence, *Experimental High-Resolution Electron Microscopy*, Oxford University Press, 2003 [330]
17. D. B. Williams and C. B. Carter, *Transmission Electron Microscopy, A Textbook for Materials Science*, Plenum Press, 1996 [380]

Some Books on Electron Optics

1. P. Grivet, *Electron Optics, Parts 1 and 2*, 2nd english edition, Pergamon, 1972 [135]
2. P. W. Hawkes and E. Kasper, *Principles of Electron Optics, Vol. 1, 2, 3*, Academic Press, 1989, 1994 [151–153]
3. P. W. Hawkes (edit.), *Aberration-corrected Electron Microscopy, Adv. Imag. and Elect. Phys., vol. 153*, Academic Press, 2008 [150]
4. O. Klemperer and M. E. Barnett, *Electron Optics, third edition*, Cambridge Univ. Press, 1971 [212]
5. A. B. El-Kareh and J. C. J. El-Kareh, *Electron Beams, Lenses, and Optics, Vol. 1, 2*, Academic Press, 1970 [91]
6. J. Orloff, *Handbook of Charged Particle Optics, 2nd edit.*, CRC Press, 2009 [276]
7. A. Septier, editor, *Applied Charged Particle Optics*, in: Adv. in Electronics and Electron Physics, Vol. 13A, B, Academic Press, 1980 [317]
8. V. K. Zworykin, G. A. Morton, E. G. Ramberg, J. Hillier, A. W. Vance, *Electron Optics and the Electron Microscope*, Wiley, 1945 [394]

Chapter 3
Linear Image Approximations

Abstract This chapter presents approximations for simple calculation of CTEM and STEM (BF, ADF, and confocal) images using linear image models. A linear image is the convolution of an object function and the point spread function (multiplication in Fourier transform space). This can provide a simple intuitive approach to interpreting the electron micrographs although it might not be particularly accurate.

At some level of approximation human vision is a linear convolution of some function of light intensity with a spatial resolution response function. This simple linear image model is not quantitatively precise but allows for an easy interpretation of everyday observations. Electron microscope images do not in general follow a simple linear-image model to any great precision. However, it is useful to try to find the conditions under which an electron micrograph can be interpreted as linear image of some physical property of the specimen.

In the linear image model the actual recorded image intensity $g(\mathbf{x})$ is related to the ideal image of the object $f(\mathbf{x})$ by a linear convolution of the object function with the point spread function $h(\mathbf{x})$ of the instrument:

$$g(x,y) = f(x,y) \otimes h(x,y) = \int f(x',y')h(x-x',y-y')\mathrm{d}x'\mathrm{d}y'$$
$$g(\mathbf{x}) = f(\mathbf{x}) \otimes h(\mathbf{x}) = \int f(\mathbf{x}')h(\mathbf{x}-\mathbf{x}')\mathrm{d}\mathbf{x}', \qquad (3.1)$$

where $\mathbf{x} = (x,y)$ is a two dimensional position vector in the image plane. The point spread function or PSF of the instrument is just the image of an isolated point in the object or specimen. The symbol $\otimes$ represents convolution. Using the Fourier convolution theorem the linear image model can also be written as a product in Fourier or reciprocal space.

$$G(\mathbf{k}) = F(\mathbf{k})H(\mathbf{k}) \qquad (3.2)$$

$H(\mathbf{k})$ is the transfer function (or modulation transfer function, MTF) and is the Fourier transform of $h(\mathbf{x})$. $G(\mathbf{k})$ and $F(\mathbf{k})$ are the Fourier transforms of $g(\mathbf{x})$

E.J. Kirkland, *Advanced Computing in Electron Microscopy*,
DOI 10.1007/978-1-4419-6533-2_3, © Springer Science+Business Media, LLC 2010

and $f(\mathbf{x})$, respectively. Capital letters will be used to denote a Fourier transformed quantity and lower case letters will denote a real space quantity. Goodman [126] has given an overview of the linear image model in light optics.

3.1 The Weak Phase Object in Bright Field

Under restricted conditions the BF image in a CTEM may be considered as a linear image model (for example Erikson [95], Hansen [145], Lenz [224], Thon [342], Hoppe [160]). It should be noted that Scherzer [311] stated most of the features of this image model but did not take the final step of writing it as a convolution. This section will discuss image formation in the context of the CTEM although BF-CTEM and BF-STEM are equivalent in this context due to the reciprocity theorem (see Sect. 2.4).

The electrons incident on the specimen have a relatively high energy (approximately 100 keV to 1000 keV) as compared to the electrons in the specimen. If the specimen is very thin, then the incident electrons pass through the specimen with only small deviations in their paths and the effect of the specimen can be modeled as a simple transmission function $t(\mathbf{x})$. The electron wave function after passing through the specimen is:

$$\psi_t(\mathbf{x}) = t(\mathbf{x})\psi_{\mathrm{inc}}(\mathbf{x}) \tag{3.3}$$

where $\psi_{\mathrm{inc}}(\mathbf{x})$ is the incident wave function. In the CTEM the incident electron wave function (see Fig. 2.3 and 3.1) is approximately a plane wave of constant intensity ($\psi_{\mathrm{inc}} \sim 1$). Later chapters will discuss the effects of specimen thickness on the transmission function (i.e., $t(\mathbf{x})$ will no longer be a simple scalar function).

The effect of the aberrations of the objective lens is to shift the phase of each frequency component by a different amount (angle and spatial frequency are proportional). If $\Psi_t(\mathbf{k})$ is the Fourier transform of $\psi_t(\mathbf{x})$ and $\Psi_i(\mathbf{k})$ is the electron wave function in the back focal plane of the objective lens then:

$$\begin{aligned}\Psi_t(\mathbf{k}) &= \mathrm{FT}[\psi_t(\mathbf{x})]\\ \Psi_i(\mathbf{k}) &= \Psi_t(\mathbf{k})\exp[-\mathrm{i}\chi(\mathbf{k})] = \Psi_t(\mathbf{k})H_0(\mathbf{k})\end{aligned} \tag{3.4}$$

where FT indicates a Fourier transform. The objective lens images this wave function into a virtual image which is equivalent to an inverse Fourier transform of the earlier equation (yielding $\psi_i(\mathbf{x})$ from $\Psi_i(\mathbf{k})$). The projector lenses further magnify this virtual image. Although magnification is the primary function of the microscope this magnification can be ignored (in the math) if the image coordinates are always referred to the dimensions on the specimen.

The actual recorded image is the intensity, or square magnitude of the image wave function. Denoting the intensity in the recorded image as $g(\mathbf{x})$ yields:

$$\begin{aligned}\psi_i(\mathbf{x}) &= \mathrm{FT}^{-1}[\Psi_i(\mathbf{k})]\\ g(\mathbf{x}) &= |\psi_i(\mathbf{x})|^2 = |\psi_t(\mathbf{x}) \otimes h_0(\mathbf{x})|^2,\end{aligned} \tag{3.5}$$

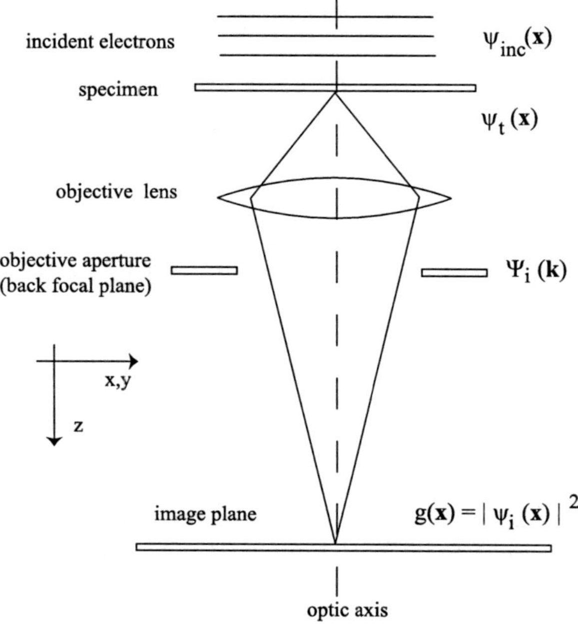

Fig. 3.1 The positions of the imaging wave functions in the CTEM column

where $h_0(\mathbf{x})$ is the complex point spread function of the objective lens (the inverse Fourier transform of $H_0(\mathbf{k})$).

In the weak phase object (WPO) approximation the specimen is assumed to be very thin and composed mainly of light atoms. The primary effect of the specimen is to produce a spatially varying phase shift in the electron wave function as it passes through the specimen. This approximation can also be referred to as the Moliere [252], WKB, or eikonal [312] approximation. If the specimen is also weak then the exponential phase factor can be expanded in a power series where only the low-order terms are important. If the incident wave function is a plane wave ($\psi_{inc} = 1$) and the specimen is a weak phase object (3.3) becomes:

$$\psi_t(\mathbf{x}) \sim t(\mathbf{x}) \sim \exp[i\sigma_e v_z(\mathbf{x})] \sim 1 + i\sigma_e v_z(\mathbf{x}) + \cdots \tag{3.6}$$

In later chapters $v_z(\mathbf{x})$ will be shown to be the projected atomic potential of the specimen and σ_e is an appropriate scaling factor (both are real). It is possible to use the opposite sign convention (reverse the sign of v_z in the exponent, with a corresponding change in $H_0(\mathbf{k})$) because only the square modulus of ψ will be important in the end. The result will be the same if all the signs are consistently changed, however this sign change can lead to some confusion when comparing different published versions of this theory in the literature (different authors use different sign conventions). The sign convention used here is consistent with the forward propagation of electrons (see Self et al. [315] for a discussion of the proper sign). Expanding the expression for $g(\mathbf{x})$ (3.5) and keeping only the lowest order terms in $v_z(\mathbf{x})$ (i.e., the weak phase object approximation) yields:

$$\begin{aligned} g(\mathbf{x}) &= \left|\left[1+\mathrm{i}\sigma_e v_z(\mathbf{x})+\mathscr{O}(v_z^2)\right]\otimes h_0(\mathbf{x})\right|^2 \\ &= |1\otimes h_0(\mathbf{x})+\mathrm{i}\sigma_e v_z(\mathbf{x})\otimes h_0(\mathbf{x})|^2+\mathscr{O}(v_z^2) \end{aligned} \tag{3.7}$$

using the Fourier convolution theorem:

$$1\otimes h_0(\mathbf{x}) = \mathrm{FT}^{-1}\left[\delta(\mathbf{k})\exp[-\mathrm{i}\chi(\mathbf{k})]\right] = 1 \tag{3.8}$$

which leaves:

$$\begin{aligned} g(\mathbf{x}) &= 1+\sigma_e v_z(\mathbf{x})\otimes\left[ih_0(\mathbf{x})-ih_0^*(\mathbf{x})\right]+\mathscr{O}(v_z^2) \\ &\approx 1+2\sigma_e v_z(\mathbf{x})\otimes h_{\mathrm{WP}}(\mathbf{x}), \end{aligned} \tag{3.9}$$

where a superscript $*$ denotes complex conjugation and $h_{\mathrm{WP}}(\mathbf{x})$ is the point spread function for BF imaging in the weak phase object approximation. It is easier to state the Fourier transform of the point spread function (i.e., the transfer function) than the PSF itself:

$$\begin{aligned} G(\mathbf{k}) &= \mathrm{FT}[g(\mathbf{x})] = \delta(\mathbf{k})+2\sigma_e V_z(\mathbf{k})H_{\mathrm{WP}}(\mathbf{k}) \\ H_{\mathrm{WP}}(\mathbf{k}) &= \mathrm{FT}\left[h_{\mathrm{WP}}(\mathbf{x})\right] \\ &= \frac{i}{2}\left\{\exp[-\mathrm{i}\chi(\mathbf{k})]-\exp[\mathrm{i}\chi(\mathbf{k})]\right\} \\ &= \sin\chi(\mathbf{k}). \end{aligned} \tag{3.10}$$

An oscillatory transfer function is about the worst thing that can happen. This means that some spacings (i.e., spatial frequencies $\mathbf{k}$) in the specimen will be transmitted as white ($H_{\mathrm{WP}}(\mathbf{k}) > 0$) at the same time that other spacings are transmitted as black ($H_{\mathrm{WP}}(\mathbf{k}) < 0$) because the transfer function is both positive and negative.

The minimum of the aberration function $\chi(\mathbf{k})$ remains approximately flat for a significant region near its minimum (see Fig. 2.10). If the defocus is adjusted so than $\sin\chi(\mathbf{k})$ is also near its minimum or maximum (± 1) when $\chi(\mathbf{k})$ is flat then the transfer function will have a significant region of uniformly transferred information (i.e., the transfer function is large and constant in a band of spatial frequencies). Allowing $\sin\chi(\mathbf{k})$ to deviate slightly from unity magnitude in the passband increases the resolution slightly. Therefore, look for conditions in which:

$$\begin{aligned} 0.7 \leq |\sin\chi(k)| &\leq 1.0 \\ \chi(k) &= -\left[\frac{2n_D-1}{2}\right]\pi \pm \frac{\pi}{4} \\ n_D &= 1,2,3,\ldots \end{aligned} \tag{3.11}$$

The minimum of the aberration function is found by setting its derivative equal to zero. Using the dimensionless form of $\chi(K)$ (2.11) and ignoring the astigmatism (assumed small) yields:

$$\begin{aligned} \frac{\partial\chi(K)}{\partial K} &= \pi(2K^3 - 2DK) = 0 \\ K^2 - D &= 0 \\ K &= \sqrt{D}. \end{aligned} \tag{3.12}$$

Next substitute this expression for K into the original expression for $\chi(K)$ (2.16) and solve for defocus D to make $\sin[\chi(K)] \sim \pm 1$:

$$\begin{aligned} \chi(K) &= \pi(0.5(\sqrt{D})^4 - D(\sqrt{D})^2) \\ &= \pi(0.5D^2 - D^2) = -0.5\pi D^2 \\ &= -\left[\frac{2n_D - 1}{2}\right]\pi - \frac{\pi}{4}. \end{aligned} \tag{3.13}$$

The optimum defocus is therefore:

$$\begin{aligned} D &= \sqrt{2n_D - 0.5} \\ \Delta f &= \sqrt{(2n_D - 0.5)C_s\lambda} \\ n_D &= 1, 2, 3, \ldots \end{aligned} \tag{3.14}$$

The special case of $n_D = 1$ is referred to as Scherzer focus. Each positive integer value of n_D produces a band of uniformly transferred spatial frequencies (Eisenhandler and Siegel [90]). The transfer function for several values of n_D is shown in Fig. 3.2. The broad passband moves to higher spatial frequencies as n_D increases but unfortunately also gets smaller so that the resolution cannot be extended dramatically using large n_D. Eisenhandler and Siegel [90] and Hoppe [160] proposed using zone plates (in the objective aperture) to select appropriate bands in the transfer function to improve resolution. Image reconstruction from a defocus series may use these wide bands to advantage (for example Kirkland et al. [209]).

Ideally, a transfer function would be flat and have the same sign over the range of spatial frequencies that are transmitted to the image. The oscillatory nature of $H_{WP}(K)$ can cause serious problems because it is not flat and changes sign. Choosing an optimum defocus produces bands of uniform sign in the transfer function but the transfer function is still oscillating. Scherzer [311] realized that it is better to limit the range of spatial frequencies so that the transfer function at least has the same sign over its allowed range. An objective aperture is placed in the back focal plane of the objective lens (see Fig. 3.1) which is conveniently the Fourier transform plane. The radius in the aperture corresponds to spatial frequency in the image. The objective aperture allows all rays within a maximum distance from the optic axis to pass. This means that the objective aperture limits the maximum spatial frequency in the image. If this maximum cutoff is made to coincide with the first zero crossing

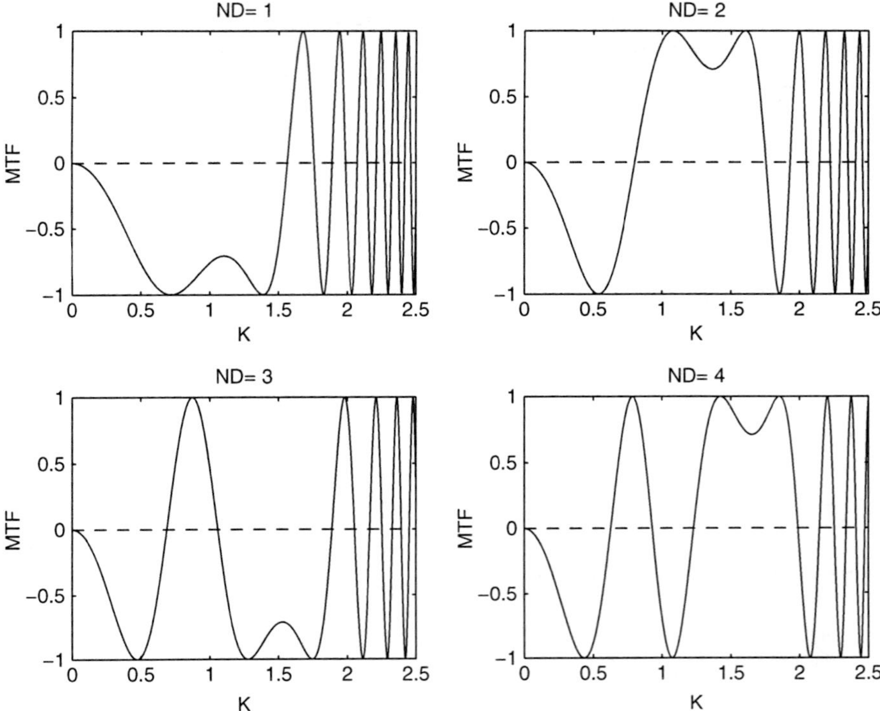

Fig. 3.2 The phase contrast transfer function H_{WP} for a weak phase object in bright field as a function of the normalized spatial frequency K for different values of the defocus index $n_d = 1,2,3,4$. Astigmatism is assumed to be zero

of the transfer function (for Scherzer focus) then the transfer function will have the same sign over its range. The first zero crossing is found in dimensionless (normalized) form from:

$$\begin{aligned} \chi(K) = \pi(0.5K_{\mathrm{max}}^4 - DK_{\mathrm{max}}^2) &= 0 \\ 0.5K_{\mathrm{max}}^2 - D &= 0 \\ K_{\mathrm{max}} &= \sqrt{2D}. \end{aligned} \tag{3.15}$$

At Scherzer focus $D = \sqrt{1.5}$. Substituting this value of D yields:

$$\begin{aligned} K_{\mathrm{max}} &= \sqrt{2\sqrt{1.5}} = k_{\mathrm{max}}(C_s\lambda^3)^{1/4} \\ k_{\mathrm{max}} &= \left(\frac{6}{C_s\lambda^3}\right)^{1/4} \\ \alpha_{\mathrm{max}} &= \lambda k_{\mathrm{max}} = \left(\frac{6\lambda}{C_s}\right)^{1/4}. \end{aligned} \tag{3.16}$$

This value of α_{max} is referred to as the Scherzer aperture. Together the Scherzer aperture and Scherzer focus are referred to as the Scherzer conditions. The corresponding resolution with Scherzer conditions is just:

$$d_s > \left(\frac{C_s\lambda^3}{6}\right)^{1/4} = 0.64(C_s\lambda^3)^{1/4} = 1/k_{max}. \tag{3.17}$$

This resolution is plotted vs. spherical aberration in Fig. 3.3. This is a lower bound because the transfer function $H_{WP} = 0$ at this spacing and there is no information transferred. There is some ambiguity in the choice of (3.11). Other choices will change the value of various constants presented earlier by small amounts. The differences appearing in the literature mainly reflect the allowed variation of $\sin\chi(K)$ in the pass band (0.7–1.0 chosen here).

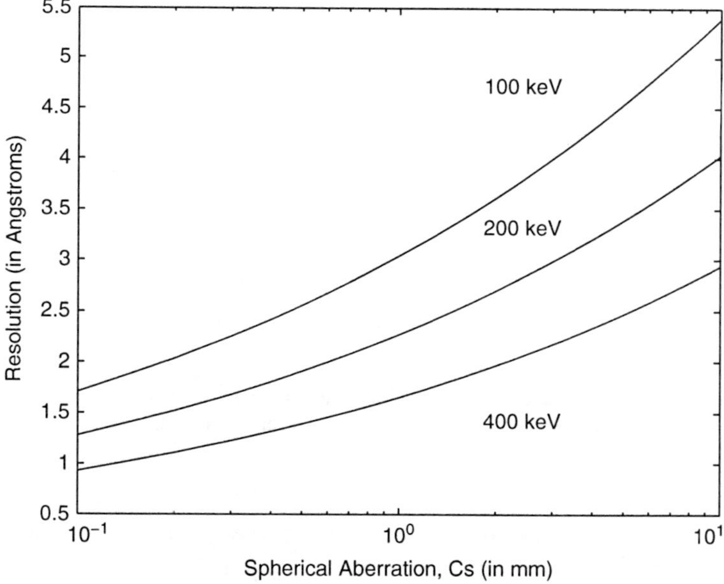

Fig. 3.3 The coherent bright field phase contrast resolution in the weak phase object approximation as a function of spherical aberration C_s for several different electron beam energies

3.2 Partial Coherence in BF-CTEM

In practice the electron microscope always has small deviations from the ideal. The incident illumination is never exactly collimated and parallel to the optic axis of the microscope. The electron energy is not completely monochromatic and the objective and condenser lens currents are never perfectly stable. The degree of collimation of the illumination incident on the specimen is related to the lateral coherence (i.e., spatial coherence) of the incident electron wave function and the stability of the beam energy and lens currents is related to the temporal coherence of the imaging process. The imaging described in the previous section assumes that these effects are negligible and that the imaging process is perfectly coherent. When the effects of a small spread in illumination angles and a small spread in beam energy and lens currents are included, the imaging process is said to be partially coherent. An analytical derivation of the effects of partial coherence on the transfer function in the weak phase object approximation has been given by Frank [111], Fejes [99], and Wade and Frank [361].

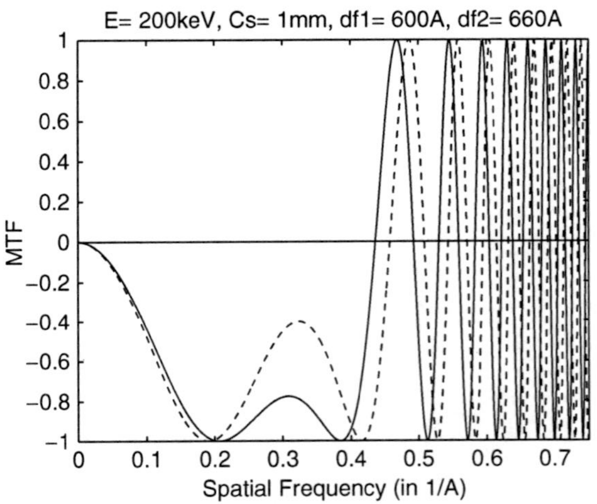

Fig. 3.4 BF-CTEM transfer function with two slightly different defocus values (*solid line* 600Å, and *dashed* 660Å) as might be caused by fluctions in the objective lens current. When these are superposed in the image lower spatial frequencies are little affected but the high spatial frequencies tend to average to zero

The transfer function oscillates both positive and negative. Partial coherence results in the superposition of adjacent portions of the transfer function (as in Fig. 3.4). When the transfer function is oscillating quickly (at high spatial frequencies) then partial coherence tends to reduce the transfer function to zero. Partial coherence limits the maximum information content of the image in the electron microscope by damping the high spatial frequency (large scattering angle) portion of the transfer function.

The electron beam illuminating the specimen (formed by the condenser system) will always have a small distribution of angles (see the right hand side of Fig. 3.5). To calculate the effect on the image first consider an illumination at a single angle as shown in the left hand side of Fig. 3.5. Previously in (3.3) the incident wave function was $\psi_{\mathrm{inc}} \sim 1$ however with the illumination at an angle β:

$$\psi_{\mathrm{inc}}(\mathbf{x}) = \exp(2\pi i \mathbf{k}_\beta \cdot \mathbf{x}), \tag{3.18}$$

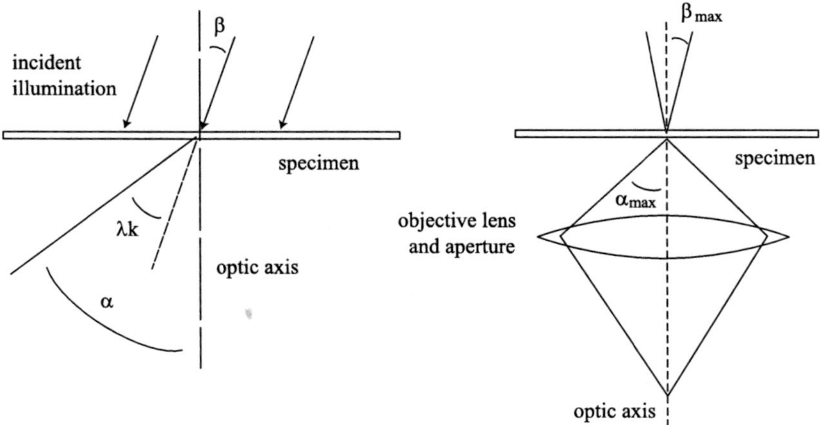

Fig. 3.5 Imaging with nonideal illumination. The incident electrons have an angle β. The specimen scatters at an angle λk and the final angle into the objective lens is α (angles measured with respect to the optic axis). A single electron trajectory is shown on the *left* and the total illumination of a single point on the image is shown on the *right*. β_{max} is typically the condenser aperture and α_{max} is the objective aperture

where $\mathbf{k}_\beta = \beta/\lambda$ and β is the angle of the incident illumination (with respect to the optic axis). Note that $\mathbf{k}_\beta$ is a two dimensional vector because β can vary in both the polar and azimuthal directions. The transmitted wave function (3.3) now becomes:

$$\psi_t(\mathbf{x}) = t(\mathbf{x})\exp(2\pi i\mathbf{k}_\beta \cdot \mathbf{x}). \tag{3.19}$$

Usually, the condenser system will deliver a small cone of illumination angles onto the specimen. Typically, each illumination angle will be incoherent with other illumination angles so the images due to each illumination angle should be summed incoherently by adding intensities $|\psi|^2$ and not amplitudes ψ. It is possible to operate a field emission gun to produce a coherent spread of illumination angles (in which case amplitudes and not intensities would be summed), however this case will not be considered here. If $p(\mathbf{k}_\beta)$ represents the (probability) distribution of (incoherent) illumination angles then (3.5) becomes:

$$\begin{aligned} g(\mathbf{x}) &= \int |\psi_i(\mathbf{x})|^2 p(\mathbf{k}_\beta) d^2k_\beta \\ &= \int \left| \left[t(\mathbf{x})\exp(2\pi i\mathbf{k}_\beta \cdot \mathbf{x}) \right] \otimes h_0(\mathbf{x}) \right|^2 p(\mathbf{k}_\beta) d^2k_\beta \end{aligned} \tag{3.20}$$

A small spread in energy of the incident electron is equivalent to a small (incoherent) spread in defocus values due to the chromatic aberration of the objective lens. Fluctuations in the focusing currents in the objective lens also produce an incoherent spread in defocus values. When this spread in defocus values is combined and included in the image:

$$g(\mathbf{x}) = \int |\psi_i(\mathbf{x})|^2 p(\mathbf{k}_\beta)p(\delta_\mathrm{f})\mathrm{d}\delta_\mathrm{f}\mathrm{d}^2k_\beta$$
$$= \int \left| \left[t(\mathbf{x})\exp(2\pi i\mathbf{k}_\beta\cdot\mathbf{x})\right] \otimes h_0(\mathbf{x},\Delta f+\delta_\mathrm{f})\right|^2 p(\mathbf{k}_\beta)p(\delta_\mathrm{f})\mathrm{d}\delta_\mathrm{f}\mathrm{d}^2k_\beta, \tag{3.21}$$

where δ_f is the fluctuation in defocus and $p(\delta_\mathrm{f})$ is the distribution of this fluctuation. Both $p(\mathbf{k}_\beta)$ and $p(\delta_\mathrm{f})$ are normalized such that their integrated value is unity. This expression for $g(\mathbf{x})$ is formidable and should be treated numerically in general (see later chapters). However, if the deviations from the ideal are assumed very small and the specimen is assumed to be a weak phase object, a modified transfer function can be obtained.

First examine the expression for the wave function in the image plane for one illumination angle and one defocus value:

$$\psi_i(\mathbf{x}) = \left[t(\mathbf{x})\exp(2\pi i\mathbf{k}_\beta\cdot\mathbf{x})\right] \otimes h_0(\mathbf{x}). \tag{3.22}$$

Figure 3.5 shows that the tilt angle can appear in either the objective lens angle α or the incident wave function ψ_{inc}. To see this mathematically take the Fourier transform of ψ_i and use the Fourier convolution theorem:

$$\mathrm{FT}\left\{ [t(\mathbf{x})\exp(2\pi i\mathbf{k}_\beta\cdot\mathbf{x})] \otimes h_0(\mathbf{x})\right\} = T(\mathbf{k}+\mathbf{k}_\beta)H_0(\mathbf{k}). \tag{3.23}$$

Now with a change of variable:

$$T(\mathbf{k}+\mathbf{k}_\beta)H_0(\mathbf{k}) = T(\mathbf{k}')H_0(\mathbf{k}'-\mathbf{k}_\beta). \tag{3.24}$$

The inverse Fourier transform is now an integration over $\mathbf{k}'$ instead of $\mathbf{k}$ (over all space) but gives the same result. This means that the image plane wave function can also be written as:

$$\psi_i(\mathbf{x}) = t(\mathbf{x}) \otimes h_0(\mathbf{x},\mathbf{k}_\beta) \tag{3.25}$$

Using $h_0(\mathbf{x},\mathbf{k}_\beta)$ will make the analytical calculation of the transfer function much simpler. This approximation would not be appropriate for thick specimens because the effect of the specimen is no longer a simple multiplicative function.

If the specimen is a weak phase object (3.6) then the expression for the final image intensity will be similar to (3.9) except that the point spread function $h_0(\mathbf{x})$ of the objective lens will be integrated over illumination angles and defocus values as in (3.21). If $\mathbf{k}_\beta$ is small then the leading background constant can also be assumed to be close to unity [as in (3.8)]:

$$1 \otimes h_0(\mathbf{x},\mathbf{k}_\beta) \sim 1. \tag{3.26}$$

The integral of the transfer function over a spread in illumination angles and defocus values using the dimensionless form of $\chi(\mathbf{K})$ is:

$$H_0(\mathbf{K}) = \int \exp\{-\mathrm{i}\chi(\mathbf{K}+\mathbf{K}_\beta, D+D_p)p(D_p)p(\mathbf{K}_\beta)\mathrm{d}D_p\mathrm{d}^2K_\beta \quad (3.27)$$

where the sign of $\mathbf{K}_\beta$ has been changed to positive for simplicity (i.e., the integrand is symmetric) and:

$$\chi(\mathbf{K}+\mathbf{K}_\beta, D+D_p) = \pi[0.5(\mathbf{K}+\mathbf{K}_\beta)^4 - (D+D_p)(\mathbf{K}+\mathbf{K}_\beta)^2] \quad (3.28)$$

$p(\mathbf{K}_\beta)$ and $p(D_p)$ are the distribution of illuminations angles $\mathbf{K}_\beta$ and defocus deviations D_p. The simplest assumption of a Gaussian distribution for each has the advantage of allowing an analytical solution.

$$p(\mathbf{K}_\beta) = \frac{1}{\pi K_s^2}\exp(-K_\beta^2/K_s^2) \quad (3.29)$$

$$p(D_p) = \frac{1}{\sqrt{\pi}D_s}\exp(-D_p^2/D_s^2), \quad (3.30)$$

where K_s is the $1/e$ width of the spread in illumination angles and D_s is the $1/e$ spread in defocus values. Taylor expanding $\chi(\mathbf{K},D)$ to lowest order in $\mathbf{K}_\beta$ and D_p:

$$\chi(\mathbf{K}+\mathbf{K}_\beta, D+D_p) \sim \chi(\mathbf{K},D) + \mathbf{K}_\beta\cdot\mathbf{W_1} + D_p\,W_2 + D_p\mathbf{K}_\beta\cdot\mathbf{W}_3 + \cdots \quad (3.31)$$

where

$$\mathbf{W}_1 = \frac{\partial\chi(\mathbf{K},D)}{\partial\mathbf{K}} = 2\pi(|\mathbf{K}|^2 - D)\mathbf{K} \quad (3.32)$$

$$W_2 = \frac{\partial\chi(\mathbf{K},D)}{\partial D} = -\pi K^2 \quad (3.33)$$

$$\mathbf{W}_3 = \frac{\partial^2\chi(\mathbf{K},D)}{\partial\mathbf{K}\partial D} = -2\pi\mathbf{K}. \quad (3.34)$$

$\mathbf{W}_1$ and $\mathbf{W}_3$ are two dimensional vector quantities. Equation (3.27) now becomes:

$$H_0(\mathbf{K}) = \frac{1}{\pi K_s^2\sqrt{\pi}D_s}\exp[-\mathrm{i}\chi(\mathbf{K},D)]\int \exp[-\mathrm{i}\mathbf{K}_\beta\cdot\mathbf{W_1} - iD_pW_2 - iD_p\mathbf{K}_\beta\cdot\mathbf{W}_3 - \mathbf{K}_\beta^2/K_s^2 - D_p^2/D_s^2]\mathrm{d}^2K_\beta\mathrm{d}D_p \quad (3.35)$$

First do the integration over $\mathbf{K}_\beta$:

$$H_0(\mathbf{K}) = \frac{1}{\sqrt{\pi}D_s}\exp[-\mathrm{i}\chi(\mathbf{K},D)]\int \exp[-0.25K_s^2|\mathbf{W}_1 + D_p\mathbf{W}_3|^2 - iD_pW_2 - D_p^2/D_s^2]\mathrm{d}D_p. \quad (3.36)$$

Next do the integral over D_p:

$$H_0(\mathbf{K}) = \frac{1}{\sqrt{1+EK^2}} \exp\left[-\mathrm{i}\chi(\mathbf{K},D) + i\frac{D_s^2 K_s^2 \mathbf{W}_1 \cdot \mathbf{W_3} W_2}{4(1+EK^2)}\right]$$
$$\times \exp\left[-0.25K_s^2|\mathbf{W}_1|^2 + \frac{D_s^2[(0.5K_s^2\mathbf{W}_1\cdot\mathbf{W_3})^2 - W_2^2]}{4(1+EK^2)}\right]. \tag{3.37}$$

Aside: if the cross term $\mathbf{W}_3 = 0$ (implying $E = 0$) is neglected a somewhat more elegant expression results:

$$H_0(\mathbf{K}) = \exp[-\mathrm{i}\chi(\mathbf{K},D) - 0.25K_s^2|\mathbf{W}_1|^2 - 0.25D_s^2 W_2^2]. \tag{3.38}$$

Next substitute for the intemediate variables:

$$H_0(\mathbf{K}) = \exp\left[-\mathrm{i}\frac{\pi K^2}{1+EK^2}\left(0.5K^2(1-EK^2) - D\right)\right]$$
$$\times \frac{1}{\sqrt{1+EK^2}} \exp\left[\frac{-\pi^2 K_s^2 (K^2-D)^2 K^2 - 0.25\pi^2 D_s^2 K^4}{1+EK^2}\right], \tag{3.39}$$

where $E = \pi^2 K_s^2 D_s^2$

Now take the imaginary part to form the transfer function for the weak phase object image approximation (3.2, 3.10) and substitute the dimensional form of the electron microscope parameters:

$$H_{\mathrm{WP}}(k) = \sin\left[\frac{\pi\lambda k^2}{1+\varepsilon k^2}\left(0.5C_s(1-\varepsilon k^2)\lambda^2 k^2 - \Delta f\right)\right]$$
$$\times \frac{1}{\sqrt{1+\varepsilon k^2}} \exp\left[-\frac{[\pi\lambda k_s k(C_s\lambda^2 k^2 - \Delta f)]^2 + 0.25(\pi\lambda\Delta_0 k^2)^2}{1+\varepsilon k^2}\right], \tag{3.40}$$

where $\varepsilon = (\pi\lambda k_s \Delta_0)^2$, $k_s = K_s(C_s\lambda^3)^{-1/4}$, and $\Delta_0 = D_s\sqrt{C_s\lambda}$. This transfer function (3.40) should be substituted into the image model (3.9 and 3.10). Apart from the extra term with ε (typically small) the oscillatory portion of the transfer functions is the same as the coherent case. The main change is the addition of a damping envelope that attenuates the transfer function at high spatial frequencies. In a practical sense $\beta = \lambda k_s$ is the condenser (illumination) semiangle and Δ_0 is approximately the rms value of all of the appropriate fluctuations multiplied by the chromatic aberration C_c.

$$\Delta_0 \sim C_\mathrm{c}\sqrt{\left(\frac{\Delta E}{E}\right)^2 + \left(2\frac{\Delta I}{I}\right)^2 + \left(\frac{\Delta V}{V}\right)^2}, \tag{3.41}$$

where E, I, and V are the electron energy, lens currents, and acceleration voltage, respectively, and ΔE, ΔI, and ΔV are the $1/e$ width of their fluctuations.

A graph of the WPO transfer function with (3.40) and without (3.10) partial coherence is shown in Fig. 3.6 for typical values of the electron optical parameters.

It is interesting to note that the cross term W_3 produced the correction ε in the transfer function which can move the zero crossings in the transfer function but seems to have a negligible effect under Scherzer conditions.

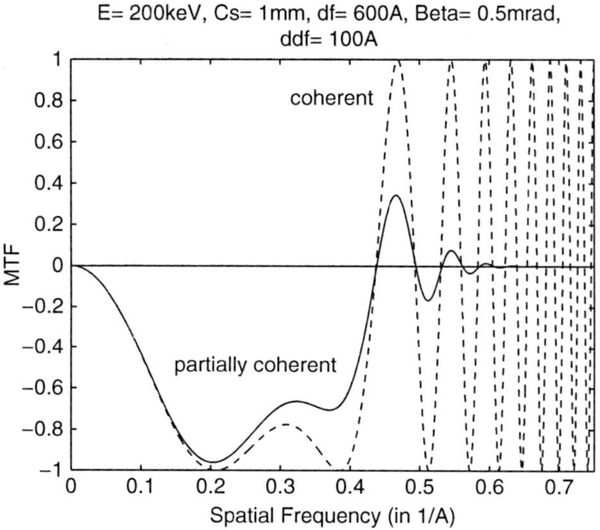

Fig. 3.6 The BF phase contrast transfer function with and without partial coherence

3.2.1 Aberration Correctors and Partial Coherence

It is straight forward but somewhat tedious to add higher order aberrations to the earlier expressions for partial coherence. The transfer function with the fifth-order spherical aberration is:

$$
\begin{aligned}
H_{\mathrm{WP}}(k) &= \frac{1}{\sqrt{1+\varepsilon k^2}} \\
&\times \sin\left[\frac{\pi\lambda k^2}{1+\varepsilon k^2}\left(\frac{1}{3}C_{S5}(1-2\varepsilon k^2)\lambda^4 k^4 + 0.5C_{S3}(1-\varepsilon k^2)\lambda^2 k^2 - \Delta f\right)\right] \\
&\times \exp\left[-\frac{[\pi\lambda k_s k(C_{S5}\lambda^4 k^4 + C_{S3}\lambda^2 k^2 - \Delta f)]^2 + 0.25(\pi\lambda\Delta_0 k^2)^2}{1+\varepsilon k^2}\right].
\end{aligned} \tag{3.42}
$$

However, terms such as the spherical aberration coefficient $C_s = C_{S3}$ are no longer really constant. The spherical aberration of the round objective lens is essentially constant because it is due to the fixed geometry of the magnetic material used to shape the focusing fields and to lowest order insensitive to small fluctuations in the lens current (which instead produces fluctuations in defocus). An aberration corrector generates a large negative C_{S3} from a combination of strong focusing multipole fields. Fluctuations in the currents in these elements can produce fluctuations in the net C_{S3} coefficients as well as all of the other dozen or so aberration coefficients. There can be of order 100 or more focusing currents (etc.) in the corrector that have there own fluctuations. These potentially cause all of the aberration coefficients to fluctuate (more study is needed here to determine how these all interact, may vary with the specific design of each corrector). Various groupings of the multipoles combine to form each aberrations (some elements may be part of more than one aberration). If the cross terms between aberrations are ignored for simplicity, the aberration phase factor becomes:

$$H_0(\mathbf{K}) = \exp\left[-i\chi(\mathbf{K},D) - \sum_k 0.25 W_k \left|\frac{\partial\chi}{\partial C_k}\right|^2\right], \tag{3.43}$$

where W_k is the width of the fluctuations for parameter C_k and the summation is over all aberration components (C_k such as defocus, spherical aberrations C_{s3}, C_{s5}, etc.) and the illumination angle. Even though the aberration corrector can effectively negate the existing aberrations in the main (round) objective lens the fluctuations in the corrector causing partial coherence will be centered about the much larger aberrations (such as C_{S3}) of the original objective lens making this effect much larger than might be expected.

It may be easier to numerically integrate over small fluctuation in all parameters. Gauss-Hermite quadrature is very efficient for integration over a Gaussian weighting. With this many parameters it might even be worthwhile performing a Monte-Carlo integration over a range of parameter values (which becomes more efficient when integrating over more than four dimensions). It is also not clear how fluctuations in the various aberration coefficient are coupled (mathematically in the aberration function and practically through common coils and currents).

3.3 Detector Influence (CTEM)

The detector that records the image may itself have a large influence on the image quality. Currently, the detector is usually a CCD camera (plus scintillator). Film (or plates) were the most common detector in the not so distant past. It was common to record the image at not very high magnification and then magnify further in the darkroom enlarger, in which case the film may produce a large effect. Film (or plates) have a transfer function of their own (for example Downing and Grano [80]). CCD detectors also have an associated transfer function which can be significant (Thust [343]). The effects of the detector can be included in the image by convolving the electron image (3.5) with detector transfer function $h_{\mathrm{DET}}(\mathbf{x})$:

$$g(\mathbf{x}) = |\psi_t(\mathbf{x}) \otimes h_0(\mathbf{x})|^2 \otimes h_{\mathrm{DET}}(\mathbf{x}) \tag{3.44}$$

The detector may reduce the contrast in the high resolution component of an image (for example lattice fringes) by a large factor (like two or three).

3.4 Incoherent Imaging of Thin Specimens (CTEM)

The electron microscope image is generated from the electrons scattered by the specimen. For the electron wave scattered from two points of the specimen to interfere the incident electron wave from the condenser system must be coherent over the distance between these points (i.e., the specimen is not self-luminous). The illuminating electrons (using a thermionic source) are produced by a spatially incoherent, quasi-monochromatic source with nonzero size. The source size is typically translated into a maximum condenser angle β_{max} (see Fig. 3.5). The electron rays incident on the specimen are not perfectly parallel but subtend a small cone of angles β_{max} at the specimen plane. The lateral coherence length perpendicular to the optic axis (see Sect. 10.4.2 of Born and Wolf [34]) of the illuminating electron wave function is approximately given by:

$$\Delta x_{coh} \sim \frac{0.16\lambda}{\beta_{max}}. \tag{3.45}$$

If the coherence length is much smaller than the resolution element of the image then the imaging will be essentially incoherent. If the coherence length is much bigger than the resolution element then the imaging process will be essentially coherent. The resolution element of the image is approximately $d \sim \lambda/\alpha_{max}$ where α_{max} is the maximum objective angle (i.e., the size of the objective aperture). Combining these two expressions yields an approximate criteria for the image coherence:

$$\begin{aligned} &\beta_{max} \ll 0.16\alpha_{max} \quad \text{coherent imaging} \\ &\beta_{max} \gg 0.16\alpha_{max} \quad \text{incoherent imaging} \end{aligned} \tag{3.46}$$

In between these two extremes the image is partially coherent.

If the image is coherent then the amplitudes ψ of the scattered electrons add and if the image is incoherent then the intensities of the scattered electrons $|\psi|^2$ add. The final image recording process is only sensitive to the intensity of the electrons and not their amplitudes. This means that phase contrast must have a coherent image process to be sensitive to the phase of the electron via some interference process. Reducing the coherence of the imaging process should also reduce the phase contrast transfer function. The ratio β_{max}/α_{max} can be used to control the coherence of the imaging process.

Because the phase contrast image will likely disappear in an incoherent image the transmission function should include the possibility of amplitude contrast. Assuming that the specimen is a weak-phase, weak-amplitude object yields a transmission function:

$$t(\mathbf{x}) \sim \exp[\mathrm{i}\sigma v_z(\mathbf{x}) - u(\mathbf{x})] \sim 1 + i\sigma_e v_z(\mathbf{x}) - u(\mathbf{x}) + \cdots \tag{3.47}$$

where $u(\mathbf{x})$ is the amplitude component of the specimen transmission function which can arise from scattering outside of the objective aperture (it is preceded by a minus sign because electrons cannot be created with elastic scattering). Alternately $u(\mathbf{x})$ can be considered as the next term in the Taylor expansion of $\exp[i\sigma v_z(\mathbf{x})]$ where $u(\mathbf{x}) \propto v_z^2(\mathbf{x})$. Both $\sigma_e v_z(\mathbf{x})$ and $u(\mathbf{x})$ are small compared to unity. The expression for the image intensity (3.21) now becomes:

$$\begin{aligned} g(\mathbf{x}) &= \int |\psi_i(\mathbf{x})|^2 p(\mathbf{k}_\beta)p(\delta_\mathrm{f})\mathrm{d}\delta_\mathrm{f}\mathrm{d}^2k_\beta \\ &= \int \left|\left[(1+\mathrm{i}\sigma_e v_z(\mathbf{x})-u(\mathbf{x}))\exp(2\pi\mathrm{i}\mathbf{k}_\beta\cdot\mathbf{x})\right]\otimes h_0(\mathbf{x},\Delta f+\delta_\mathrm{f})\right|^2 \\ &\qquad p(\mathbf{k}_\beta)p(\delta_\mathrm{f})\mathrm{d}\delta_\mathrm{f}\mathrm{d}^2k_\beta. \end{aligned} \tag{3.48}$$

The objective aperture will play a significant role in the following derivation so it should be included with the point spread function $h_0(\mathbf{x})$ or equivalently the transfer function:

$$H_0(\mathbf{k}) = \exp[-\mathrm{i}\chi(\mathbf{k},\Delta f+\delta_\mathrm{f})]A(\mathbf{k}), \tag{3.49}$$

where $A(\mathbf{k})$ is the aperture function:

$$\begin{aligned} A(\mathbf{k}) &= 1 \quad ; \quad \lambda|\mathbf{k}| = \alpha < \alpha_\mathrm{max} \\ &= 0 \quad ; \quad \text{otherwise} \end{aligned} \tag{3.50}$$

and α_max is the maximum semiangle allowed by the objective aperture.

Now expand the integrand keeping only the terms of lowest order in $\sigma_e v_z(\mathbf{x})$ and $u(\mathbf{x})$ and drop the explicit reference to the independent arguments for simplicity:

$$\begin{aligned} g(\mathbf{x}) = \int \{ &|\exp(2\pi\mathrm{i}\mathbf{k}_\beta\cdot\mathbf{x})\otimes h_0|^2 \\ &+ [\exp(2\pi\mathrm{i}\mathbf{k}_\beta\cdot\mathbf{x})\otimes h_0]^*[i\sigma_e v_z\exp(2\pi\mathrm{i}\mathbf{k}_\beta\cdot\mathbf{x})\otimes h_0] \\ &- [\exp(2\pi\mathrm{i}\mathbf{k}_\beta\cdot\mathbf{x})\otimes h_0]^*[u\exp(2\pi\mathrm{i}\mathbf{k}_\beta\cdot\mathbf{x})\otimes h_0] \\ &+ [\text{complex conjugate}]\}p(\mathbf{k}_\beta)p(\delta_\mathrm{f})\mathrm{d}\delta_\mathrm{f}\mathrm{d}^2k_\beta. \end{aligned} \tag{3.51}$$

The first term is a constant of order unity that does not vary with position in the image. The rest of the right hand side is rather unpleasant but can be simplified a little. Because $h_0(\mathbf{x})$ is simplest to write in reciprocal space, look at the Fourier transform of the term containing $v_z(\mathbf{x})$, and for simplicity drop explicit reference to the defocus Δf and its fluctuations δ_f:

$$\begin{aligned} &\mathrm{FT}\{[\exp(2\pi\mathrm{i}\mathbf{k}_\beta\cdot\mathbf{x})\otimes h_0]^*[i\sigma_e v_z\exp(2\pi\mathrm{i}\mathbf{k}_\beta\cdot\mathbf{x})\otimes h_0]\} \\ &\quad = [\delta(\mathbf{k}-\mathbf{k}_\beta)H_0(-\mathbf{k})]^*\otimes[\mathrm{i}\sigma_e V(\mathbf{k}+\mathbf{k}_\beta)H_0(\mathbf{k})] \\ &\quad = \mathrm{i}\sigma_e\int\delta(\mathbf{k}'-\mathbf{k}_\beta)H_0^*(-\mathbf{k}')V(\mathbf{k}+\mathbf{k}_\beta-\mathbf{k}')H_0(\mathbf{k}-\mathbf{k}')\mathrm{d}\mathbf{k}' \\ &\quad = \mathrm{i}\sigma_e V(\mathbf{k})H_0^*(-\mathbf{k}_\beta)H_0(\mathbf{k}-\mathbf{k}_\beta), \end{aligned} \tag{3.52}$$

where $\delta(\mathbf{k})$ is the Dirac delta function.

This expression is now a simple product so when inverse Fourier transformed back into real space $\sigma_e v_z(\mathbf{x})$ is convolved with a new form of the point-spread function. Repeating this procedure with the amplitude component and combining the complex conjugate terms yield:

$$g(\mathbf{x}) \sim C_0 + 2\sigma_e v_z(\mathbf{x}) \otimes h_{\mathrm{WP}}(\mathbf{x}) - 2u(\mathbf{x}) \otimes h_{\mathrm{WA}}(\mathbf{x}), \tag{3.53}$$

where C_0 is a background constant of order unity:

$$C_0 = \int |\exp(2\pi i \mathbf{k}_\beta \cdot \mathbf{x}) \otimes h_0|^2 p(\mathbf{k}_\beta) p(\delta_{\mathrm{f}}) \mathrm{d}\delta_{\mathrm{f}} \mathrm{d}^2 k_\beta \tag{3.54}$$

and the transfer function for a weak phase object is:

$$\begin{aligned} \mathrm{FT}[h_{\mathrm{WP}}(\mathbf{x})] &= H_{\mathrm{WP}}(\mathbf{k}) \\ &= \mathrm{Imag} \int H_0(\mathbf{k}_\beta - \mathbf{k}) H_0^*(-\mathbf{k}_\beta) p(\mathbf{k}_\beta) p(\delta_{\mathrm{f}}) \mathrm{d}\delta_{\mathrm{f}} \mathrm{d}^2 k_\beta \end{aligned} \tag{3.55}$$

and the transfer function for a weak amplitude object is:

$$\begin{aligned} &\mathrm{FT}[h_{\mathrm{WA}}(\mathbf{x})] = H_{\mathrm{WA}}(\mathbf{k}) \\ &= \mathrm{Real} \int H_0(\mathbf{k}_\beta - \mathbf{k}) H_0^*(-\mathbf{k}_\beta) p(\mathbf{k}_\beta) p(\delta_{\mathrm{f}}) \mathrm{d}\delta_{\mathrm{f}} \mathrm{d}^2 k_\beta. \end{aligned} \tag{3.56}$$

With the assumption of a small Gaussian spread of defocus [see (3.30)] the integral over defocus can be done analytically yielding an expression for the transfer function for a weak phase object:

$$\begin{aligned} H_{\mathrm{WP}}(\mathbf{k}) = &\int A(\mathbf{k}_\beta) A(\mathbf{k}_\beta - \mathbf{k}) \sin[\chi(\mathbf{k}_\beta - \mathbf{k}) - \chi(\mathbf{k}_\beta)] \\ &\times \exp\{-0.25\pi^2\lambda^2\Delta_0^2[|\mathbf{k}_\beta - \mathbf{k}|^2 - |\mathbf{k}_\beta|^2]^2\} p(\mathbf{k}_\beta) \mathrm{d}^2 k_\beta \end{aligned} \tag{3.57}$$

and an expression for the transfer function for a weak amplitude object:

$$\begin{aligned} H_{\mathrm{WA}}(\mathbf{k}) = &\int A(\mathbf{k}_\beta) A(\mathbf{k}_\beta - \mathbf{k}) \cos[\chi(\mathbf{k}_\beta - \mathbf{k}) - \chi(\mathbf{k}_\beta)] \\ &\times \exp\{-0.25\pi^2\lambda^2\Delta_0^2[|\mathbf{k}_\beta - \mathbf{k}|^2 - |\mathbf{k}_\beta|^2]^2\} p(\mathbf{k}_\beta) \mathrm{d}^2 k_\beta. \end{aligned} \tag{3.58}$$

The only difference between these two transfer functions is the switch between sin and cos. If $\mathbf{k}_\beta$ is large then neither of these integrals can be done analytically and both must be done numerically (in two dimensions).

The phase and amplitude transfer functions are shown in Figs. 3.7 and 3.8, respectively, using Scherzer defocus and the Scherzer aperture. The phase contrast transfer function is initially similar to the coherent transfer (see Fig. 3.2) function but decays to zero as the condenser angle β_{max} is increased as expected (i.e., when the image is incoherent there cannot be any interference to produce phase contrast). The amplitude contrast transfer function however transforms from an oscillatory function into a smoothly falling function similar to the transfer function in normal light

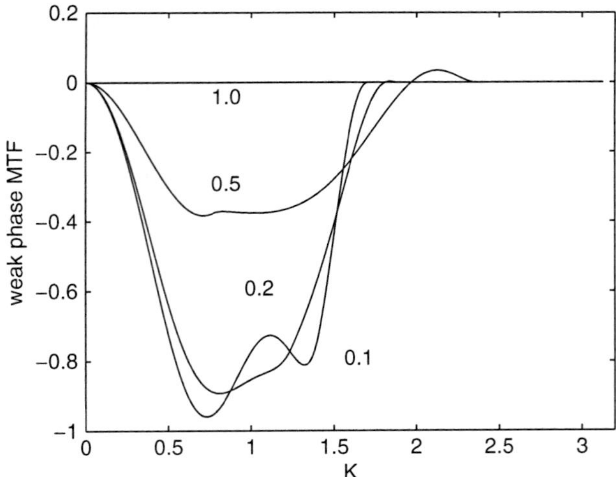

Fig. 3.7 BF-CTEM transfer function for weak phase objects with increasing condenser angle β_{max} as a function of dimensionless spatial frequency K (Scherzer defocus and aperture α_{max}). $\beta_{\text{max}}/\alpha_{\text{max}} = 0.1, 0.2, 0.5, 1.0$. The phase contrast transfer function is zero when the image process is incoherent $\beta_{\text{max}}/\alpha_{\text{max}} \geq 1$

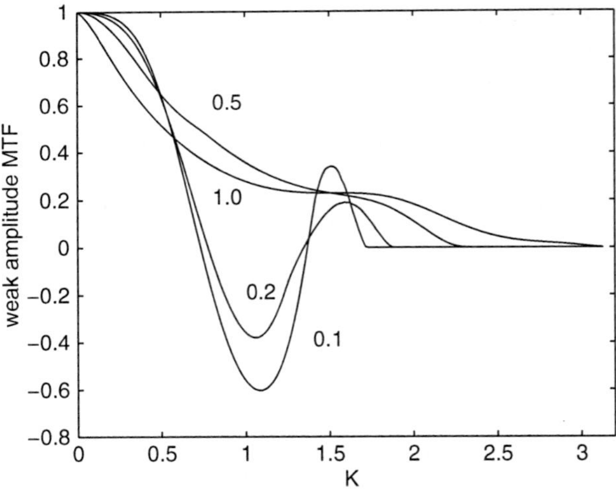

Fig. 3.8 BF-CTEM transfer function for weak amplitude objects with increasing condenser angle β_{max} as a function of dimensionless spatial frequency K (Scherzer defocus and aperture α_{max}). $\beta_{\text{max}}/\alpha_{\text{max}} = 0.1, 0.2, 0.5, 1.0$. The amplitude contrast transfer function is similar to normal light optics when the image process is incoherent $\beta_{\text{max}}/\alpha_{\text{max}} \geq 1$

optics (Black and Linfoot [30]). Note also that the first zero of the transfer function has moved by almost a factor of two in the incoherent ($\beta_{\text{max}}/\alpha_{\text{max}} = 1$) case. The improvement in resolution from incoherent imaging over coherent imaging has been known for a long time (Rayleigh [292] and Goodman [126]). The essential features of this section were discussed by Hanszen [145], Hanszen and Trepte [146], and Thomson [341].

3.5 Annular Dark Field STEM

The order of the optical components of the STEM (Fig. 2.4) is reversed from that of the CTEM (Fig. 2.3). The objective lens is before the specimen and forms a focused probe on the specimen. The portion of the electrons transmitted through the specimen that fall on the detector form the image brightness at one point in the image. A whole image is built up by scanning the focused probe over the specimen and recording the transmitted intensity at each position of the probe. In bright field the detector integrates (incoherently) over a very small angle centered about zero scattering angle, and in annular dark field (ADF) the detector integrates everything except the center regions. The ADF-STEM image would be equivalent to the BF-CTEM image using incoherent hollow cone illumination by the reciprocity theorem (Engel [92]). In practice a CTEM condenser system may not be able to handle the large angles (typically 100 to 300 mrad at 100 keV) equivalent to the angles in the ADF-detector. Alternately, an image similar to an ADF-STEM image but with reversed contrast may be obtained using BF-CTEM with a large solid cone that is mirror image of hollow cone illumination (Kirkland [204]). The simplified image model discussed in this section can be referred to as the incoherent image model.

A focused probe is calculated by integrating the aberration wave function $\exp[-i\chi(k)]$ over the objective aperture with translation to a particular point in the image. Figure 3.9 shows the relative placement of the wave functions in the STEM column. The aberrated electron probe wave function in the plane of the specimen when deflected to position $\mathbf{x}_p$ is:

$$\psi_p(\mathbf{x},\mathbf{x}_p) = A_p \int_0^{k_{\max}} \exp[-i\chi(\mathbf{k}) - 2\pi i\mathbf{k}\cdot(\mathbf{x}-\mathbf{x}_p)]\mathrm{d}^2\mathbf{k}, \tag{3.59}$$

where $\lambda k_{\max} = \alpha_{\max}$ is the maximum angle in the objective aperture and A_p is a normalization constant chosen to yield;

$$\int |\psi_p(\mathbf{x},\mathbf{x}_p)|^2 \mathrm{d}^2\mathbf{x} = 1. \tag{3.60}$$

With this normalization the total incident intensity in the electron probe is also unity (alternately the probe integral could be scaled to yield the actual value of the beam current in some appropriate choice of units). The probe is really just the demagnified image of the source and this expression assumes that the electron source (the electron gun demagnified by the condenser system) has a negligible size in the plane of the specimen (discussed later in Sect. 3.5.2). The probe size is limited instead by the aberrations of the objective lens. A dedicated STEM would usually use a high brightness field emission gun with a small virtual source size so that there is enough current left after demagnification to get a large enough signal to detect. Deflecting the beam to different positions on the specimen changes the angles through the objective lens. At high resolution this angle is of order 100 Å divided by the focal length of the objective lens (of order 1 mm) producing a deflection angle of order 0.01 mrad. which is a negligible compared to a typical angle of 10 mrad. in the objective aperture (large deflection angles would change the apparent aberrations but only occur at low magnification).

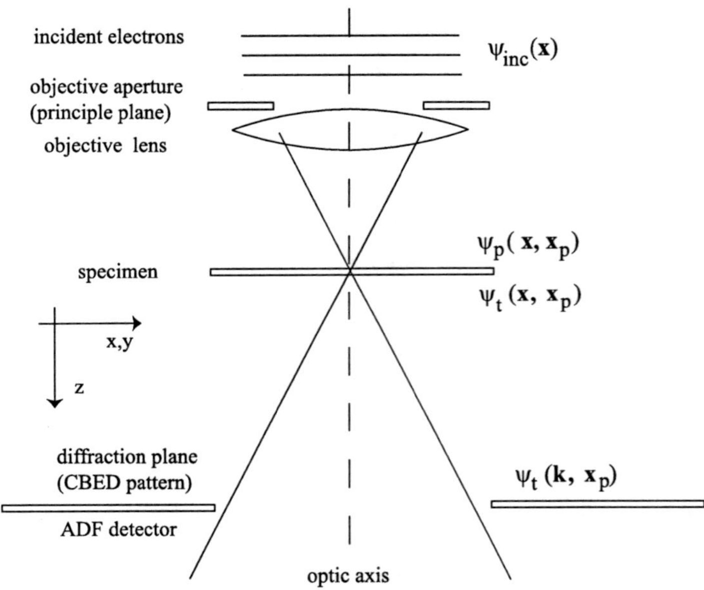

Fig. 3.9 Electron wave functions in the STEM column

The electron probe passes through the specimen and is modulated by the specimen transmission function $t(\mathbf{x})$ (identical to the transmission function for the CTEM). $t(\mathbf{x})$ is in general a complex valued function for a weak phase, weak amplitude object [see (3.47)]. The transmitted wave function is:

$$\psi_t(\mathbf{x}, \mathbf{x}_p) = t(\mathbf{x})\psi_p(\mathbf{x}, \mathbf{x}_p). \tag{3.61}$$

Again a discussion of the effects of specimen thickness on the transmission function will be deferred to later chapters. This wave function is then diffracted onto the detector plane (represented by a Fourier transform).

$$\Psi_t(\mathbf{k}, \mathbf{x}_p) = \mathrm{FT}[\psi_t(\mathbf{x}, \mathbf{x}_p)] = \int \exp(2\pi i\mathbf{k}\cdot\mathbf{x})\psi_t(\mathbf{x}, \mathbf{x}_p)\mathrm{d}^2\mathbf{x}. \tag{3.62}$$

The intensity of this wave function $|\Psi_t(\mathbf{k}, \mathbf{x}_p)|^2$ as a function of scattering angle $\lambda \mathbf{k}$ is the Convergent Beam Electron Diffraction CBED pattern.

The CBED pattern is incoherently integrated over the detector geometry and the result is the final STEM image signal $g(\mathbf{x_p})$ for one probe position $\mathbf{x_p}$.

$$g(\mathbf{x}_p) = \int |\Psi_t(\mathbf{k}, \mathbf{x}_p)|^2 D(\mathbf{k}) \mathrm{d}^2\mathbf{k}, \tag{3.63}$$

where $D(\mathbf{k})$ is the detector function.

$$D(\mathbf{k}) = 1 \quad \text{for } k_{D\min} \leq k \leq k_{D\max} \tag{3.64}$$

$$= 0 \text{ otherwise}, \tag{3.65}$$

where $\lambda k_{D\min}$ and $\lambda k_{D\max}$ are the inner and outer angles of the ADF detector. This process is repeated for each probe position $\mathbf{x}_p$.

Equation (3.63) is difficult to intuitively relate to any specific structure in the specimen (later chapters will treat this equation more exactly using numerical calculations). However with some approximations a linear image model can be derived that is easier to understand. The ADF detector should go to very large angles (of order 300 mrad or more at 100 keV) so that all electrons that are incident on the specimen either fall on the ADF detector or go through the central hole in the ADF detector. This means that the signal formed by integrating all of the electron in the central hole in the ADF detector is just one minus the ADF signal (where the total incident intensity is assumed to be unity). This effective signal from the central hole in the ADF detector is also equivalent to the BF-CTEM signal (via the reciprocity theorem) with a very large condenser illumination angle $\beta_{\max}/\alpha_{\max} >> 1$. Therefore, the ADF image process is incoherent if $k_{D\min} >> k_{\max}$ (see Sect. 3.4). If the outer dimension of the ADF detector is essentially infinite then the large diameter of the central hole in the ADF detector produces an incoherent image. This means that phase contrast should be negligible and the image should be predominately amplitude contrast. Simple imaging approximations for ADF-STEM have been considered by Misell et al. [242], Cowley [60], Spence [332], Jesson and Pennycook [187], Treacy and Gibson [347], and Loane et al. [228].

An approximate linear image model for thin specimens assuming an incoherent image process is:

$$g(\mathbf{x}) = f(\mathbf{x}) \otimes h_{\mathrm{ADF}}(\mathbf{x}), \tag{3.66}$$

where the specimen function $f(\mathbf{x})$ is approximately the probability for scattering to the large angles of the ADF detector.

$$f(\mathbf{x}) \sim \int D(\mathbf{k}) \frac{\partial \sigma(\mathbf{x})}{\partial k_s} \mathrm{d}^2 k_s = \int_{k_{D\min}}^{k_{D\max}} \frac{\partial \sigma(\mathbf{x})}{\partial k_s} \mathrm{d}^2 k_s \tag{3.67}$$

$\partial\sigma/\partial k_s$ is the partial cross section (the square of the scattering factor) for scattering to angle k_s at position $\mathbf{x}$ of the specimen. With the incoherent image assumption an ADF-STEM image of a very thin specimen is essentially a mass thickness map of the specimen. The calculation with this approximation is simple enough to perform interactively inside a web browser and is useful to quickly build intuition on the imaging process although it may not be quantitatively accurate. This simple incoherent image model also is close to the assumed image models for a variety of sophisticated image restoration algorithms such as the Richardson-Lucy [229, 302] and maximum entropy methods (for example [165]) so these methods can then be easily applied to ADF-STEM images (for example Kirkland [201]).

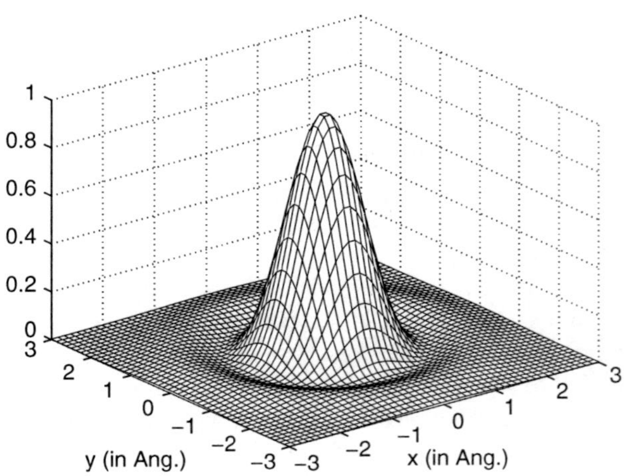

Fig. 3.10 The STEM probe intensity (approximate point spread function) when astigmatism and source size are negligible vs. position (200 keV, $C_s = 1.3$ mm, $\Delta f = 571$Å, $\alpha_{\max} = 9.4$ mrad)

The point spread function (with the assumption of an incoherent image model) is just the intensity distribution in the focused probe. The aberration limited probe is:

$$\begin{aligned} h_{\mathrm{ADF}}(\mathbf{x}) &= |\psi_p(\mathbf{x})|^2 \\ &= A_p \left| \int_0^{k_{\max}} \exp[-\mathrm{i}\chi(\mathbf{k}) - 2\pi \mathrm{i}\mathbf{k}\cdot\mathbf{x}]\mathrm{d}^2\mathbf{k} \right|^2, \end{aligned} \tag{3.68}$$

The constant A_p is chosen to normalize the point spread function to have a total integrated value of unity. Furthermore, if the astigmatism is negligible the azimuthal integral can be done analytically leaving a one dimensional integral for the probe intensity:

$$h_{\mathrm{ADF}}(r) = A_p \left| \int_0^{k_{\max}} \exp[-\mathrm{i}\chi(k)] J_0(2\pi kr) k\mathrm{d}k \right|^2, \tag{3.69}$$

where $J_0(x)$ is the zeroth order Bessel function of the first kind and r is the radial coordinate. $\chi(k)$ is a function of only the magnitude of k when astigmatism is not present. This integral cannot be done analytically and must be done numerically. The point spread function is plotted in Fig. 3.10 vs. position (on the specimen). Figure 3.11 shows a more condensed form of this graph for various values of defocus

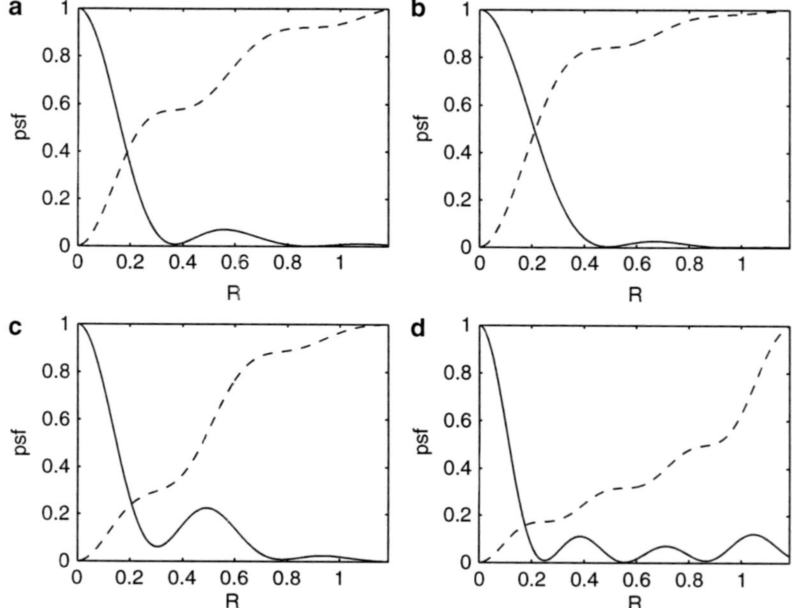

Fig. 3.11 The STEM probe intensity or approximate point spread function (*solid line*) when astigmatism and source size are negligible vs. normalized radius $R = r(C_s\lambda^3)^{-1/4}$ for various values of the normalized defocus $D = \Delta f(C_s\lambda)^{-1/2}$ and objective aperture $K_{\max} = k_{\max}(C_s\lambda^3)^{1/4}$. (a) $D = 1.2$, $K_{\max} = 1.56$ (Scherzer conditions), (b) $D = 0.80$, $K_{\max} = 1.22$, (c) $D = 1.5$, $K_{\max} = 1.5$, (d) $D = 2.5$, $K_{\max} = 2.5$. The total integrated current is shown as a *dashed line*

and objective aperture. The graph has been scaled to make $h_{\mathrm{ADF}}(0) = 1$. Figure 3.11a shows the probe profile for Scherzer defocus and aperture with a full width at half max of approximately $0.43(C_s\lambda^3)^{1/4}$.

The transfer function is just the inverse Fourier transform of the point spread function. With azimuthal symmetry (i.e., no astigmatism), and neglecting the source size the transfer function is:

$$H_{\mathrm{ADF}}(k) = A_p \left| \int_0^{\infty} h_{\mathrm{ADF}}(r) J_0(2\pi kr) r\mathrm{d}r \right|^2 \tag{3.70}$$

The transfer function is plotted in Fig. 3.12 and BF-CTEM and ADF-STEM are compared in Fig. 3.13.

It is interesting to compare an aberration corrected probe and an uncorrected probe [232]. Figure 3.14 shows a probe at 100 keV with and without an aberration corrector. The curves are normalized to have the same integrated current. It is very surprising that the corrected probe is dramatically higher than the uncorrected probe. The tails of the probe at large radius have a very large contribution to the total current even though they appear small in this type of graph. An aberration corrector can increase the image contrast a lot more than might be expected.

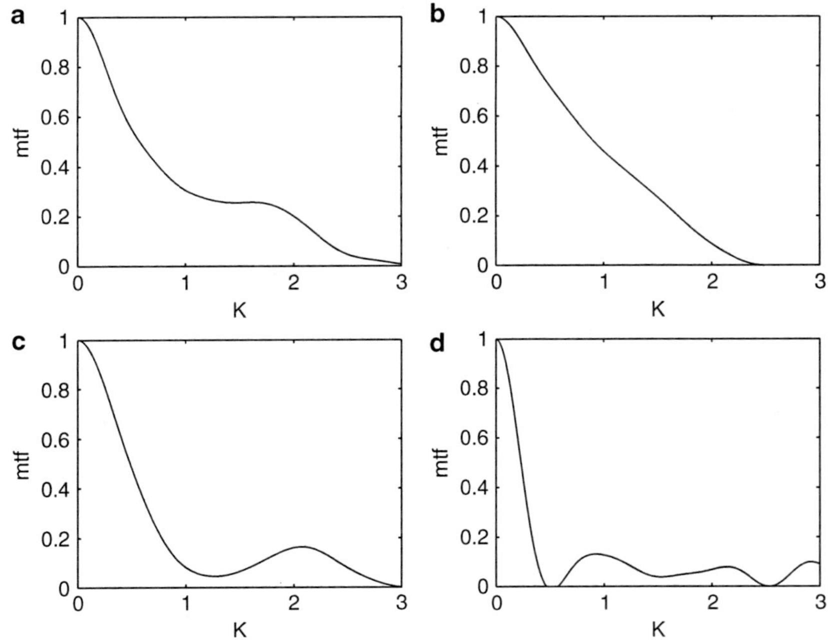

Fig. 3.12 The approximate STEM transfer function corresponding to the defocus values and objective apertures (a) $D = 1.2$, $K_{max} = 1.56$ (Scherzer conditions), (b) $D = 0.80$, $K_{max} = 1.22$, (c) $D = 1.5$, $K_{max} = 1.5$, (d) $D = 2.5$, $K_{max} = 2.5$ vs. the normalized spatial frequency $K = k(C_s\lambda^3)^{1/4}$. Source size is assumed negligible

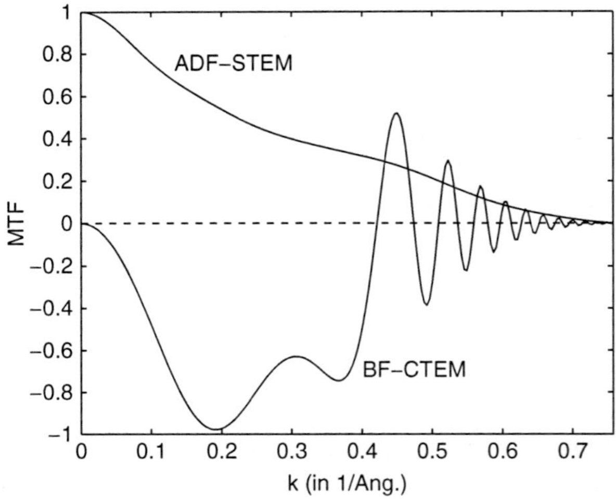

Fig. 3.13 Comparison of ADF-STEM and BF-CTEM transfer functions for the same spherical aberration, $C_s = 1.2$ mm at 200 keV. $\Delta f = 550$ Å and $\alpha_{max} = 9.5$ mrad for STEM and $\Delta f = 670$ Å for BF-CTEM with a condenser half angle of 0.1 mrad and a defocus spread of 100Å

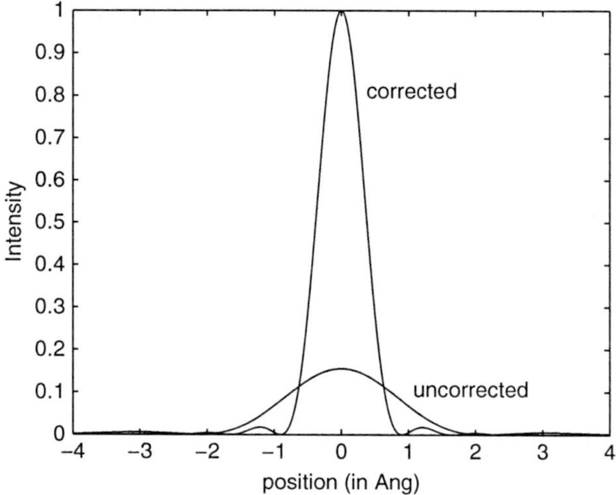

Fig. 3.14 Comparison of ADF-STEM probe with and without an aberration corrector at 100 keV. The uncorrected probe has $C_s = 1.3$ mm, $\Delta f = 694$ Å and $\alpha_{\text{max}} = 10.3$ mrad. The corrected probe has $C_s = \Delta f = 0$ and $\alpha_{\text{max}} = 25$ mrad, Both curves are normalized to have the same total (integrated) current

3.5.1 Minimum Probe Conditions

It is difficult to define an optimum defocus and aperture to produce the best probe for the highest resolution. In Fig. 3.11d the full width at half maximum (FWHM) is clearly much smaller than the FWHM for Scherzer conditions (Fig. 3.11a), however the tails (small wiggles at large radius) are dramatically increased in size. A small value at large radius can produce a large signal when integrated over a larger circumference. The total current inside a radius r in the probe is:

$$I(r) = 2\pi \int_0^r |\psi_p(\mathbf{x}')|^2 r' \mathrm{d}r' \tag{3.71}$$

This current is shown in Fig. 3.11 as a dashed line. The extra factor of r inside the integral (3.71) can shift the signal to a larger radius in a significant manner. The radius containing half of the current is a reasonable definition of the probe size (many other definitions are possible). This FWHM of the integrated intensity is a good measure of where the signal comes from.

A plot of this FWHM radius vs. both defocus and objective aperture size is shown in Fig. 3.15. The minimum probe rms radius appears to be at defocus $D = 0.87$ and objective aperture $K_{\text{max}} = 1.34$ (Fig. 3.11b). This produces the smallest tails but increases the FWHM (minimum rms radius approx. $0.43(C_s\lambda^3)^{1/4}$) to about twice that of the probe with Scherzer conditions. Scherzer conditions seem to be a compromise between a small full width half maximum and large tails. Mory et al. [254] have also considered the optimum probe defocus for STEM imaging and microanalysis. Intaraprasonk et al. [171] have discussed optimizing the probe with spherical aberration through fifth order. There is also a long history of balancing aberrations in incoherent light optics similar to the treatment here, starting with Maréchal [234] (see also Sect. 4.4 of O'Neil [275] for balancing 3rd and 5th order spherical aberrations) and Black and Linfoot [30] and including annular objective apertures (for example Barakt and Houston [18]).

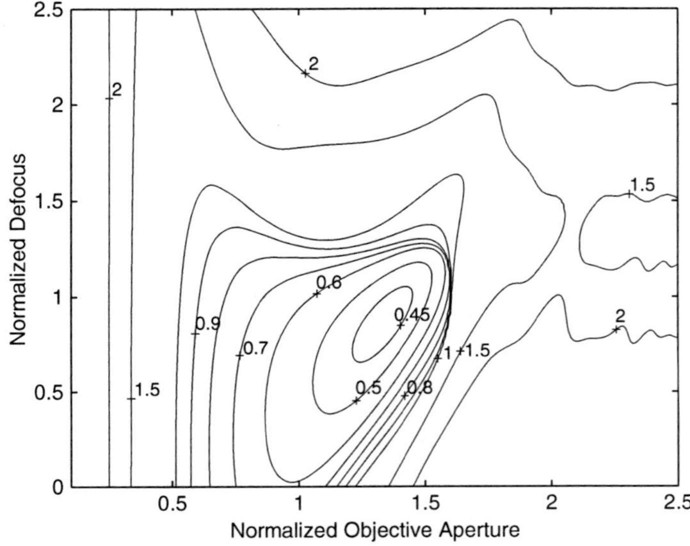

Fig. 3.15 The normalized rms radius $r_{\mathrm{rms}}(C_s\lambda^3)^{-1/4}$ of the STEM probe as a function of the normalized objective aperture $k_{\max}(C_s\lambda^3)^{1/4}$ and the normalized defocus $\Delta f(C_s\lambda)^{-1/2}$ with a large $C_S = C_{S3}$ (no aberration corrector)

3.5.2 Source Size

The probe is just the image of the electron source, which can also contribute to the probe size. The brightness of a source is defined as:

$$\beta = \frac{j}{\pi\alpha^2} \tag{3.72}$$

where j is the current density in the probe and α is the convergence half angle ($\pi\alpha^2$ is the solid angle). Brightness is conserved in a magnetic lens but may vary with beam energy. Various condenser lenses and the objective lens are used to demagnify the image of the source onto the specimen. More source demagnification produces less current (and a smaller source contribution to the probe size) in a predictable manner. If the probe is approximated as a disk of diameter d_S (the source size) and uniform intensity then the current in the probe $I_P = j(\pi d_S^2/4)$ and the probe size are related as:

$$I_P = \frac{1}{4}\pi^2\alpha^2\beta d_S^2. \tag{3.73}$$

The probe current is shown in Fig. 3.16 for two different types of electron sources. A probe current of about 10–100 pAmp is needed for practical imaging. Only a field emission source can produce enough current to produce a source size of about 1 Angstrom (or smaller) and is preferred. This simple approximation is a convenient means of estimating the source contribution from just a measurement of the probe current and aperture size although it may not be that accurate.

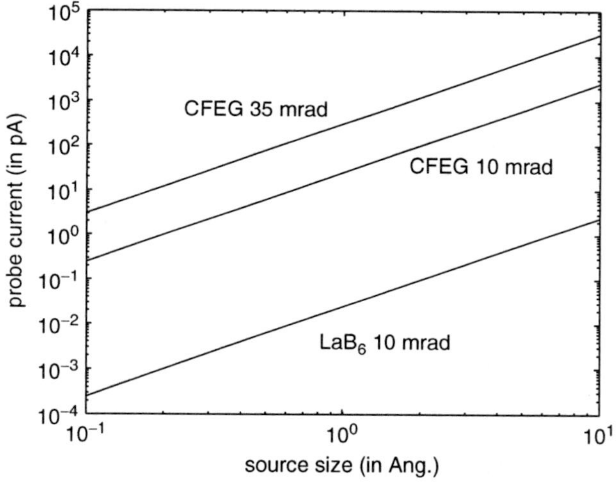

Fig. 3.16 Approximate probe current vs. source size from brightness for a cold field emitter (CFEG) with brightness $\beta = 10^9$ amp/(cm^{2}sr) for two different aperture sizes and a typical LaB_6 source with $\beta = 10^6$ amp/(cm^{2}sr)

Each part of the electron source can emit electrons that travel through the microscope and form an image of their own in some way. Each of these images is offset in position equivalent to the shift in position on the source demagnified by the lenses in the microscope. This can be summarized as a further convolution with an effective source size in the specimen plane (including appropriate demagnification).

$$g(\mathbf{x}) = f(\mathbf{x}) \otimes h_{\mathrm{ADF}}(\mathbf{x}) \otimes h_{\mathrm{source}}(\mathbf{x}) \tag{3.74}$$

The source contribution is typically taken as a Gaussian, and may become the dominate factor at high resolution. If there are other nondirectional instabilities in the microscope then some of these may be treated in a similar manner. For example if the stage has some small random vibrations convolving the final image with a small Gaussian may be an appropriate way to model this effect (stage vibrations likely have a preferred direction so the convolution kernel should match this asymmetry).

3.5.3 Defocus Spread

As in the CTEM, small fluctuations in the accelerating voltage, lens current and the thermal energy spread in the source itself produce a small spread in defocus values. The ADF-STEM transfer function does not vary as dramatically with defocus as the BF-CTEM transfer function so the effect of defocus spread is small on the ADF-STEM transfer function and can frequently be ignored. In a simple approximation the ADF-STEM transfer function (3.70) can be integrated over a small range of defocus values to approx. the effects of these small fluctuations, leaving an effective transfer function:

$$H_{\mathrm{ADF}}^{\mathrm{eff}}(k) = \int_{-\infty}^{\infty} H_{\mathrm{ADF}}(k,\Delta f)p(\Delta f)d(\Delta f), \tag{3.75}$$

where $p(\Delta f)$ is the probability distribution of defocus values. This integration may also be done in real space in this linear imaging approximation. Typically $p(\Delta f)$ is a Gaussian distribution about its mean value, which is easily performed numerically using a Gauss-Hermite quadrature formula (5–9 points are probably sufficient in most cases), which conveniently includes a Gaussian weighting of the integrand (for example Sect. 4.6 of Press et al. [289], or Chap. 25 of Abramowitz and Stegun [1]). Sheppard and Wilson [319] have considered partial coherence in scanning microscopy in a more general manner.

3.6 Confocal Mode for Weak Phase Objects

The light optical scanning transmission microscope (in confocal mode) has exhibited improved performance relative to a conventional microscope, but what happens in the electron microscope? A simple imaging theory is similar to that of the light optical microscope (Wilson and Sheppard [383]). Nellist et al. [262] have also described the theory of imaging in a double corrected instrument.

Confocal mode is a combination of STEM and CTEM (Fig. 2.8). A focused electron probe is raster scanned across the specimen (much like a STEM). The electrons transmitted through the specimen are imaged by a collector lens (much like CTEM) onto a detector (usually a small point like detectors).

The incident probe focused onto the specimen has a complex wavefunction as given by (3.59). The wave transmitted through the specimen is given by (3.61). The probe is scanned across the specimen, and in confocal mode there is also a collection lens that images the transmitted probe onto a detector. The collection lens adds a second complex points spread function. If the probe is scanned (moved) across the specimen the transmitted beam must be inversely scanned so that there is no net motion on the detector. Alternately the electron beam may be fixed and the specimen moved in a raster fashion. The theory is slightly less complicated in this alternate mode (used here). The transmitted wave function is:

$$\psi_T(\mathbf{x},\mathbf{x}_{\mathrm{P}}) = h_{\mathrm{P}}(\mathbf{x})t(\mathbf{x}+\mathbf{x}_{\mathrm{P}}), \tag{3.76}$$

where $\mathbf{x}$ is position in the specimen plane and $\mathbf{x}_\mathrm{P}$ is the position of the specimen (or probe). $t(\mathbf{x})$ is the complex transmission function of the specimen, and $h_\mathrm{P}(\mathbf{x}) = \psi_P(\mathbf{x})$ is the point spread function of the probe forming lens (3.59). The wave function incident on the detector is:

$$\begin{aligned}\psi_D(\mathbf{x},\mathbf{x}_\mathrm{P}) &= h_\mathrm{C}(\mathbf{x}) \otimes [h_\mathrm{P}(\mathbf{x})t(\mathbf{x}+\mathbf{x}_\mathrm{P})] \\ &= \int h_\mathrm{C}(\mathbf{x}')h_\mathrm{P}(\mathbf{x}'-\mathbf{x})t(\mathbf{x}'-\mathbf{x}-\mathbf{x}_\mathrm{P})\mathrm{d}^2\mathbf{x}' \qquad (3.77)\end{aligned}$$

where $h_\mathrm{C}(\mathbf{x})$ is the point spread function of the collector lens. The subscript P or C refer to the probe or collector lens, respectively. If the detector is a point at position $\mathbf{x} = 0$ then:

$$\begin{aligned}\psi_D(\mathbf{x}=0,\mathbf{x}_\mathrm{P}) &= \int h_\mathrm{C}(\mathbf{x}')h_\mathrm{P}(\mathbf{x}')t(\mathbf{x}'-\mathbf{x}_\mathrm{P})\mathrm{d}^2\mathbf{x}' \\ &= [h_\mathrm{C}(\mathbf{x})h_\mathrm{P}(\mathbf{x})] \otimes t(\mathbf{x}). \qquad (3.78)\end{aligned}$$

In practice the detector must have a nonzero size so this is an unphysical approximation to simplify the mathematics. The transmitted intensity for a given probe position with this approximation is therefore:

$$g(\mathbf{x}) = \left|[h_p(\mathbf{x})h_c(\mathbf{x})] \otimes t(\mathbf{x})\right|^2, \qquad (3.79)$$

where the distinction between $\mathbf{x}$ and $\mathbf{x}_\mathrm{P}$ has been dropped for simplicity.

A transfer function is only defined for a linear system but this is still rather nonlinear. Approximate the transmission function of the specimen as a weak-phase object as in (3.6) to obtain a simplified linear theory:

$$t(\mathbf{x}) = \exp[\mathrm{i}\sigma v_z(\mathbf{x})] \sim 1 + \mathrm{i}\sigma v_z(\mathbf{x}) \qquad (3.80)$$

or weak-amplitude object (3.47):

$$t(\mathbf{x}) = \exp[u(\mathbf{x})] \sim 1 + u(\mathbf{x}) \qquad (3.81)$$

Using the weak phase object approximation, and keeping only lowest terms in v_z, the image model becomes:

$$\begin{aligned}g(\mathbf{x}) &\sim \left|[1+\mathrm{i}\sigma v_z(\mathbf{x})] \otimes (h_p(\mathbf{x})h_c(\mathbf{x}))\right|^2 &(3.82)\\ &\sim 1 \otimes (h_p h_c) + 2Re[i\sigma v_z \otimes (h_p h_c)] &(3.83)\\ &\sim 1 + 2\sigma v_z \otimes h_\mathrm{WPA} &(3.84)\\ h_\mathrm{WPA}(\mathbf{x}) &= Re[ih_p(\mathbf{x})h_c(\mathbf{x})]. &(3.85)\end{aligned}$$

The transfer function is just the 2D Fourier transform of h_{WPA}. Using azimuthal symmetry the transfer function for weak phase objects in confocal mode is:

$$H_{\mathrm{WPA}}(k) = \int_0^{r_{\mathrm{max}}} \mathrm{Re}[ih_p(r)h_c(r)]J_0(2\pi kr)r\mathrm{d}r \tag{3.86}$$

Remember that both $h_p(\mathbf{x})$ and $h_c(\mathbf{x})$ are complex valued functions. To repeat this derivation in the weak amplitude approximation just remove the factor of i, to obtain the transfer function for weak amplitude objects:

$$H_{\mathrm{WAA}}(k) = \int_0^{r_{\mathrm{max}}} \mathrm{Re}[h_p(r)h_c(r)]J_0(2\pi kr)r\mathrm{d}r \tag{3.87}$$

Confocal mode has twice as many optical parameters and can produce a larger variety of features in the transfer function. Some examples are shown in Fig. 3.17 for the same set of parameters as ADF-STEM in Fig. 3.12. The confocal transfer function is compared to the BF-CTEM transfer function (both in the weak phase approximation) in Fig. 3.18. This particular choice of parameters produces a transfer function similar to ADF-STEM except that it is reversed in contrast (negative instead of positive). There is most likely a difference in the signal strength as well which is not apparent in this representation.

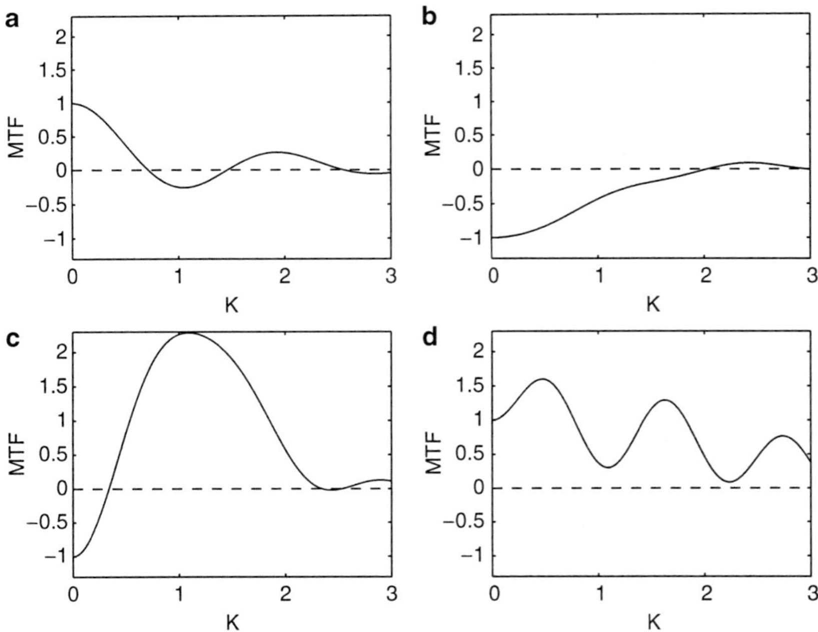

Fig. 3.17 Confocal transfer functions in the weak phase approximation (3.86) for the same aberrations as ADF-STEM (a) $D = 1.2$, $K_{\mathrm{max}} = 1.56$ (Scherzer conditions), (b) $D = 0.80$, $K_{\mathrm{max}} = 1.22$, (c) $D = 1.5$, $K_{\mathrm{max}} = 1.5$, (d) $D = 2.5$, $K_{\mathrm{max}} = 2.5$. Both the probe and collectors lens aberration are the same in each graph

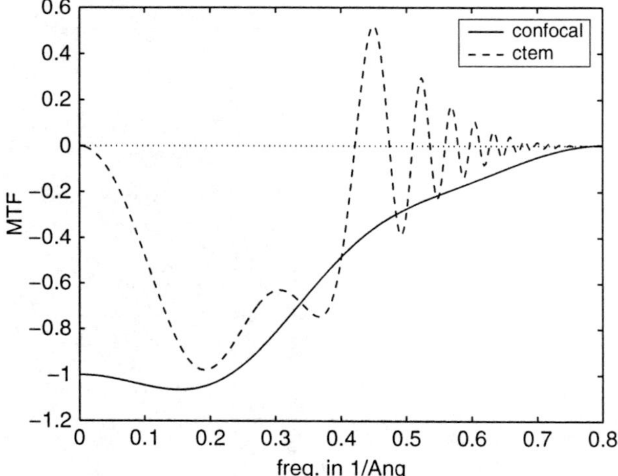

Fig. 3.18 Comparison of confocal and CTEM transfer functions in the weak phase approximation for the same spherical aberration, $C_s = 1.2$ mm, at 200 keV. $\Delta f = 550$ Å and $\alpha_{\max} = 9.5$ mrad for confocal. Both the probe and collectors lens aberration in confocal were the same. $\Delta f = 670$ Å for BF-CTEM with a condenser half angle of 0.1 mrad and a defocus spread of 100Å (Scherzer conditions)

3.7 Phase and Amplitude Contrast Revisited

The theory presented earlier is a traditional view and seems plausible, but it is worth checking a little. Figure 3.19 shows how a pure amplitude and pure phase object (combined into a single image) are imaged in BF-CTEM and ADF-STEM, using methods that will be discussed later. The letters "AMP." are a amplitude object and the letters "PHASE" are a phase object. The image transmission function is given by:

$$t(\mathbf{x}) = \exp(\mathrm{i}p - a), \tag{3.88}$$

where a and p are the amplitude and phase components (respect.) with a value of 0.05 on the appropriate letters and zero otherwise. The BF-CTEM and the ADF-STEM images contain both amplitude and phase contrast features. It is more the specimen that determines which form of contrast will be in the image. The ADF-STEM image only renders the high frequency components from the edges (scattered at high angle onto the ADF detectors). However, every specimen is composed of many small atoms, each of which scatters to high angle so this edge effect is not in a real specimen composed of many individual atoms (sharp edges, or points are everywhere). The simple theory is very helpful in developing an intuitive understanding of electron microscope image but there is still a need for more detailed simulations to more completely understand what the images mean. The property of coherence and incoherence also plays a large role in image formation.

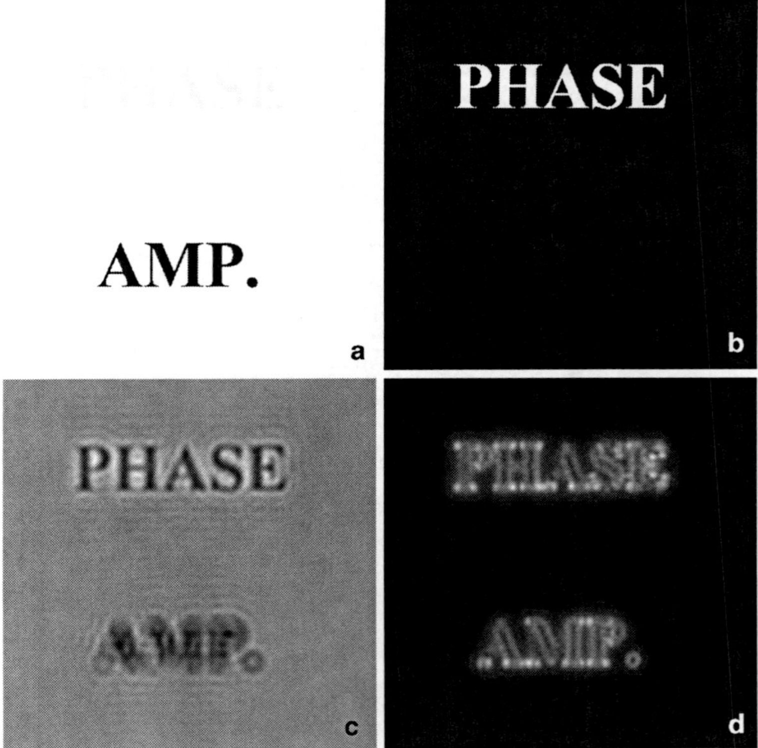

Fig. 3.19 Phase and amplitude contrast in BF-CTEM and ADF-STEM. (**a**) Pure amplitude object, (**b**) pure phase object. The object was formed from the superposition of (**a**) and (**b**) and imaged as a (**c**) BF-CTEM image and (**d**) ADF-STEM image

Chapter 4
Sampling and the Fast Fourier Transform

Abstract This chapter takes a small detour and discusses some numerical approximations that will be necessary to theoretically compute electron microscopy images. Specifically, digital sampling (pixels and levels) and the all important fast Fourier transform are introduced. The FFT will be the principle tool to speed up later calculation. If you are familiar with these topics, this chapter may be skipped.

To go further with image calculations requires a detailed numerical calculation using a computer program. The mathematics gets too complicated (or long) to perform analytically with pencil and paper. This chapter gives some necessary computer background prior to calculating images in later chapters. The computer imposes its own set of rules that must be understood and dealt with to perform these calculation. Experienced computer user may prefer to skip this chapter.

Image simulation or image processing with the computer presumes that the image is somehow represented inside the computer. A digital computer naturally operates on numerical data. Therefore, an image must be represented as a two dimensional array of numbers inside the computer. Each number is one pixel or spot in the image whose intensity is proportional to its numerical value. The trick is to have a sufficiently large number of pixels so that when they are displayed as an image the individual numbers or pixels are not individually distinguishable. Sampling the image in this manner leads to some specific rules and limitation that are summarized in this chapter.

The fast Fourier transform or FFT is one of the most efficient computer algorithms available. The FFT computes the Fourier transform of discretely sampled data in a minimum amount of computer time. Image simulation, such as the multislice method, are usually organized around the FFT to reduce the computer time required for simulations. The mechanism of the FFT is closely coupled to discretely sampled data and is also included in this chapter.

E.J. Kirkland, *Advanced Computing in Electron Microscopy*,
DOI 10.1007/978-1-4419-6533-2_4, © Springer Science+Business Media, LLC 2010

4.1 Sampling

In practice each image is calculated in a rectangular grid of $N_x \times N_y$ pixels or picture elements as shown in Fig. 4.1. The images are sampled at N_x discrete points along x and N_y discrete points along y and form a supercell with dimensions of $a \times b$ in real space. Figure 4.2 shows the visual effects of changing the number of pixels in an image for the special case of $N_x = N_y$ and $a = b$. Each pixel has dimensions $a/N_x \times b/N_y$ in real space and has a single value associated with it that is the average over the area of the pixel (or in the case of a complex wave function two values representing the real and imaginary parts). The real space coordinates take on only discrete values of:

$$x = i\Delta x \quad i = 0, 1, 2, \cdots, (N_x - 1) \tag{4.1}$$

$$y = j\Delta y \quad j = 0, 1, 2, \cdots, (N_y - 1), \tag{4.2}$$

where $\Delta x = a/N_x$ and $\Delta y = b/N_y$.

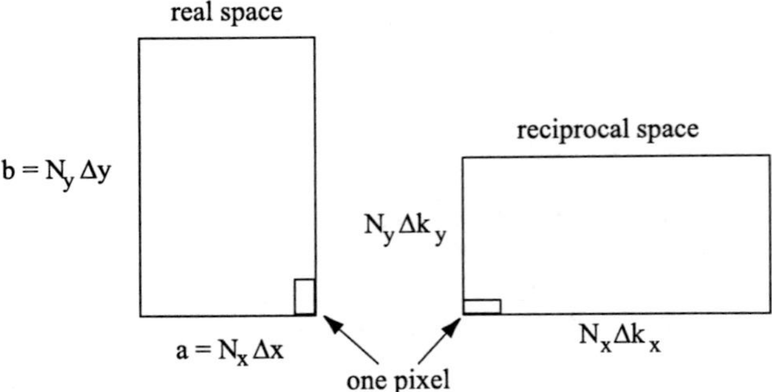

Fig. 4.1 Sampling of an image of size $a \times b$ in a plane perpendicular to the optic axis of the microscope. There are $N_x \times N_y$ pixels with a size of $a/N_x \times b/N_y$ in real space and $1/a \times 1/b$ in reciprocal or Fourier transform space. Neither the pixels or the image have to be *square*

The Fourier transform of this image will also have $N_x \times N_y$ pixels but the dimensions of each pixels change to $1/a \times 1/b$ and the reciprocal space coordinates take on values:

$$k_x = i\Delta k_x \quad i = 0, 1, 2, \cdots, (N_x - 1) \tag{4.3}$$

$$k_y = j\Delta k_y \quad j = 0, 1, 2, \cdots, (N_y - 1), \tag{4.4}$$

where $\Delta k_x = 1/a$ and $\Delta k_y = 1/b$. The supercell does not have to be square and there may be a different number of pixels in x and y although this is usually less efficient. It is interesting to note that the longest dimension of the supercell will be reversed in real and reciprocal space (i.e., a tall narrow supercell in real space becomes short and wide in reciprocal space).

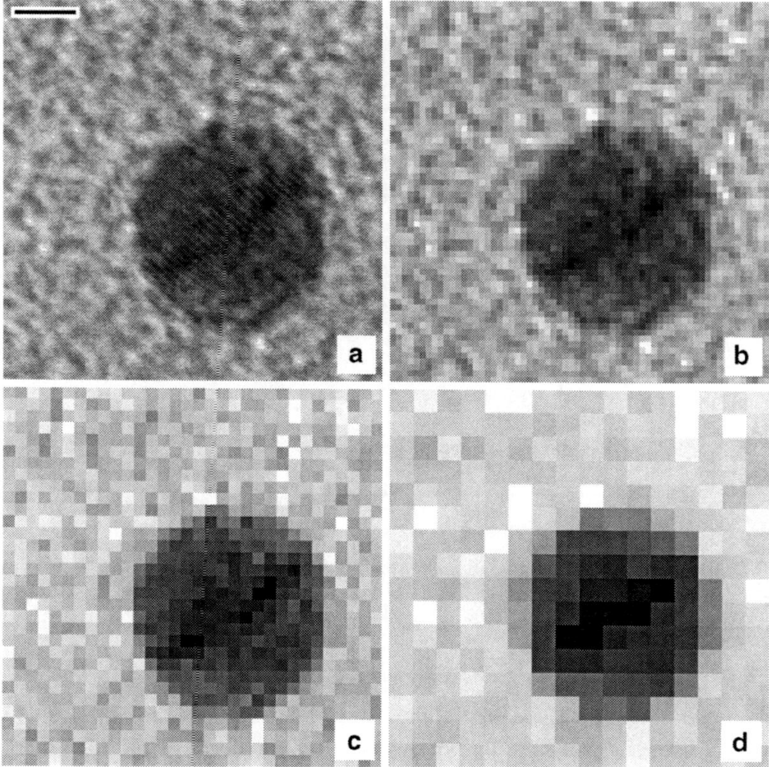

Fig. 4.2 The visual effects of changing the number of pixel in an image. (a) 256×256 pixels, (b) 64×64 pixels, (c) 32×32 pixels, and (d) 16×16 pixels. Each pixel has eight bits. The image is a bright field STEM image of a *gold particle* on an amorphous carbon film recorded on a VG HB-501 STEM ($C_s = 1.3$ mm, 100 keV). The 2.35Å gold lattice fringes are barely visible in the center of the particle. The scale bar in (a) is approximately 20Å

Discrete sampling also imposes a limit on the maximum spatial frequency in the image of:

$$|k_x| < \frac{1}{2\Delta x}$$
$$|k_y| < \frac{1}{2\Delta y} \tag{4.5}$$

This is referred to as the Nyquist limit. In principle this limit may be different in each direction, however in practice the larger limit should be reduced to the smaller so that the effective resolution is isotropic in the image (i.e., to avoid sampling artifacts in the image). If the sampling size Δx or Δy is too large then the signal is under sampled and aliasing occurs. Figure 4.3 shows the effect of under sampling a sine wave. The high frequency sine wave appears to be a low frequency sine wave if it is under sampled. This should be avoided by making the sampling size smaller or explicitly limiting the bandwidth of the original signal.

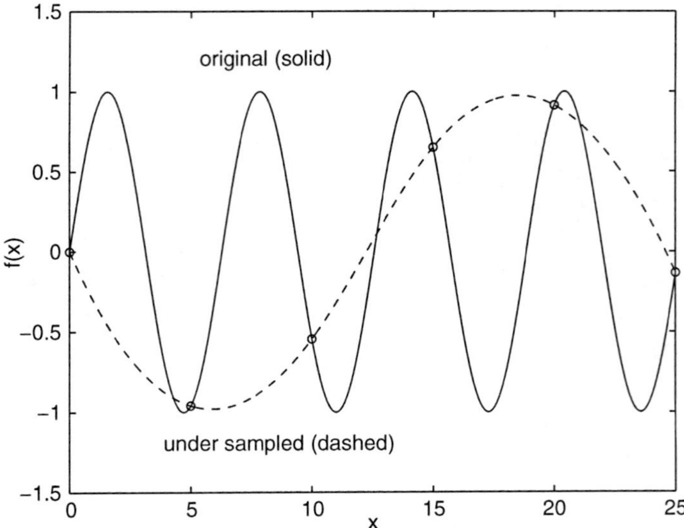

Fig. 4.3 The affect of under sampling a function. The original function is shown as a *solid line*. It is under sampled at regular intervals (*open circles*). The apparent signal is shown as a *dashed line*. Under sampling causes high frequencies to be improperly aliased as low frequencies

The numerical value associated with each pixel is itself discretely sampled. This value must be encoded as a finite number of bits (one bit has a value of 0 or 1) for each pixel in the computer. The human eye can distinguish at best about 30–50 different shades of grey. Figure 4.4 shows the effects on human perception of varying the number of bits in each pixel. The computer hardware is usually capable of conveniently handling data in units as small as 8 bits at a time. Therefore, about 8 bits of information is needed for each pixel to display the image. Any substantial amount of calculation with only 8 bits per pixels will however quickly get into a lot of trouble. Rounding each value to 8 bits introduces an error of at least one in $2^8 = 256$. Even worse, dividing two eight bit integer numbers truncates the result to the lowest integer (for example $1/2 = 0$, $128/50 = 2$, with integer arithmetic) which can introduce very large errors for each arithmetic operation between pixels. An image simulation may require thousands or millions of operations on each pixel. Therefore, during image simulation each pixel should be represented as a floating point number with much more than 8 bits per pixel. Most computer hardware is equipped to handle 32 bit single precision (typically there is one sign bit, 8 exponent bits and 23 mantissa bits) and 64 bit double precision floating point arithmetic. Single precision (32 bits) gives about six decimal digits of accuracy per pixel and is usually sufficient for most image calculations. Each arithmetic operation between two single precision floating point numbers can be thought of as adding an error of about ± 1 in 10^6 or 1 in 6 digits. This error is sometimes referred to as round-off error, and calculations with a finite number of bits is sometimes referred to as finite precision arithmetic. Although this error may seem insignificant it may be necessary to perform a million operations on some numbers so the errors can add up to be significant even with single precision (32 bit) floating point arithmetic. A good computer program should be organized in such a way as to minimize the effects of round-off error. Storing several images of sizes of 512×512 pixels or more requires a lot of memory so double precision is usually not used to keep the memory requirements to a reasonable level. During numerical simulation each pixels should be stored as a 32 bit (or more) floating point number and for the final displayed result 8 bits is probably sufficient.

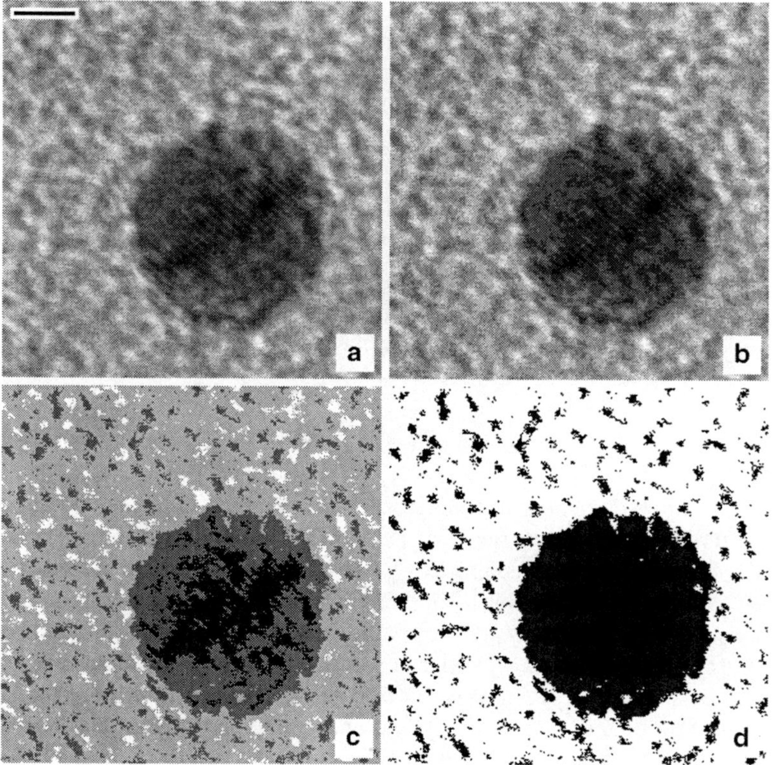

Fig. 4.4 The effect of limiting the number of bits in each pixel. (a) 4 bits/pixel, (b) 3 bits/pixel, (c) 2 bits/pixel, and (d) 1 bits/pixel. Each image was normalized to fill the available greyscale. The image is a bright field STEM image of a gold particle on an amorphous carbon film recorded on a VG HB-501 STEM ($C_s = 1.3$ mm, 100 keV). The 2.35Å gold lattice fringes are barely visible in the center of the particle. The scale bar in (a) is approximately 20Å. (Fig. 4.2 a is the same image with 8 bits per pixel)

4.2 Discrete Fourier Transform

Image simulation and image processing frequently uses a discrete Fourier transform (DFT) to perform convolutions or to convert from reciprocal space to real space and vice versa. There are several different ways to define a DFT that differ mainly in the placement of the minus signs and normalization constants. The specific definition of the Fourier transform FT and its inverse FT^{-1} that will be used here is:

$$\mathrm{FT}[f(\mathbf{x})] = F(\mathbf{k}) = \sum_{x,y} \exp(2\pi i \mathbf{k}\cdot\mathbf{x}) f(\mathbf{x}) \tag{4.6}$$

$$\mathrm{FT}^{-1}[F(\mathbf{k})] = f(\mathbf{x}) = \frac{1}{N_x N_y} \sum_{k_x,k_y} \exp(-2\pi i \mathbf{k}\cdot\mathbf{x}) F(\mathbf{k}) \tag{4.7}$$

where $\mathbf{x} = (x,y)$ and $\mathbf{k} = (k_x,k_y)$. The inverse Fourier transform can be written as the complex conjugate of the forward transform of the complex conjugate of $F(\mathbf{k})$.

$$\mathrm{FT}^{-1}[F(\mathbf{k})] = \frac{1}{N_x N_y} \{\mathrm{FT}[F^*(\mathbf{k})]\}^*. \tag{4.8}$$

It is only necessary to program one or the other transform (forward or inverse) and the other can be obtained with suitable complex conjugation and scaling.

When expressed in (x,y) Cartesian coordinates the Fourier transform is separable in x and y.

$$F(k_x,k_y) = \sum_x \exp(2\pi i k_x x) \left[\sum_y \exp(2\pi i k_y y) f(x,y)\right] \tag{4.9}$$

A two dimensional transform (forward or inverse) may be implemented by successive one-dimensional transforms. First perform a one dimensional transform along all of the columns and then along all of the rows (or vice versa) of the sampled image (row-column decomposition). Therefore it is only necessary to program a single one-dimensional transform to perform both forward and inverse transforms on two dimensional images. It is much easier in practice to arrange the images to be sampled in a rectangular (x,y) Cartesian grid to exploit the separability of the Fourier transform. Also note that the spacing in x and y may be different but it is usually advisable that the x and y spacings be within a factor of two or so of each other.

4.3 The Fast Fourier Transform or FFT

The one-dimensional discrete Fourier transform is:

$$F(n\Delta k) = F_n = \sum_j f(j\Delta x) \exp[2\pi i (n\Delta k)(j\Delta x)] \tag{4.10}$$

$$k = n\Delta k; \quad \Delta k = 1/a; \quad n = 0,1,2,\cdots,(N-1)$$

$$x = j\Delta x; \quad \Delta x = a/N; \quad j = 0,1,2,\cdots,(N-1)$$

with $f_j = f(j\Delta x)$ this simplifies to:

$$F_n = \sum_j f_j \exp[2\pi i(nj/N)] \tag{4.11}$$

The amount of computer time this requires typically scales as the number of floating point operations such as add, subtract, multiply, and divide. A one-dimensional DFT of length N requires N sums each with N terms and thus requires a computer time that is proportional to N^2.

The computer time may be greatly reduced by use of the fast Fourier transform or FFT algorithm (Cooley and Tukey [53], Brigham [39], and Bracewell [37]). The FFT requires that the length N of the transform be a highly composite number (i.e., be factorable into many smaller integer prime factors). Usually factors of two are a little more efficient although this is not a strict requirement. If the length of the data array is some power of two, $N = 2^m$ (N and m integer) then the data array index j may be written as:

$$j = j_0 + j_1 2 + j_2 2^2 + j_3 2^3 + \cdots + j_{m-1} 2^{m-1}, \tag{4.12}$$

where each of the $j_0, j_1, \ldots$ takes on values of 0 or 1. Equation (4.11) can then be written as:

$$\begin{aligned} F_n &= \sum_{j_0}\sum_{j_1}\sum_{j_2}\cdots\sum_{j_{m-1}} f_{j_0 j_1 j_2 \cdots j_{m-1}} \\ &\quad \exp[2\pi i n(j_0 + j_1 2 + j_2 2^2 + \cdots + j_{m-1} 2^{m-1})/N]. \end{aligned} \tag{4.13}$$

The sums can be rearranged as:

$$\begin{aligned} F_n &= \sum_{j_0} \exp[2\pi i n j_0/N] \sum_{j_1} \exp[2\pi i n j_1 2/N] \\ &\quad \cdots \sum_{j_{m-1}} \exp[2\pi i n j_{m-1} 2^{m-1}/N] f_{j_0 j_1 j_2 \cdots j_{m-1}}. \end{aligned} \tag{4.14}$$

At first glance it seems as if this has just gotten a lot more complicated without any gain. However a close inspection of the arithmetic reveals that there are $m = \log_2 N$ sums each with two terms. Each of the N Fourier components F_n requires m sums of length 2 so the total computer time becomes proportional to $Nm = N \log_2 N$. This is a huge savings in computer time for large N.

When N is decomposed into factors of 2 the FFT is said to be a radix-2 FFT. The radix-2 FFT requires frequent multiplication by sine and cosine of 0 and π (equals 0 and ± 1) which can be hand coded to avoid some floating point arithmetic operations. Curiously, if factors of 4 are also treated as prime factors, sine and cosine of 0, $\pi/2$, $3\pi/2$ and π appear (equals 0, ± 1 and $\pm i$). These can be hand coded without floating point arithmetic to produce an FFT that is a little faster than using only factors of 2. Factors of 8 also give a very small improvement in speed but significantly increase the code size so only factors of 2 and 4 are commonly treated. When factors of four are used the FFT is said to be a radix-4 FFT. When N is decomposed into both factors of 2 and 4 the FFT is said to be a mixed radix FFT. It would first do all of the factors of 4 and then at most one factor of 2 to get any length that is a power of 2. If a two dimensional (or higher) FFT is needed, an additional improvement in performance can also be obtained by using a look up table for the sines and cosines because they only need to be calculated once. Higher radix FFT's generally trade integer operations for floating point operation. Until recently floating point operations were almost always much slower than integer operation and higher radix FFT's run faster. Some new computer architectures close the gap (in speed) between integer and floating and it is possible that radix-4 or radix-8 may not be significantly faster than radix-2 on some specific types of computers if floating point and integer operations are equally fast. Multidimensional transforms can also be limited by the memory bandwidth, and may benefit from careful attension to the order in which the data is accessed. The FFT is still a DFT in some sense (the FFT also satisfies (4.6) but just does it faster) but the two names will be used to distinguish the simple sum from the fast form of the sum.

The computer time for a radix-2 two dimensional transform of length $N_x \times N_y$ for the DFT and the FFT (4.6) is approximately proportional to:

$$\text{CPU time for simple DFT} \propto N_y N_x^2 + N_x N_y^2 \tag{4.15}$$

$$\text{CPU time for FFT} \propto N_x N_y \log_2(N_x N_y). \tag{4.16}$$

The constant of proportionally is of order unity in both cases. The advantage of the FFT over the DFT is very large as N gets bigger. Table 4.1 illustrates the relative CPU time of the DFT vs. the FFT for some typical lengths of the data array. In two dimensions the ratio of the FFT to the DFT remains the same as in the table if $N_x = N_y = N$.

Table 4.1 Comparison of the relative CPU time required for a simple discrete Fourier transform (DFT) and a fast Fourier transform (FFT) for different lengths N

Number of data points N	$\log_2(N)$	DFT	FFT	Ratio
32	5	1024	160	6.4
64	6	4096	384	10.7
128	7	16,384	896	18.3
256	8	65,536	2048	32
512	9	262,144	4608	56.9
1024	10	1,048,576	10240	102.4

The FFT is the workhorse of image simulation and image processing. It is worthwhile to optimize the code for efficient execution in the computer because it gets used over and over again. There are a variety of other tricks that can be incorporated into an actual program. The book by Brigham [39] gives an excellent discussion of strategies for implementing an efficient FFT. Working code for the FFT has been given by Press et al. [288] and Gonzalez and Wintz [124, 125]. A version of a one dimensional FFT using a mixed radix-2 and radix-4 approach in C is given at the end of this chapter. There are many freely available FFT subroutines available to download. The currently popular FFTW package (www.fftw.org [115]) seems to be one of the fastest (if not the fastest) available code. Multidimensional FFTs can be implemented by successively applying a 1D FFT to each dimension. In a 2D FFT each row and then each column. This procedure is easy to adapt to a multiply CPU (multithreaded) computer. Each row (or subset of rows) is independent so can be done simultaneously on a different CPU (many rows at the same time) and then each column (or subset of columns) on a different CPU. The openMP syntax for multithreading is relatively easy to use and has become commonly implemented in several different compilers and computer (and is mostly platform independent).

4.4 Wrap Around Error and Rearrangement

A consequence of a discrete Fourier transform (DFT or FFT) is that the sampled data is repeated indefinitely in a periodic array. An identical copy of the image appears on all four sides of the image. The image is usually not drawn this way but is implied by the use of discrete sampling and the discrete Fourier transform. In practice this periodic repetition means that the left and right edges of the image are effectively adjacent to one another and can interfere ($x = 0$ is equivalent to $x =$ a, $x =$ 2a, etc.). The same is true of the top and bottom of the image. This effect is called the wrap-around error because the left and right or top and bottom edges effectively wrap around and touch each other. If the underlying specimen does not share this periodicity then some rather dramatic artifacts that have nothing to do with the specimen can be produced in simulated images. The underlying specimen periodicity should match the periodicity of the supercell (the supercell dimensions should be an integer multiple of the specimen unit cell size if the specimen is crystalline) or there should be a buffer zone around the edge of the image that can be discarded later. The width of this buffer zone varies with the application and may not always be obvious.

The wrap around effect applies in both real space and reciprocal space. This causes a strange distribution of spatial frequencies in the FFT. Large positive frequencies are the same as small negative frequencies. The sampled frequencies in reciprocal space (4.4) may be written in order from left to right (in x) and bottom to top (in y) as:

$$k_x = 0, \Delta k_x, 2\Delta k_x, 3\Delta k_x, \cdots, (N_x - 1)\Delta k_x \tag{4.17}$$

$$k_y = 0, \Delta k_y, 2\Delta k_y, 3\Delta k_y, \cdots, (N_y - 1)\Delta k_y. \tag{4.18}$$

With wrap around the $k_x = (N_x - 1)\Delta k_x$ position is touching the $k_x = 0$ position on the left. This means that the $N_x - 1$ position is the same as the -1 position (likewise in the y direction). A circle with constant magnitude of spatial is drawn in FFT space on the left side of Fig. 4.5 (labeled default). The origin is in the lower left corner.

What is normally the first quadrant is in the lower left corner, the second quadrant is in the lower right, the third quadrant in the upper right and the fourth quadrant in the upper left. The distribution of spatial frequencies is normally used in this order during calculation because it would waste computer time to rearrange it. However when displayed the FFT will be rearranged (as on the right hand side of Fig. 4.5) to conform to a normal diffraction pattern with zero spatial frequency in the center.

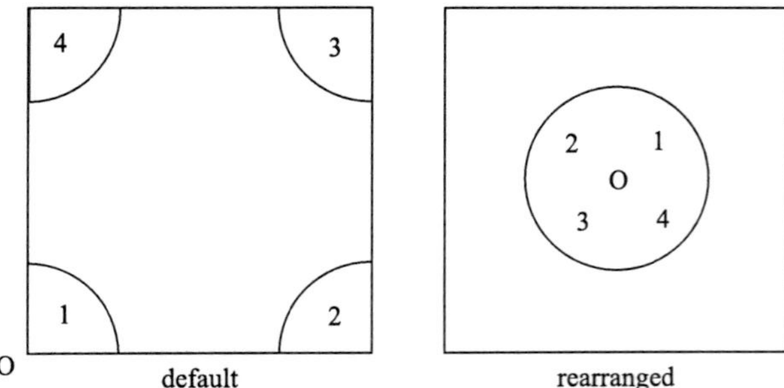

Fig. 4.5 Geometry of the FFT (or DFT). A *circle* of constant spatial frequency is drawn in the default configuration (*left*) and the rearranged configuration (*right*). The origin is at the lower left in the default configuration and in the center in the configuration at right rearranged for display purposes

4.5 Fourier Transforming Real Valued Data

The FFT discussed in Sect. 4.3 assumed a general case in which the data $f(x)$ is complex valued with a real and imaginary part (as in an electron wave function). Sometimes the data will be explicitly real valued (as in the potential $V(\mathbf{r})$ of the specimen). The special case of Fourier transforming real data allows a further reduction in computer time of about a factor of two. There are two possible methods to use. One is to transform a single real valued array more efficiently and the other is to transform more than one real array at a time with a single complex FFT. Both of these methods are discussed in Press et al. [288]. Performing a two dimensional FFT on real valued data is easiest using the second of these methods.

Consider two one dimensional arrays of N real values, $f_a(x)$ and $f_b(x)$, linearly combined into a single complex valued array $f_c(x)$ as:

$$f_c(x) = f_a(x) + if_b(x) \tag{4.19}$$

$f_c(x)$ is now in the form that can be used in a single FFT to produce N complex valued Fourier coefficients. The trick is to untangle the Fourier transform of $f_a(x)$ from that of $f_b(x)$. In real space $f_a(x)$ and $f_b(x)$ can be retrieved from $f_c(x)$ using (4.19) as:

$$f_a(x) = [f_c(x) + f_c^*(x)]/2 \tag{4.20}$$

$$f_b(x) = [f_c(x) - f_c^*(x)]/(2i) \tag{4.21}$$

In Fourier or reciprocal space:

$$F_c(k) = FT[f_c(x)] = \int f_c(x)\exp(2\pi ikx)dx \tag{4.22}$$

$$F_c^*(-k) = FT[f_c^*(x)] = \int f_c^*(x)\exp(2\pi ikx)dx \tag{4.23}$$

Therefore, the transforms of $f_a(x)$ and $f_b(x)$ may be extracted from the transform of $f_c(x)$ [using (4.21)] as:

$$FT[f_a(x)] = F_a(k) = [F_c(k) + F_c^*(-k)]/2 \tag{4.24}$$

$$FT[f_b(x)] = F_b(k) = [F_c(k) - F_c^*(-k)]/(2i) \tag{4.25}$$

If there are N real values in each of $f_a(x)$ and $f_b(x)$ then there are N complex values in $f_c(x)$ and $F_c(k)$. It follows that there are $N/2$ complex valued Fourier coefficients in each of $F_a(k)$ and $F_b(k)$ because:

$$F_a(k) = F_a^*(-k) \tag{4.26}$$

$$F_b(k) = F_b^*(-k) \tag{4.27}$$

To calculate a two dimensional Fourier transform of $N_x \times N_y$ real valued data points, first calculate the Fourier transform of all of the N_y columns two at a time, yielding N_y complex arrays of length $N_x/2$. Next Fourier transform N_x rows of $N_x/2$ complex values. This is referred to as a real to complex FFT and results in a net sped up of about a factor of two.

4.6 Displaying Diffraction Patterns

The square modulus of the Fourier transform of a function is called its power spectra. The power spectra of the wave function transmitted through the specimen is also equivalent to the electron diffraction pattern of the specimen. A diffraction pattern typically has a very large dynamic range in its intensity. The low spatial frequency information (low scattering angle) has a large amplitude but the high spatial frequency information (high scattering angle) has a much lower amplitude. A normal image display device (computer screen or printed paper) does not have a sufficient dynamic range to display both sets of information. The high spatial frequency information (which is frequently the interesting part) is not visible if the diffraction pattern is normalized to fill the available grey scale in a linear manner. In practice when a diffraction pattern is recorded the film or other detector may be saturated near the central beam to produce a nonlinear scale or multiple patterns may be recorded at different exposures to accommodate this large dynamic range. Gonzalez and Wintz [124, 125] and Pratt [287] suggest that a numerically calculated power spectra be displayed on a logarithmic scale to compress the dynamic range so that the entire diffraction pattern is visible. A simple logarithm will not work because some points of the power spectra (diffraction pattern) may be identically zero. The computer program must somehow limit the negative extent of the image scale. One method is to simply clip all negative values (of the logarithm) to some minimum value. For example, the minimum grey scale can be set to the average value in some region of reciprocal space about half way between the minimum and maximum spatial frequencies. An alternative (similar to that proposed by Gonzalez and Wintz [124, 125] and Pratt [287]) is to transform the intensities as:

$$D(k_x, k_y) = \log(1 + c|F(k_x, k_y)|^2), \tag{4.28}$$

where $F(k_x, k_y)$ is the Fourier transform of the image, $D(k_x, k_y)$ is the actual value displayed and c is a scaling constant that can be varied to adjust the contrast. Figure 4.6 shows the power spectra of the electron wave function transmitted through approximately 100Å of silicon in the 110 orientation at an electron energy of 100 keV (simulated using the multislice method discussed in later chapters and a super cell size of 27×27Å with 128×128 pixels). The linear grey scale in Fig. 4.6a shows only the zero order beam in the center (rearranged as in Sect. 4.4) plus a few of the low order reflections. When displayed using (4.28) with c = 0.1 the higher order diffraction spots become visible in Fig. 4.6b. Most diffraction patterns shown in this book will use a compressed scale something like this.

4.7 An FFT Subroutine in C

The FFT is one of the most efficient (fast computation) algorithms available and is one of the primary drivers of the multislice algorithm to be considered in later chapters. There are many FFT subroutines available for downloading or purchase alone or as part of standard libraries. Below is the code for a simple one dimensional FFT. Multidimensional transforms are typically performed as successive 1D FFT's in each direction, and can be easily implemented in parallel in a shared memory multiprocessor computer

```
/*------------------------ fft42 -------------------------

    fft42( fr[], fi[], n )        radix-4,2 FFT in C

    fr[], fi[] = (float) real and imag. array with input data
    n          = (long) size of array

   calculate the complex fast Fourier transform of (fr,fi)
   input array fr,fi (real,imaginary) indexed from 0 to (n-1)
   on output fr,fi contains the transform
```

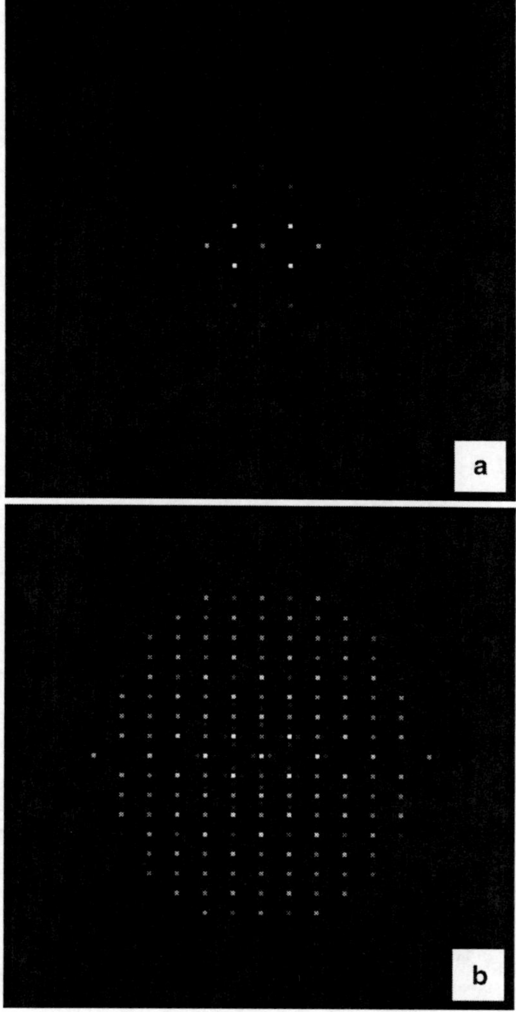

Fig. 4.6 The calculated diffraction pattern or power spectra of 110 silicon. (a) Is shown on a linear scale and (b) is a logarithmic scale as in (4.28) White is a larger positive value

```
*/

void fft42 ( float *fr, float *fi, long n  )
{
#define TWOPI   6.283185307

    long i, j, nv2, nm1, k, k0, k1, k2, k3, kinc, kinc2;
    float qr, qi, rr, ri, sr, si, tr, ti, ur, ui;
    double x1, w0r, w0i, w1r, w1i, w2r, w2i, w3r, w3i;
```

```
kinc = n;

while( kinc >= 4 ) {      /* start radix-4 section */

    kinc2 = kinc;
    kinc = kinc / 4;

    for( k0=0; k0<n; k0+=kinc2) {
        k1 = k0 + kinc;
        k2 = k1 + kinc;
        k3 = k2 + kinc;

        rr =  fr[k0] + fr[k2];     ri = fi[k0] + fi[k2];
        sr =  fr[k0] - fr[k2];     si = fi[k0] - fi[k2];
        tr =  fr[k1] + fr[k3];     ti = fi[k1] + fi[k3];
        ur = -fi[k1] + fi[k3];     ui = fr[k1] - fr[k3];

        fr[k0] = rr + tr;     fi[k0] = ri + ti;
        fr[k2] = sr + ur;     fi[k2] = si + ui;
        fr[k1] = rr - tr;     fi[k1] = ri - ti;
        fr[k3] = sr - ur;     fi[k3] = si - ui;
    }

    x1 = TWOPI/( (double) kinc2 );
    w0r = cos( x1 );    w0i = sin( x1 );
    w1r = 1.0;    w1i = 0.0;

    for( i=1; i<kinc; i++) {
        x1 = w0r*w1r - w0i*w1i;     w1i = w0r*w1i + w0i*w1r;
        w1r = x1;
        w2r = w1r*w1r - w1i*w1i;    w2i = w1r*w1i + w1i*w1r;
        w3r = w2r*w1r - w2i*w1i;    w3i = w2r*w1i + w2i*w1r;

        for( k0=i; k0<n; k0+=kinc2) {
            k1 = k0 + kinc;
            k2 = k1 + kinc;
            k3 = k2 + kinc;

            rr =  fr[k0] + fr[k2];     ri = fi[k0] + fi[k2];
            sr =  fr[k0] - fr[k2];     si = fi[k0] - fi[k2];
            tr =  fr[k1] + fr[k3];     ti = fi[k1] + fi[k3];
            ur = -fi[k1] + fi[k3];     ui = fr[k1] - fr[k3];

            fr[k0] = rr + tr;     fi[k0] = ri + ti;

            qr = sr + ur;     qi = si + ui;
            fr[k2] = (float) (qr*w1r - qi*w1i);
            fi[k2] = (float) (qr*w1i + qi*w1r);

            qr = rr - tr;     qi = ri - ti;
            fr[k1] = (float) (qr*w2r - qi*w2i);
            fi[k1] = (float) (qr*w2i + qi*w2r);

            qr = sr - ur;     qi = si - ui;
```

```
                fr[k3] = (float) (qr*w3r - qi*w3i);
                fi[k3] = (float) (qr*w3i + qi*w3r);
            }
        }

    } /*  end radix-4 section */

    while( kinc >= 2 ) {     /* start radix-2 section */

        kinc2 = kinc;
        kinc = kinc /2 ;

        x1 = TWOPI/( (double) kinc2 );
        w0r = cos( x1 );    w0i = sin( x1 );
        w1r = 1.0;    w1i = 0.0;

        for( k0=0; k0<n; k0+=kinc2 ){
            k1 = k0 + kinc;
            tr = fr[k0] - fr[k1];  ti = fi[k0] - fi[k1];
            fr[k0] = fr[k0] + fr[k1];
            fi[k0] = fi[k0] + fi[k1];
            fr[k1] = tr;  fi[k1] = ti;
        }

        for( i=1; i<kinc; i++) {
            x1 = w0r*w1r - w0i*w1i;  w1i = w0r*w1i + w0i*w1r;
            w1r = x1;
            for( k0=i; k0<n; k0+=kinc2 ){
                k1 = k0 + kinc;
                tr = fr[k0] - fr[k1];    ti = fi[k0] - fi[k1];
                fr[k0] = fr[k0] + fr[k1];
                fi[k0] = fi[k0] + fi[k1];
                fr[k1] = (float) (tr*w1r - ti*w1i);
                fi[k1] = (float) (tr*w1i + ti*w1r);
            }
        }

    } /* end radix-2 section */

    nv2 = n / 2;
    nm1 = n - 1;
    j = 0;

    for (i=0; i< nm1; i++) {  /* reorder in bit rev. order */
        if( i < j ){
            tr = fr[j]; ti = fi[j];
            fr[j] = fr[i];  fi[j] = fi[i];
            fr[i] = tr; fi[i] = ti; }
        k = nv2;
        while ( k <= j ) { j -=  k;  k = k>>1; }
        /* while ( k <= j ) {j=j-k;  k= k /2; }  is slower */
        j += k;
    }
```

```
#undef TWOPI
}  /* end fft42() */
```

4.8 Further Reading

Some Books on Computer Image Processing

1. K. R. Castleman, *Digital Image Processing*, Prentice Hall, 1979 [46]
2. R.C. Gonzalez and R.E. Woods, *Digital Image Processing, 3nd edition*, Prentice-Hall, 2008 [125]
3. E.L. Hall, *Computer Image Processing and Recognition*, Academic Press, 1979 [143]
4. B. Jahne, *Digital Image Processing, 3rd edition*, Springer, 1995 [183]
5. A. Jain, *Fundamentals of Digital Image Processing*, Prentice Hall, 1989 [184]
6. D.L. Missel, *Image Analysis, Enhancement annd Interpretation*, North Holland, 1978 [243]
7. W.K. Pratt, *Digital Image Processing*, Wiley, 1978 [287]
8. A. Rosenfeld and A.C. Kak,*Digital Picture Processing*, Academic Press, 1976 [305]
9. W.O. Saxton, *Computer Techniques for Image Processing in Electron Microscopy, Adv. in Electronics and Electron Physics, Supplement 10*, Academic Press, 1978 [309]

Some Books on Fourier Transforms and Fourier Optics

1. E.O. Brigham, *The Fast Fourier Transform*, Prentice-Hall, 1974 [39]
2. J.W. Goodman, *Intro. to Fourier Optics, 3rd. edit.*, Roberts and Co., 2005 [126]
3. James S. Walker, *Fast Fourier Transforms, 2nd edit.*, CRC Press, 1996 [362]

Chapter 5
Calculation of Images of Thin Specimens

Abstract This chapter presents approximate methods of calculating transmission electron microscope images of thin specimens. The thickness of the specimen is ignored, which may be appropriate for very thin specimens. Multiple scattering is also generally ignored. This approach is intermediate between the transfer function (in previous chapters) and the multislice and Bloch wave methods (discussed in later chapters) and has the advantage of requiring much less computer time.

This chapter discusses the calculation of an electron microscope image neglecting the geometrical thickness of the specimen (i.e., very thin specimens). Many practical specimens are too thick for this type of calculation to be quantitatively correct. However, this approach can provide a qualitative insight into the structure in the image and it requires much less computer time. This type of image simulation is sometimes referred to as a phase grating approximation or a kinematical image approximation because it does not properly include the effects of multiple or plural scattering within the specimen. Calculation of the transmission function of thin specimens is also a necessary part of more advanced calculations including a realistic specimen thickness that will be considered in later chapters. In particular the calculation presented in this chapter will form a single slice of the multislice algorithm.

The kinetic energy of the imaging electrons in the electron microscope approaches their rest mass energy. A detailed quantum mechanical calculation of the motion of these electrons should properly be calculated using relativistic quantum mechanics (the Dirac equation with spin). As discussed in Sect. 2.3 the relativistic effects can be approximated by using the nonrelativistic Schrödinger equation (neglecting electron spin) with the relativistically correct wavelength and mass of the electron. This approximation is probably accurate enough at 100 keV but may be less accurate at 1 MeV. Nonrelativistic quantum mechanics is however dramatically easier to work with, and this approximation will be used here.

E.J. Kirkland, *Advanced Computing in Electron Microscopy*,
DOI 10.1007/978-1-4419-6533-2_5, © Springer Science+Business Media, LLC 2010

5.1 The Weak Phase Object

The primary interaction between the specimen and the imaging electrons is between the electrostatic potential of the specimen and the charge on the electron. The electrons traveling down the column of the microscope (before hitting the specimen) are a superposition of one or more plane waves. In the CTEM the incident electrons are primarily in a single plane wave and the STEM probe is a superposition of many plane waves (i.e., a spherically convergent probe). It suffices to consider the effect of the specimen on one plane wave. The wave function ψ for one plane wave traveling along the optic axis in the z direction is:

$$\psi(\mathbf{x}) = \exp(2\pi i k_z z) = \exp(2\pi i z/\lambda), \tag{5.1}$$

where λ is the wavelength of the electron and $k_z = 1/\lambda$ is the propagation wave vector. The relativistic expression for the reciprocal of the electron wavelength in vacuum [see (2.5)] is:

$$k_z = \frac{1}{\lambda} = \frac{\sqrt{eV(2m_0c^2 + eV)}}{hc}, \tag{5.2}$$

where m_0 is the rest mass of the electron, c the speed of light in vacuum, h Planck's constant, and eV is the kinetic energy of the electron in vacuum.

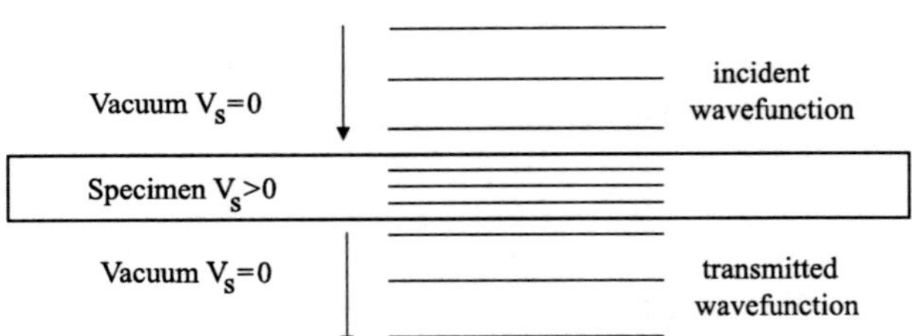

Fig. 5.1 An incident (high energy) electron plane wave passing through the electrostatic potential V_s of the specimen. The wave function is drawn as lines of constant phase, and the specimen is assumed to have a uniform constant potential. The electron wavelength is reduced by the positive potential inside the specimen. This drawing is not to scale

The imaging electrons typically have a much higher energy than the electrons in the specimen. If the specimen is thin the imaging electrons pass through the specimen with only a small deviation in their path. This deviation can be approximated as a small change in wavelength of the electrons as they pass through the specimen (see Fig. 5.1). The specimen has a small electrostatic potential which influences the electron wavelength. If the potential inside the specimen is positive then the imaging electrons are accelerated inside the specimen giving them a smaller wavelength. If eV_s is the additional electrostatic potential energy of the imaging electrons while in the specimen and λ_s is their wavelength then while inside the specimen:

$$\begin{aligned}\frac{1}{\lambda_s} &= \frac{[(eV+eV_s)(2m_0c^2+eV+eV_s)]^{1/2}}{hc} \\ &= \frac{[eV(2m_0c^2+eV)+eV_s(2m_0c^2+2eV+eV_s)]^{1/2}}{hc} \\ &= \frac{1}{\lambda}\left[1+\frac{eV_s(2m_0c^2+2eV+eV_s)}{eV(2m_0c^2+eV)}\right]^{1/2}. \end{aligned} \tag{5.3}$$

Expanding this equation and keeping only the lowest order terms in V_s/V yields:

$$\begin{aligned}\frac{1}{\lambda_s} &\sim \frac{1}{\lambda}\left[1+\frac{eV_s(2m_0c^2+2eV)}{2eV(2m_0c^2+eV)}+\cdots\right] \\ &\sim k_z+\frac{V_s(m_0c^2+eV)}{\lambda V(2m_0c^2+eV)}+\cdots \end{aligned} \tag{5.4}$$

Changing the wavelength is equivalent to shifting the phase of the electron as it passes through the specimen (i.e., changing the wave vector k_z). Therefore, the electron wave function while passing through the specimen is:

$$\psi(\mathbf{x}) \sim \exp(2\pi i k_z z)\exp(i\sigma V_s z), \tag{5.5}$$

where the interaction parameter σ is:

$$\sigma = \frac{2\pi}{\lambda V}\left(\frac{m_0c^2+eV}{2m_0c^2+eV}\right) = \frac{2\pi m e\lambda}{h^2} \tag{5.6}$$

$m=\gamma m_0$ is the relativistic mass. This expression assumes that the specimen potential V_s is much smaller than the beam energy ($V_s/V \ll 1$). Also remember that the specimen potential varies with position although not explicitly written earlier. It is traditional to use the same symbol for both the interaction parameter and the scattering cross section, but the meaning should usually be clear from the context in which each is used. The interaction parameter is plotted vs. the electron kinetic energy in Fig. 5.2. The interaction parameter decreases rapidly with increasing electron energy at low electron energy but is nearly constant for electron energies above about 300 keV.

If the specimen is very thin then the electron wave function accumulates a total phase change while passing through the specimen that is just the integral of the potential of the specimen. The incident electrons pass through the specimen and the effect of the specimen is to multiply the incident wave function (5.1) by the specimen transmission function $t(\mathbf{x})$. The wave function transmitted through the specimen is:

$$\begin{aligned}\psi_t(\mathbf{x}) &= t(\mathbf{x})\exp(2\pi i k_z z) \\ t(\mathbf{x}) &= \exp[i\sigma v_z(\mathbf{x})] \end{aligned} \tag{5.7}$$

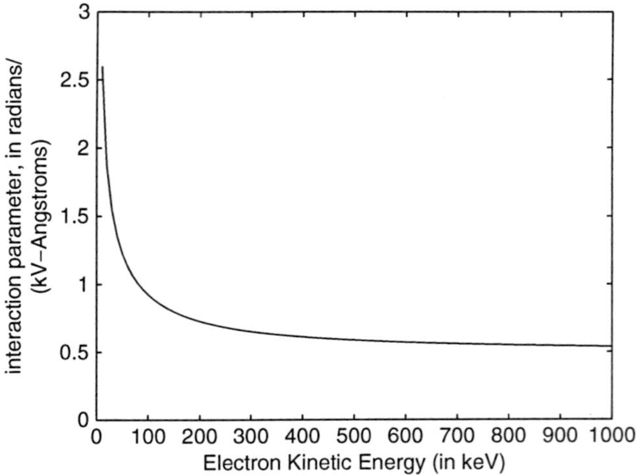

Fig. 5.2 The interaction parameter σ vs. the electron kinetic energy

and the projected atomic potential $v_z(\mathbf{x})$ is the integral along the optic axis, z of the specimen:

$$v_z(\mathbf{x}) = v_z(x,y) = \int V_s(x,y,z)\mathrm{d}z \tag{5.8}$$

This is the so-called weak phase object approximation (Cowley and Iijima [62]). There are really two assumptions in this approximation. One is that the potential inside the specimen is very small and the other is that the accumulated effect of the specimen can be replaced with a simple integral along z.

5.2 Single Atom Properties

Single atoms are a reasonable starting point to discuss the simulation of electron microscope images in the weak phase object approximation. Single isolated atoms with low to medium atomic number satisfy the thin specimen requirement and can actually be seen in some microscopes under the appropriate conditions. Furthermore the potential and charge distribution of single atoms can be calculated from first principles using relativistic Hartree-Fock theory in a reasonably well defined although tedious manner. To find the potential and charge distribution in a single atom requires finding the wave function of all electrons in the atom (to a good approximation the atomic nucleus may be regarded as a fixed point charge at the origin). Hydrogen is the only atom which can be solved analytically and a derivation is usually given in most quantum mechanics text books (see for example Eisberg [88] or Schiff [312]). Unfortunately, atoms with more than one electron (that is, the rest of the periodic chart) must be solved numerically with some approximations. The Hartree-Fock method forms an effective many-electron wave function obeying the Pauli exclusion principle that also satisfies the Schrödinger wave equation (or in the case of the relativistic Hartree-Fock method the Dirac wave equation). The method reduces to repeatedly solving a single particle wave equation for each electron orbital that is moving in an effective potential due to the charge on the nucleus and the average interactions with all other electrons in the atoms. Each electron orbital is calculated using the current distribution of the other electrons in the atoms and the process is repeated for all orbitals until the electron wave functions for all electrons converge to a final self-consistent result. Relativistic effects are probably negligible for low atomic numbers (like carbon) but are significant for high atomic numbers (like gold) because the core electrons near the nucleus experience a very large electric field and have a large kinetic energy (i.e., velocity). A detailed discussion of the Hartree-Fock method is beyond the scope of this book but may be found in the books by Froese-Fischer [117], Szasz [339] and Froese-Fischer et al. [106]. Appendix C gives a detailed description of using a relativistic Hartree-Fock program to calculate a complete set of atomic potentials for the whole periodic chart (atomic number $Z = 2$ through $Z = 103$).

5.2.1 Radial Charge Distribution

The radial electron charge distribution $\rho(r)$ of each atom is generated as part of the Hartree-Fock atomic structure calculation. The calculated radial charge distribution for a few selected atoms is shown in Fig. 5.3. The peaks in the charge distribution correspond to the atomic orbitals (or electron shells) of each atom. It is interesting to note that although the total number of electrons increases with atomic number Z the actual size of the atoms does not change dramatically with atomic number. The increasing charge of the nucleus (with increasing Z) causes the electrons to be attracted more strongly to the nucleus roughly keeping the actual atomic size relatively constant at about one Angstrom in diameter.

5.2.2 Potential

The atomic potential is a more interesting quantity for electron microscopy because the imaging electrons in the microscope interact directly with the atomic potential [see (5.5)]. The charge distribution and potential are related via Poisson's equation from electromagnetic theory. The Mott-Bethe [25, 27, 255, 256] formula (C.7) is equivalent to Poisson's equation, however it is stated in reciprocal space. Figure 5.3 only includes the electron charge distribution, however the large point charge on the atomic nucleus probably has the strongest interaction with the imaging electrons. The electron cloud in Fig. 5.3 mainly serves to shield the atomic nucleus (remember that the nucleus has a positive charge but the electron cloud has a negative charge). When the charge on the atomic nucleus is added to the atomic electron charge

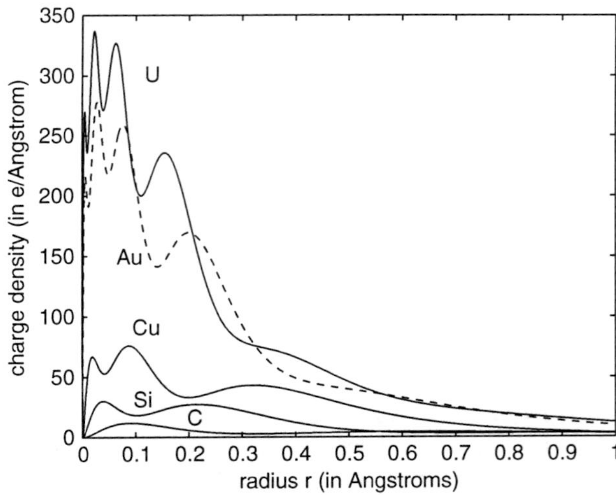

Fig. 5.3 The calculated electron charge distribution $4\pi r^2\rho(r)$ (in electrons per Å) for isolated single atoms at the origin vs. the three dimensional radius r. The atoms are carbon (Z6), silicon (Z = 14), copper (Z = 29), gold (Z = 79), and uranium (Z = 92)

distribution and the resulting charge distribution is transformed into an atomic potential the total potential is much more strongly peaked near the nucleus (at the origin). The Hartree-Fock procedure of necessity ends up with a large table of numbers. Appendix C details how to parameterize the tabulated Hartree-Fock results. To a reasonably good approximation the atomic potential (including the nucleus) may be written as:

$$\begin{aligned} V_a(x,y,z) &= 2\pi^2 a_0 e \sum_{i=1}^{3} \frac{a_i}{r} \exp(-2\pi r\sqrt{b_i}) + \\ & 2\pi^{5/2} a_0 e \sum_{i=1}^{3} c_i d_i^{-3/2} \exp(-\pi^2 r^2/d_i) \\ & \text{with} \quad r^2 = x^2 + y^2 + z^2, \end{aligned} \tag{5.9}$$

where a_0 is the Bohr radius and the a_i, b_i, c_i, and d_i coefficients are tabulated in Appendix C. There is a different set of coefficients for each element. A graph of the atomic potential vs. radius in three dimensions is shown in Fig. 5.4 for a few selected elements.

When the atomic potential is integrated along the z direction (i.e., the optic axis of the microscope) the result is the projected atomic potential:

$$\begin{aligned} v_z(x,y) &= \int_{-\infty}^{+\infty} V_a(x,y,z)\mathrm{d}z \\ &= 4\pi^2 a_0 e \sum_{i=1}^{3} a_i K_0(2\pi r\sqrt{b_i}) + 2\pi^2 a_0 e \sum_{i=1}^{3} \frac{c_i}{d_i} \exp(-\pi^2 r^2/d_i) \\ & \text{with} \quad r^2 = x^2 + y^2, \end{aligned} \tag{5.10}$$

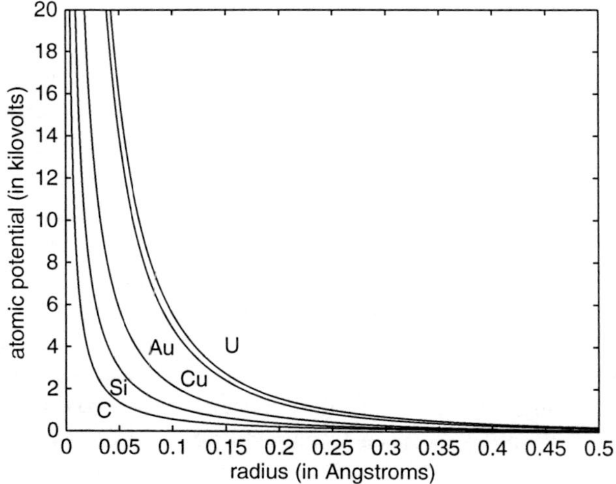

Fig. 5.4 Calculated atomic potential vs. the three dimensional radius r for isolated single atoms at $r = 0$. The atoms are carbon (Z = 6), silicon (Z = 14), copper (Z = 29), gold (Z = 79), and uranium (Z = 92)

where $K_0(x)$ is the modified Bessel function of zeroth order. The projected atomic potential is shown in Fig. 5.5 for the same elements as in Fig. 5.4.

The atomic nucleus is essentially a point charge on this scale (typical nuclear sizes are of order a few times 10^{-5}Å). This causes a singularity in the projected

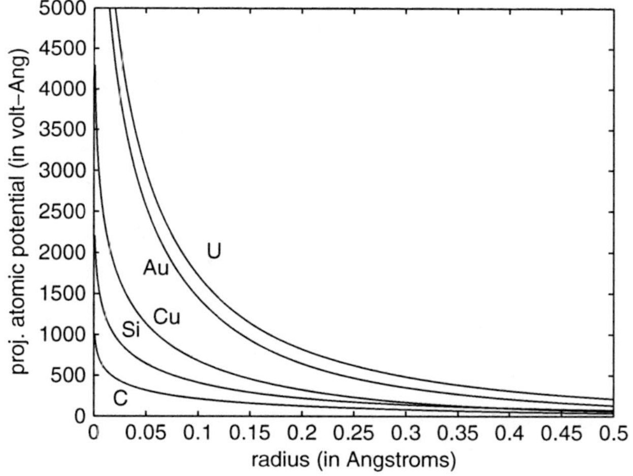

Fig. 5.5 Calculated projected atomic potential vs. the two dimensional radius r for isolated single atoms at r = 0. The atoms are carbon (Z = 6), silicon (Z = 14), copper (Z = 29), gold (Z = 79), and uranium (Z = 92)

atomic potential at $r = 0$ (there is also a singularity in the 3D atomic potential.) In reality the finite size of the nucleus removes the singularity so there is no real problem. No electron microscope that currently exists has enough resolution to see this strong singularity either, so the limited resolution of the microscope will further smear out this singularity in practice. In practical computer simulations the singularity will also be removed by finite sampling requirements because only the projected atomic potential averaged over a nonzero sized sampling element or pixel is used. The projected atomic potential (Fig. 5.5) diverges less strongly than the atomic potential (Fig. 5.4).

At a radius of 0.1Å the projected atomic potential of silicon (Si) is 0.41 kV-Å and the projected atomic potential of gold (Au) is 1.45 kV-Å (see Fig. 5.5). The interaction parameter σ (5.6) is 0.92 radians/(kV-Å) at a beam energy of 100 keV. This means that a single silicon atom will produce a total phase shift of 0.38 radians and a single gold atom will produce a phase shift of 1.34 radians (both at a beam energy of 100 keV and a radius of 0.1 Å). A single gold atom is not a weak phase object [in the sense of (3.6)] but a single silicon atom is a reasonable weak phase object (at 100 keV). The situation improves slightly at beam energies of 300 keV or above because σ decreases by almost a factor of two. All atoms have a near singularity at a radius of zero so no single atom is truly a weak phase object in a strict sense.

5.2.3 Atomic Size

There is no single unambiguous method to calculate the effective size of single atoms. The rms (root-mean-square) radius is as good as any other method. The three dimensional mean square (rms) radius of the charge distribution can be defined as:

$$r_q = \left[\frac{\int_0^\infty r^2 \rho(r) r^2 \mathrm{d}r}{\int_0^\infty \rho(r) r^2 \mathrm{d}r} \right]^{1/2} \quad \text{where} \quad r = x^2 + y^2 + z^2, \tag{5.11}$$

where $\rho(r)$ is the radial charge density. In a similar manner the two dimensional mean square radius of the projected atomic potential is:

$$r_v = \left[\frac{\int_0^\infty r^2 v_z(r) r \mathrm{d}r}{\int_0^\infty v_z(r) r \mathrm{d}r} \right]^{1/2} \quad \text{where} \quad r = x^2 + y^2. \tag{5.12}$$

Portions of the potential at large radius contribute more to this calculation due to their increased effective area. For example if the projected atomic potential is multiplied by r the result is shown in Fig. 5.6. The rms atomic radii for all atoms as calculated from (5.11) and (5.12) is shown in Fig. 5.7. The large tails of the potential increase the apparent size of the atoms. The effective rms size of the atoms as determined by the projected atomic potential is about two Angstroms in diameter. Note that the effective full-width-half-maximum of a single atom image may be smaller because of the strong potential near the nucleus of each atom.

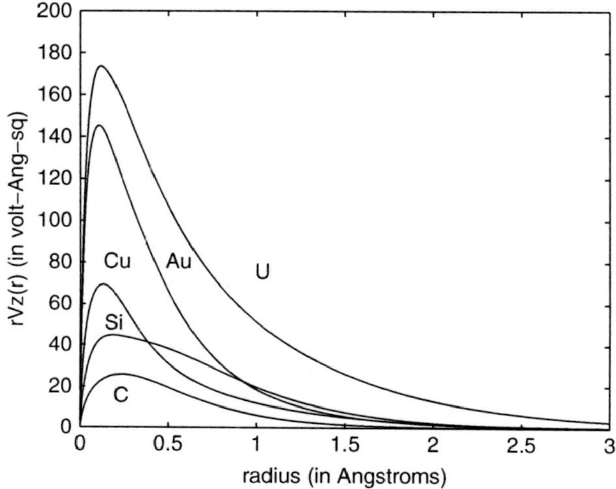

Fig. 5.6 Projected atomic potential multiplied by the radius r to illustrate the relative contribution to an image. Each curve is a different atom; carbon (Z = 6), silicon (Z = 14), copper (Z = 29), gold (Z = 79), and uranium (Z = 92)

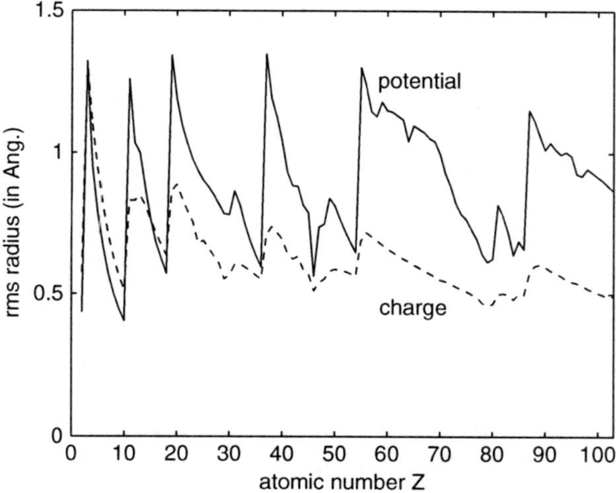

Fig. 5.7 The rms radius of isolated single atoms as determined from the (3D) electron charge and the (2D) projected atomic potential

5.2.4 Scattering Factors

The traditional physics view of electron scattering starts with a plane wave incident on the atom which gives rise to an outgoing plane wave plus an outgoing spherical wave (the atom has spherical symmetry) with amplitude $f_e(q)$.

$$\psi(\mathbf{x}) = \exp(2\pi i k_z z) \quad \text{incident} \tag{5.13}$$
$$= \exp(2\pi i k_z z) + f_e(q)\frac{\exp(2\pi i \mathbf{q}\cdot\mathbf{r})}{r} \quad \text{scattered,} \tag{5.14}$$

where $\mathbf{q}$ is the difference between the incident and scattered wave vectors (three dimensional vector). The scattering amplitude $f_e(q)$ can also be referred to as the scattering factor. There are several methods of calculating the scattering amplitudes. The most popular approximation to the scattering amplitude is the first Born approximation that is simply the three dimensional Fourier transform of the atomic potential (see for example Sect. 38 of Schiff [312]):

$$f_e(\mathbf{q}) = \frac{2\pi m_0 e}{h^2}\int V_a(\mathbf{r})\exp(2\pi i \mathbf{q}\cdot\mathbf{r})d^3r$$
$$= \frac{1}{2\pi e a_0}\int V_a(\mathbf{r})\exp(2\pi i \mathbf{q}\cdot\mathbf{r})d^3r, \tag{5.15}$$

where $V_a(\mathbf{r})$ is the 3D atomic potential of the atom, m_0 is the rest mass of the electron, e is the magnitude of the charge of the electron, h is Planck's constant, and $a_0 = \hbar^2/m_0e^2 = 0.5292$Å is the Bohr radius. $f_e(q)$ is in units of Å and must be multiplied by the relativistic mass ratio m/m_0 for different incident electron energies. For the case where the atom is spherically symmetric this reduces to:

$$f_e(q) = \frac{1}{\pi e a_0 q}\int_0^\infty V_a(r)\sin(2\pi q r) r dr. \tag{5.16}$$

The scattering factor is the amplitude for scattering of a single electron by a single atom. The first Born approximation is totally inadequate for directly calculating electron scattering in the electron microscope image (Zeitler and Olsen [388, 389], Glauber and Shoemaker [121]). In general $f_e(q)$ should be a complex valued quantity but the first Born approximation only yields a real valued quantity. However the first Born approximation is convenient because it is also the Fourier transform of the atomic potential. Image simulation will eventually use the specimen potential directly and not the scattering factors, so the Born approximation is still useful as a means of calculating the specimen potential. Combined with the Fourier projection theorem (Appendix B) the Born approximation provides a convenient method of calculating the projected atomic potential of thin specimens. It is also independent of the incident electron energy so that is easy to tabulate. As given in Appendix C the scattering amplitude in the first Born approximation can be written as:

$$f_e(q) = \sum_{i=1}^{3} \frac{a_i}{q^2 + b_i} + \sum_{i=1}^{3} c_i \exp(-d_i q^2), \tag{5.17}$$

where the a_i, b_i, c_i, and d_i coefficients are tabulated in Appendix C for each element and are found by fitting the results of the relativistic Hartree-Fock program. The scattering factors for several atoms are plotted in Fig. 5.8. Note that the scattering factor for carbon (Z = 6) and silicon (Z = 14) cross at low angle. The low angle scattering factor is dependent on the state of the outer valence shell which fills periodically with increasing atomic number in the periodic chart. The scattering factor at large angles is primarily due to the scattering from the nucleus which is a monotonic function of atomic number Z. Doyle and Turner [82] have given the most popular tabulation currently in use. Many other authors have tabulated relevant parameters for single atoms and Table C.1 of appendix C gives a more complete listing of the data available in the literature.

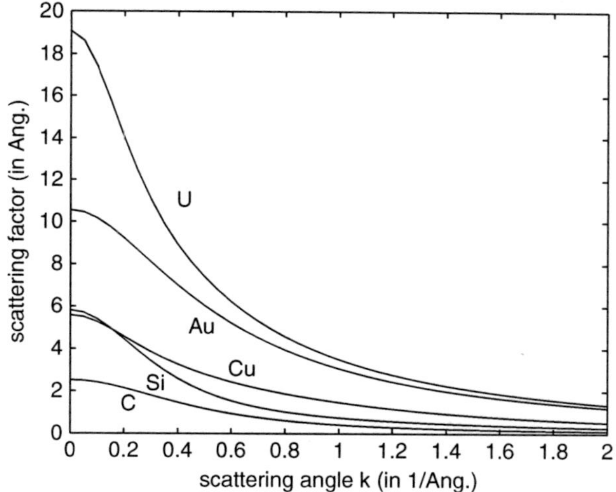

Fig. 5.8 The electron scattering factor in the first Born approximation vs. scattering angle $k = \alpha/\lambda$ of isolated single atoms. Each curve is a different atom which are carbon (Z = 6), silicon (Z = 14), copper (Z = 29), gold (Z = 79), and uranium (Z = 92)

A more detailed derivation of the scattering factor in the Moliere [252] or eikonal [312] approximation yields an improved expression of:

$$f_e(q) = \frac{2\pi i}{\lambda} \int_0^\infty J_0(2\pi qr) \left\{ 1 - \exp\left[i\sigma \int V(x,y,z) dz \right] \right\} r dr \tag{5.18}$$

where $r = \sqrt{x^2 + y^2}$ can also be interpreted as the impact parameter, $J_0(x)$ is the Bessel function of order zero which arises from the azimuthal integration of a spherically symmetric function and σ is the interaction parameter (5.6). This expression for the electron scattering factor (5.18) is a complex quantity (Zeitler and Olsen

[388, 389], Frank [110], Reimer and Gilde [297], Ferwerda and Visser [103]) and can be recognized as the Fourier transform of the weak phase object approximation (5.5). The scattering processes discussed here are purely elastic so electrons should not be created or destroyed. The optical theorem (for example Schiff [312]) requires that the scattering factor $f_e(q)$ be complex to preserve the total number of electrons. A complex valued scattering factor does not imply that the atomic potential is complex. The atomic potential is a real valued function but a complete elastic scattering factor should be complex to preserve the total number of particles. The Moliere approximation for $f_e(q)$ (5.18) is a different value for each incident electron energy and is therefore difficult to tabulate for a general incident electron energy. The first Born approximation can be tabulated independent of energy which probably accounts for its popularity. The scattering amplitude is plotted in Fig. 5.9 in both the Moliere and Born approximations. The first Born approximation gets the magnitude about right but gets the phase of the scattering amplitude completely wrong. The phase of $f_e(q)$ increases dramatically at high angles whereas the phase of the first Born approximation is identically zero for all angles.

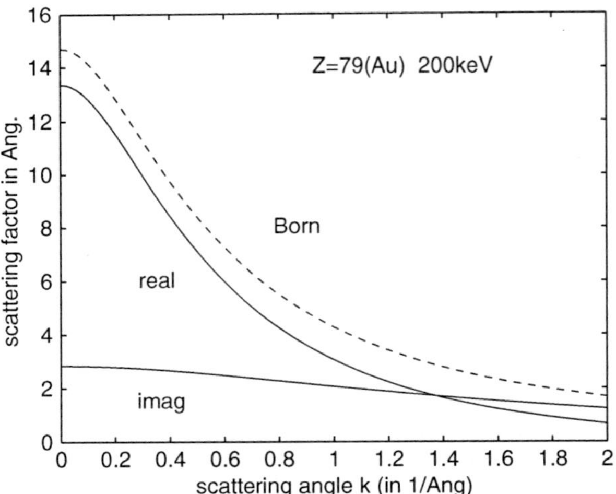

Fig. 5.9 The electron scattering factor for gold at 200 keV. The curves labeled real and imag. are the complex scattering factor in the Moliere approximation (5.18) and the curve labeled Born is the scattering factor in the first Born approximation scaled by the relativistic mass ratio m/m_0

5.3 Total Specimen Potential

The imaging electrons in the microscope interact with the effective potential of the specimen as a whole. Simulation of a whole electron microscope image requires a knowledge of the position of all atoms in the specimen. The main question is how to combine the single atom potentials to get the potential of the whole specimen. The simplest approach is to form the potential of the specimen by a linear superposition of the potentials of each atom in the specimen.

$$v_z(\mathbf{x}) = \sum_{j=1}^{N} v_{zj}(\mathbf{x} - \mathbf{x}_j) \tag{5.19}$$

where $\mathbf{x}_j = (x_j, y_j)$ is the position of atom j in a plane perpendicular to the optical axis of the microscope and $v_{zj}(\mathbf{x})$ is its projected atomic potential. This linear superposition approximation would be exact for single atoms separated by a distance that is large compared to the size of the atom. However, in an actual solid specimen the atoms are bound together and their outer valence electrons will have been rearranged slightly. This electronic rearrangement will also alter the projected atomic potential (and low angle scattering) slightly. The principle interaction causing high angle scattering is the interaction between the imaging electrons (in the microscope) and the large point charge at the atomic nucleus. Because the nucleus is unaffected by bonding in the solid the high angle scattering (as in ADF-STEM) should be well represented by a simple linear superposition of atomic potentials. The bonding in the solid should primarily affect the low angle scattering such as in bright field phase contrast (STEM and CTEM).

Inherent in the discussion of Sect. 5.2 is the assumption that each atom is spherically symmetric. This is only true for atoms whose valence shells are in the $l = 0$ or s angular momentum state. Early work by McWeeny [238, 239] and Freeman [112, 113] showed that the X-ray scattering from aspherical atoms (p-state valence shells) may vary by approximately 5%–10% with azimuthal angle in low Z atoms at small scattering angles. Electronic rearrangement in the solid should produce a similar magnitude of error in the projected atomic potential of the specimen. The error introduced with the linear superposition of atomic potentials approximation should be regarded as about 5% to 10% in the low angle scattering. This is an acceptable error for many calculations but caution should be taken when trying to extract precise quantitative information out of a simulation.

It is straight forward to write a general purpose computer program to calculate images of thin specimens with any element in the periodic chart that runs in a reasonable amount of computer time using the linear superposition of tabulated single atom potentials from appendix C (or other similar tabulations). Unfortunately, computing the electronic structure of solids including bonding is still a demanding task. It is not yet easy to write a general purpose computer program to calculate the electronic structure of an arbitrary solid, both because the theory is still under development to some extent and because of the large amount of computer time required. With the rapid advances in computational science this situation will likely improve in the near future, but for now a simple linear superposition (5.19) will have to suffice.

Previous tabulations of single atom properties usually were stated in terms of the electron scattering factors in the first Born approximation (5.15). The first Born approximation is totally inadequate to describe the scattering process but has a simple relationship to the atomic potential (as well as the X-ray scattering factors, see appendix C). The scattering factors in the first Born approximation naturally lend themselves to a calculation of the projected atomic potential in reciprocal (or Fourier) space. Once the potential has been calculated then a more correct scattering factor (5.18) or specimen transmission function can be calculated (5.7) from the specimen potential. The Fourier transform of (5.19) is:

$$V_z(\mathbf{k}) = \sum_{j=1}^{N} V_{zj}(\mathbf{k}) \exp(2\pi i \mathbf{k} \cdot \mathbf{x}_j) \tag{5.20}$$

substituting the electron scattering factor in the first Born approximation yields:

$$\sigma V_z(\mathbf{k}) = \lambda \frac{m}{m_0} \frac{1}{ab} \sum_j f_{ej}(k_x, k_y, 0) \exp(2\pi i \mathbf{k} \cdot \mathbf{x}_j), \tag{5.21}$$

where $f_{ej}(\mathbf{q})$ is the electron scattering factor in the first Born approximation for atom j. The total area of the specimen being simulated (with supercell dimensions $a \times b$) is ab and is required to normalize the Fourier transform properly. This constant may vary with different implementations of the Fourier transform. Note that $f_{ej}(\mathbf{q})$ is a function of a three dimensional wave vector $\mathbf{q} = (k_x, k_y, k_z)$ but the inverse transform is with respect to only two dimensions ($\mathbf{k} = (k_x, k_y)$) with zero for the third coordinate k_z. This results in the projected atomic potential (i.e., integration along the z axis) because of the Fourier projection theorem (see appendix B). The real space atomic potential can be found by inverse Fourier transforming $V_z(\mathbf{k})$ as:

$$v_z(\mathbf{x}) = \mathrm{FT}^{-1}[V_z(\mathbf{k})]. \tag{5.22}$$

This $v_z(\mathbf{x})$ can be used in (5.18) or (5.7). The summation on the right hand side of (5.21) is formally called the structure factor of the specimen given by:

$$F(\mathbf{q}) = \sum_j f_{ej}(\mathbf{q}) \exp(2\pi i \mathbf{q} \cdot \mathbf{x}_j). \tag{5.23}$$

If the specimen is a crystal then there will be a few discrete points $\mathbf{k}$ at which the Fourier component of the potential $V_z(\mathbf{k})$ is significantly larger than its neighboring values. These correspond to the Bragg reflections for this particular projection of the crystal. The Fourier components of the potential are zero at points in between the Bragg reflections for crystalline specimens. Thermal vibration of the atoms in the specimen may cause a small diffuse background in between the Bragg peaks (thermal diffuse scattering, TDS). If the specimen is amorphous then there will be a nearly continuous distribution of values at all points $\mathbf{k}$ in reciprocal space. Crystals containing defects or interfaces should be treated as if they are amorphous and all Fourier components of the potential should be calculated (i.e., not just the Bragg reflections).

5.4 BF Phase Contrast Image Calculation

This section will be discussed in the context of BF-CTEM image formation although it also applies to BF-STEM via the reciprocity theorem. In simple coherent BF-CTEM image formation the electron wave function incident on the specimen is a single plane wave of unit intensity. This wave function will pass through a thin specimen and experience a position dependent phase shift modeled as the specimen transmission function. Once the projected atomic potential of the specimen has been calculated (5.19 and 5.21) it is relatively straight forward to calculate the actual electron microscope image. From the projected atomic potential the wave function transmitted through the specimen is:

$$\psi_t(\mathbf{x}) = t(\mathbf{x})\exp(2\pi i k_z z) \sim t(\mathbf{x}), \tag{5.24}$$

where a common factor of $\exp(2\pi i k_z z)$ has been ignored (it will drop out because only the intensity matters in the end and it is unaffected by the transfer function). The specimen transmission function is:

$$t(\mathbf{x}) = \exp\left[i\sigma v_z(\mathbf{x})\right], \tag{5.25}$$

where $v_z(\mathbf{x})$ is the total projected atomic potential of the specimen (5.8, 5.22) and σ is the interaction parameter (5.6).

The electron microscope image is fundamentally cylindrically symmetric. It is essential that any simulation preserve this symmetry. Figure 5.10 shows a view of a sampled image in reciprocal space. The real space dimensions are $a \times b$ (in x and y). The general case in which a and b are very different is shown. There are two things that may alter the basic symmetry of the image and produce odd artifacts in the image. First, if the real space image size is not the same in x and y then the maximum spatial frequency may be different in each direction. The second related problem is the few Fourier coefficient in the four corners. If the entire area of the Fourier transform is just filled completely then various artifacts with rectangular symmetry can creep into the image. It is usually advisable to bandwidth limit the image with cylindrical symmetry. Only the spatially frequencies inside the largest inscribed circle as shown in (5.10) should be allowed to contribute to the final image. Although this may limit the resolution a little it at least does it with the proper symmetry and introduces fewer nonphysical artifacts in the image. This also means that it is best to make the image size square, otherwise a large percentage of the pixels will have to be set to zero and the calculation will not be very efficient. This bandwidth limit should be applied to both the projected atomic potential and the transmission function because the calculation in (5.25) is nonlinear and will generate many higher frequencies even though $v_z(\mathbf{x})$ may be properly bandwidth limited.

The transmitted wave function (5.24) is imaged by the objective lens of the microscope. The effects of the aberrations of the objective lens are easiest

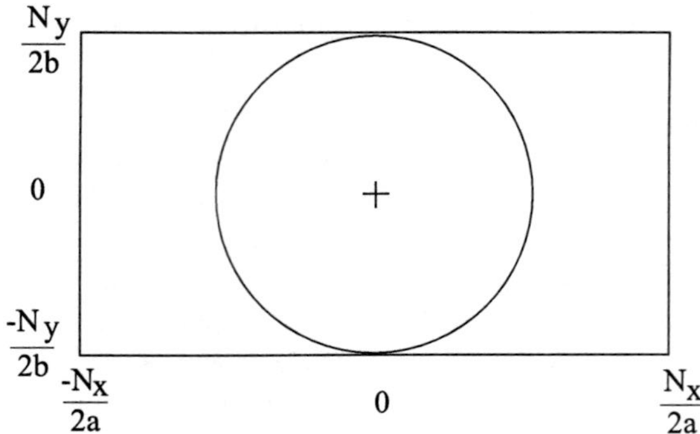

Fig. 5.10 Symmetrical bandwidth limiting the image in reciprocal space. The real space image has dimensions of $a \times b$ (in x and y) and is sampled with $N_x \times N_y$ pixels. Only those Fourier components inside the *largest inscribed circle* should be allowed to contribute to the image to avoid artifacts with an incorrect symmetry

to calculate by Fourier transforming the transmitted wave function and then multiplying by the transfer function of the objective lens.

$$\begin{aligned} \Psi_t(k) &= \mathrm{FT}[\psi_t(\mathbf{x})] \\ \Psi_i(k) &= \Psi_t(\mathbf{k})H_0(\mathbf{k}), \end{aligned} \tag{5.26}$$

where $\Psi_i(k)$ is the image wave function in the back focal plane of the objective lens and $H_0(k)$ is the transfer function of the objective lens.

$$\begin{aligned} H_0(\mathbf{k}) &= \exp[-\mathrm{i}\chi(\mathbf{k})]A(\mathbf{k}) \\ \chi(k) &= \pi\lambda k^2(0.5C_s\lambda^2k^2 - \Delta f), \end{aligned} \tag{5.27}$$

where Δf is defocus, C_s is the coefficient of spherical aberration and $A(\mathbf{k})$ is the aperture function:

$$\begin{aligned} A(\mathbf{k}) &= 1; \quad \lambda k = \alpha < \alpha_{\max} \\ &= 0; \quad \text{otherwise} \end{aligned} \tag{5.28}$$

$\alpha_{\max}$ is the maximum semiangle allowed by the objective aperture. Other aberrations may be included but only the simplest few are given here for simplicity. The objective lens images this wave function to form the final electron microscope image. The actual magnification of the objective lens may be ignored in the calculation (but NOT in practice) if the image coordinates are always referred to the actual specimen dimensions.

The actual recorded image $g(\mathbf{x})$ is the magnitude squared of the image wave function after inverse Fourier transforming back into real space.

$$\begin{aligned} \psi_i(\mathbf{x}) &= \mathrm{FT}^{-1}[\Psi_i(\mathbf{k})] \\ g(\mathbf{x}) &= |\psi_i(\mathbf{x})|^2 = |\psi_t(\mathbf{x}) \otimes h_0(\mathbf{x})|^2, \end{aligned} \tag{5.29}$$

where $h_0(\mathbf{x})$ is the complex point spread function of the objective lens (the inverse Fourier transform of $H_0(\mathbf{k})$). The steps required to calculate the image of a thin specimen are summarized in algorithmic form in Table 5.1.

Table 5.1 Steps in the simulation of CTEM images of thin specimens (a phase grating calculation)

Step 1	Calculate the projected atomic potential $v_z(\mathbf{x})$ from (5.19) or (5.21).
Step 2	Calculate the transmission function $t(\mathbf{x}) = \exp[i\sigma v_z(\mathbf{x})]$ (5.25) and symmetrically bandwidth limit it. The incident wave function is a plane wave so the transmitted wave function is equal to the transmission function.
Step 3	Fourier transform the transmission function $T(\mathbf{k}) = \mathrm{FT}[t(\mathbf{x})]$.
Step 4	Multiply the Fourier transform of the transmission function by the transfer function of the objective lens, $H_0(k)$ (5.27) to get the image wave function in the back focal plane $\Psi_i(\mathbf{k}) = H_0(k)T(\mathbf{k})$.
Step 5	Inverse Fourier transform the image wave function $\psi_t(\mathbf{k}) = \mathrm{FT}^{-1}[\Psi_i(\mathbf{k})]$.
Step 6	Calculate the square modulus of the image wave function (in real space) to get the final image intensity $g(\mathbf{x}) = |\psi_t(\mathbf{x}) \otimes h_o(\mathbf{x})|^2$.

5.4.1 Single Atom Images

Figure 5.11 shows the specimen transmission function (5.25) for the five single atoms plotted in Fig. 5.5. The atoms are arranged in a row 10 Å apart. The atomic potential was calculated in an image size of 50 Å on a side and 512×512 pixels using (5.21). The slight ringing near each atom is due to the finite bandwidth of the sampled image. There is also a slight asymmetry in some of the atoms. This occurs because there will always be some atom positions that are not exactly integer multiples of the pixel spacing. The atom potential is then spread across neighboring pixels in a nonsymmetrical manner. This asymmetry should be small and vanish in the final images. Also note that the real part of the transmission function has a stronger dependence on atomic number Z than the imaginary part.

The actual height of the peak for each atom is mainly a function of the sampling size. The actual potential has a singularity at the center of each atom (see Fig. 5.5). The value at the center of the atom is the average over one pixel. As this pixel gets smaller this value is closer to the singular value at the center of the atom. The integrated value should come through to the image properly with different pixel sizes although some care is required in choosing an appropriate pixel sampling size.

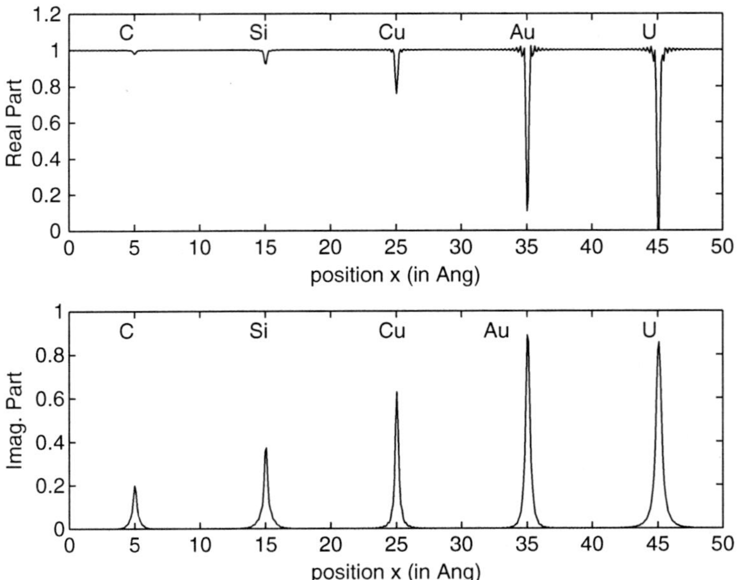

Fig. 5.11 Line scan of the complex transmission function for five isolated single atoms and an incident electron beam energy of 200 keV. This was calculated with a sampled image of 512×512 pixels. The scan goes through the center of each atom (atomic number $Z = 6,14,29,79,92$)

Figure 5.12 shows the image intensity (5.29) of the five isolated single atoms calculated from Fig. 5.11 with Scherzer conditions. The incident electron intensity is assumed to be unity for this simulation so that the image intensity in between the atoms (i.e., vacuum) should be one. The rings surrounding each atom are part of the so-called Airy disk caused by a sharp cut off in reciprocal space due to the objective aperture. The rings on the right wrap around to interfere with the atom on the left (the wrap-around effect). Figure 5.13 show a line scan through the center of each atom in Fig. 5.12. Eisenhandler and Siegel [89] and Reimer and Gilde [297] have also plotted single atom image profiles.

In the special case of isolated single atoms the image intensity $g(\mathbf{x})$ in bright field phase contrast may be written as:

$$g(\mathbf{x}) = \left| 1 + 2\pi i \int_0^{k_{\max}} f_e(k) \exp[-i\chi(k)] J_0(2\pi k r) k dk \right|^2 \tag{5.30}$$

where $f_e(k)$ is the electron scattering factor in the Moliere approximation (found by numerical integration of (5.18) using the projected atomic potential from appendix C), $\chi(k)$ is the aberration function, $k_{\max} = \alpha_{\max}/\lambda$ is the maximum spatial frequency in the objective aperture and $J_0(x)$ is the Bessel function of order zero. Note that the first Born approximation for the scattering factor should not be used in this expression. A graph of the peak single atom signal $1 - g(0)$ in coherent bright

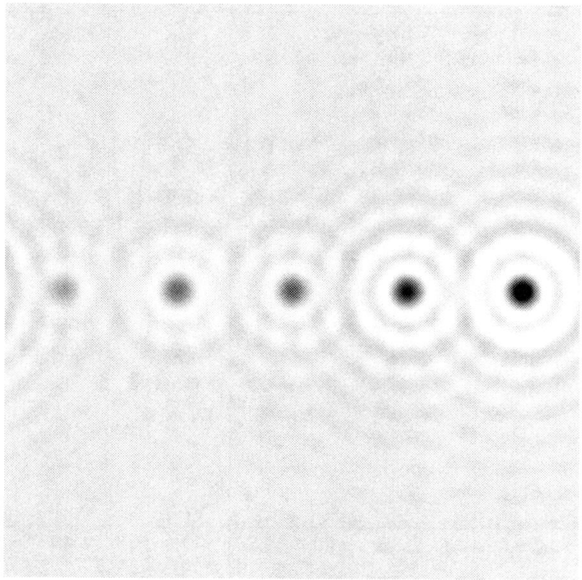

Fig. 5.12 The coherent bright field phase contrast image of five isolated single atoms and an incident electron beam energy of 200 keV. The electron optical parameters are spherical aberration $C_s = 1.3$ mm, defocus $\Delta f = 700$ Å and an objective aperture of 10.37 mrad (Scherzer conditions). This was calculated with a sampled image of 512×512 pixels. Atomic number Z = 6(C),14(Si),29(Cu),79(Au),92(U) (left to right). The image ranges from 0.72 (*black*) to 1.03 (*white*)

field phase contrast found by numerical integration of (5.30) is shown in Fig. 5.14. The BF signal varies weakly with atomic number. The overall trend approximately follows $Z^{0.6}$ to $Z^{0.7}$ however there is a significant variation reflecting the filling of different valence shells as in the periodic chart. Note that different atomic numbers can have the same signal and heavier atoms can have a smaller signal than lighter atoms (the signal is not monotonic in atomic number Z). It is possible to distinguish heavy atoms and light atoms but it would not be possible to precisely identify any specific atom by its bright field phase contrast signal alone.

5.4.2 Thin Specimen Images

Silicon is a reasonable starting point to begin discussing image simulation. It is a low enough atomic number that it is reasonable to approximate it as a weak phase object and it has a relatively simple structure. A precise simulation should still include thickness effects as covered in later chapters but for now the geometrical effects of specimen thickness will be ignored. Thickness can be included in some sense by

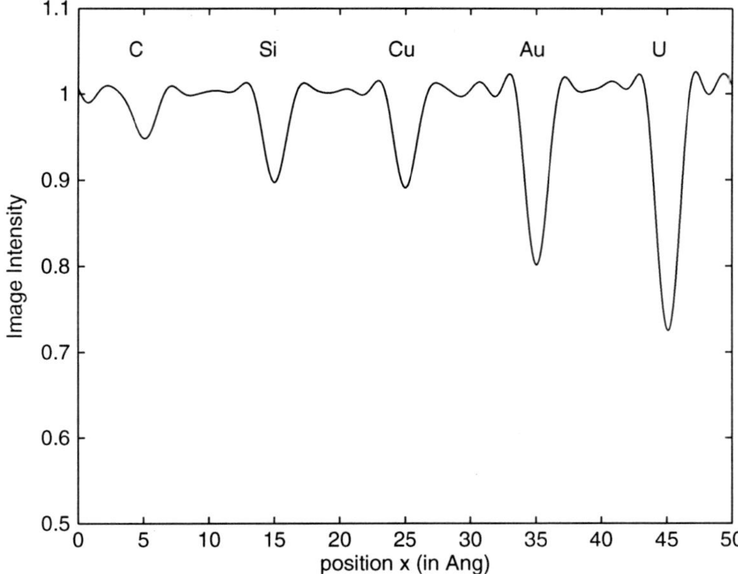

Fig. 5.13 Line scan through the center of the atoms in the coherent bright field phase contrast image in Fig. 5.12

superimposing the atoms along the optic axis in the specimen but the position of each atom along the beam direction will be ignored until later.

Silicon is a technologically important material and is the basic building block of most electronic devices. Its common crystalline form is the diamond structure (two interpenetrating face centered cubic or fcc lattices). When viewed in the electron microscope the three-dimensional lattice is projected into a two dimensional image. The three common projections of the diamond lattice with a cubic cell dimension of a are shown in Fig. 5.15. The (100) projection has a repeat length of a along the beam direction with one atom per repeat length. The (110) projection repeats every $a\sqrt{2}$ with two atoms per repeat length and two different types of atomic layers. The (111) projection repeats every $a\sqrt{3}$ with two atoms per site and three different layers.

To calculate an image of silicon in the weak phase object approximation requires sampling the atomic potential of silicon in a rectangular grid with the atoms placed at their respective positions in the grid. The discrete Fourier transform (DFT or FFT) requires that the sampled image obey periodic boundary conditions so the boundaries of the sampled image should match the natural periodicity of the actual specimen. This means that the sampling grid size should be an integer number of unit cells shown in Fig. 5.15 or equivalent. The full scale dimensions of the sampling grid will be labeled a, b in the horizontal and vertical directions, respectively. This nomenclature is drawn from crystallography and should not be confused with the real physical dimensions of the crystal (which are also frequently labeled a, b, c).

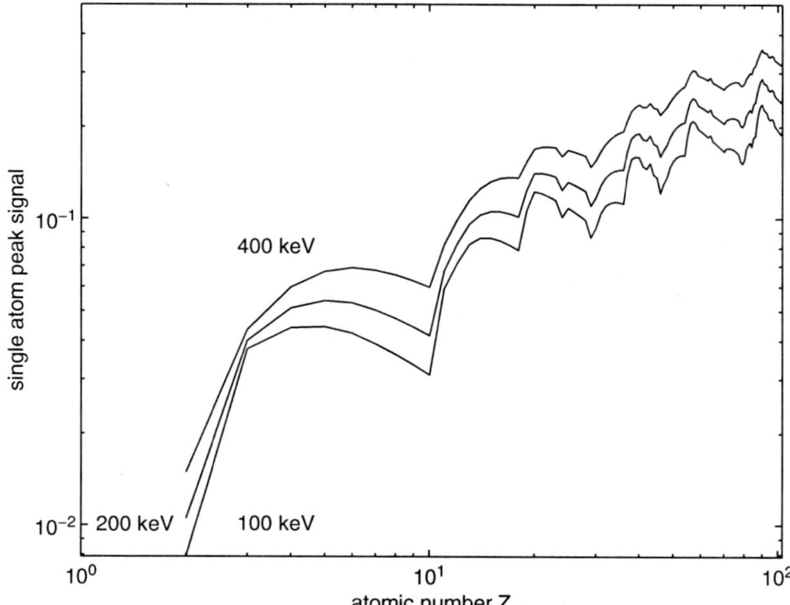

Fig. 5.14 Single atom peak signal in coherent bright field phase contrast vs. atomic number Z for incident electron energies of 100 keV, 200 keV, and 400 keV. Spherical aberration was fixed at $C_s = 1.3$ mm. Scherzer conditions were used for defocus and objective aperture

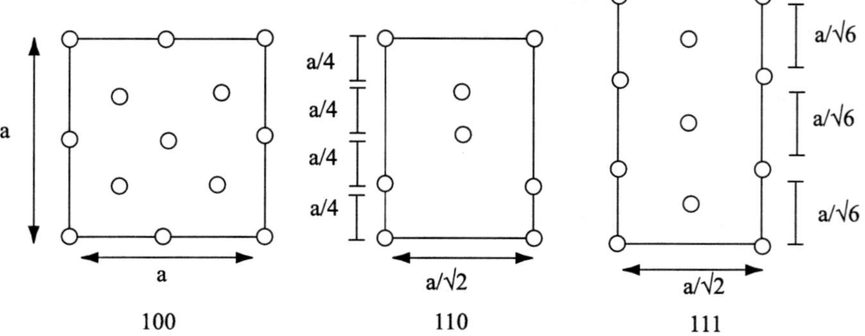

Fig. 5.15 Common projections of the silicon lattice (*diamond structure* with cubic cell size a = 5.43 Å). The *circles* represent the projected position of the silicon atoms. The indicated crystal direction (100, 110, or 111) is along the optic axis of the microscope and perpendicular to the plane of the paper

The meaning will usually be clear from the context in which they are used. As an example an image of the (110) projection of silicon will be calculated. The full scale dimensions of the (110) unit cell in Fig. 5.15 will be called $a_0 = a_{Si}/\sqrt{2}$ in the horizontal direction and $b_0 = a_{Si}$ in the vertical direction. a_{Si} is the the real physical cubic cell size of silicon.

It is tempting to make the real space sampling grid the same size as the two dimensional unit cell in Fig. 5.15 but this is not a good idea. A typical electron microscope has a resolution of no better than about 1.5–2Å. A good rule of thumb is to keep the real space sampling resolution ($\Delta x, \Delta y$) to be no bigger than about 1/3–1/4 of the final resolution. This way atoms still look round and not rectangular. With a_0 =3.84Å and b_0 =5.43Å this would mean that $N_x = a_0/\Delta x \sim 7.7$ pixels and $N_y = b_0/\Delta y \sim 10.9$ pixels. Rounding up to the nearest power of two would give an image of 8 by 16 pixels. Although the real space sampling size is adequate the reciprocal space sampling is very wrong in this simple argument. There are actually two sampling requirements that must be meet. Both real space and reciprocal space are important. With a single unit cell of size $a_0 \times b_0$ the resolution in reciprocal space is $\Delta k_x = 1/a_0 = 0.26\text{Å}^{-1}$ and $\Delta k_y = 1/b_0 = 0.18\text{Å}^{-1}$. At 200 keV the electron wavelength is 0.02508Å. Using $\alpha = \lambda k$ means that this choice has an angular resolution of about 6.5 mrad and 4.6 mrad. In reciprocal space a typical objective aperture semiangle α is about 10 mrad. There would only be two or three pixels inside the objective aperture in the final image. This would be totally inadequate to sample the transfer function or the specimen. In practice it is a good idea to have at least 5–10 pixels in the radius of the objective aperture (much more is better).

The solution to this problem is slightly nonintuitive. The size of the sampled image should really be several times the size of the minimum unit cell of the crystal specimen. As long as a and b are integer multiples of a_0 and b_0 the periodic boundary conditions are satisfied, and the reciprocal space resolution can be improved (i.e., Δk_x and Δk_y made smaller). To get a reciprocal space sampling of about 1 mrad at 200 keV requires a supercell size of 7×5 unit cells (26.89×27.15 Å).

For this particular problem (a perfect crystal) expanding the number of unit cell in real space simply moves the Bragg peaks further apart in Fourier space with zero in between. The transfer function is not strictly needed at these in between zeros so there is no real improvement for this particular problem. However, in later examples with imperfect crystals and STEM probes this sampling will be important so this example will continue as if it were a more general problem, although there is some argument that this is not the most efficient approach (this is a quick calculation so CPU time is not much of a concern) for this particular problem.

The normalized coordinates for one projected unit cell of (110) silicon is shown in Table 5.2. The projected atomic potential for a four atom thick specimen of (110) silicon is shown in Fig. 5.16 a) with a super cell size of 7×5 silicon unit cells. Note the characteristic "dumb-bell" shape of pairs of silicon atoms. Figure 5.16 b,c) is the complex transmission function at a beam energy of 200 keV for this specimen and Fig. 5.17 a,b are the bright field images that would result at electron beam energies of 200 keV and 400 keV, respectively. The lattice is not easily visible at an energy of 100 keV under Scherzer conditions at this value of C_s. Deviations from Scherzer conditions can however produce some interesting effects (see for example Izui [180], Hutchison and Waddington [166]). When the image is calculated at 100 keV the result has a structure that resembles a crystal lattice but has a range that is of order 10^{-5} to 10^{-6} of its average value. This is just the roundoff error of the numerical calculation and has no physical significance. This can happen often in simulation and you have to be careful to look at the actual numerical range of the image as well as its structure. The computer will just scale whatever numerical range is there to fill the available grey scale and produce an image. Although the dumb-bell shape is not resolved at either energy the higher voltage does start to get an elongated shape for the atom pairs. Spence et al. [333] and Desseaux et al. [78] have given a similar discussion of 110 germanium.

Table 5.2 The normalized coordinates for one projected unit cell of the silicon (Z=14) lattice in the (110) projection

Atom	Occupancy	x/a	y/b
1	1	0	0
2	1	0.5	0.75
3	1	0	0.25
4	1	0.5	0.5

The two dimensional unit cell dimensions are $a \times b$ where $a = a_{Si}/\sqrt{2} = 3.8396$Å and $b = a_{Si} = 5.43$Å

The apparent resolution of the simulated images in Fig. 5.17 is consistent with the linear image transfer function as shown in Fig. 5.18. The lowest order allowed reflection in the 110 projected unit cell (Fig. 5.15) are the 111 type beams with a spacing of $a_{Si}/\sqrt{3}$ =3.13Å, and the distance between the two atoms in the pair in the center of the unit cell is $a_{Si}/4$ =1.36Å (Edington [87]). The larger of these spacing should be clearly resolved in all but the 100 keV case, but the smaller of these two should not be resolved at all, consistent with the simulation in Fig. 5.17.

5.4.3 Partial Coherence and the Transmission Cross Coefficient

The BF image intensity distribution is the square modulus of the transmitted wave function $\psi_t(\mathbf{x})$ convolved with point spread function $h_0(\mathbf{x})$ of the objective lens:

$$g(\mathbf{x}) = |\psi_t(\mathbf{x}) \otimes h_0(\mathbf{x})|^2 = [\psi_t(\mathbf{x}) \otimes h_o(\mathbf{x})]\,[\psi_t^*(\mathbf{x}) \otimes h_o^*(\mathbf{x})]\,. \tag{5.31}$$

The Fourier transform of this equation is:

$$\begin{aligned} G(\mathbf{k}) &= [\Psi_t(\mathbf{k})H_0(\mathbf{k})] \otimes [\Psi_t^*(-\mathbf{k})H_0^*(-\mathbf{k})] \\ &= \int \Psi_t(\mathbf{k}')H_0(\mathbf{k}')\Psi_t^*(\mathbf{k}'+\mathbf{k})H_0^*(\mathbf{k}'+\mathbf{k})\mathrm{d}^2\mathbf{k}' \\ &= \int T_{\mathrm{cc}}(\mathbf{k}',\mathbf{k}'+\mathbf{k})\Psi_t(\mathbf{k}')\Psi_t^*(\mathbf{k}'+\mathbf{k})\mathrm{d}^2\mathbf{k}', \end{aligned} \tag{5.32}$$

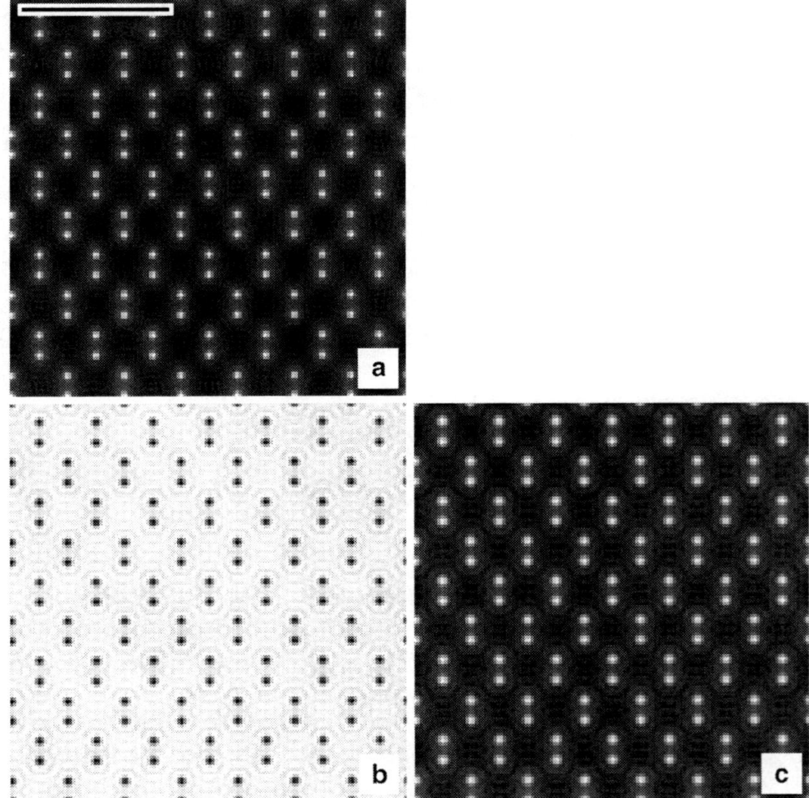

Fig. 5.16 Steps in simulating an image of 110 silicon in the weak phase object approximation. (a) Projected atomic potential (4 atoms thick, 128×128 pixels). (b,c) Real and imaginary part of the specimen transmission function at 200 keV. The scale bar in (a) is 10 Å. The image ranges are (b) 0.96–1.00, (c) 0.00–0.28. *White* is a larger positive number and atoms should appear white in (a)

where $T_{cc}(\mathbf{k}', \mathbf{k}' + \mathbf{k})$ is the transmission cross coefficient that is similar to the function of the same name in light optics (for example Sect. 10.5.3 of Born and Wolf [34]). In the special case of perfectly coherent image formation the transmission cross coefficient (distinguished by an additional superscript coh) is:

$$T_{cc}^{coh}(\mathbf{k}', \mathbf{k}' + \mathbf{k}) = \exp\left[-i\chi(\mathbf{k}') + i\chi(\mathbf{k}' + \mathbf{k})\right] A(\mathbf{k}')A(\mathbf{k}' + \mathbf{k}), \tag{5.33}$$

where $\chi(\mathbf{k})$ is the aberration function of the objective lens (5.27) and $A(\mathbf{k})$ is the aperture function (5.28). The portion of the transmitted wave function that passes through the objective aperture is combined with itself as illustrated in Fig. 5.19. $\Psi_t(\mathbf{k})$ is duplicated and offset by a vector $\mathbf{k}$. The overlap region is multiplied by the transmission cross coefficient and the integrated value in the intersection of the two apertures forms a single Fourier component at position $\mathbf{k}$. The direct implementation (in a program) of the transmission cross coefficient for coherent imaging is rather inefficient. It would require the calculation of a two dimensional convolution

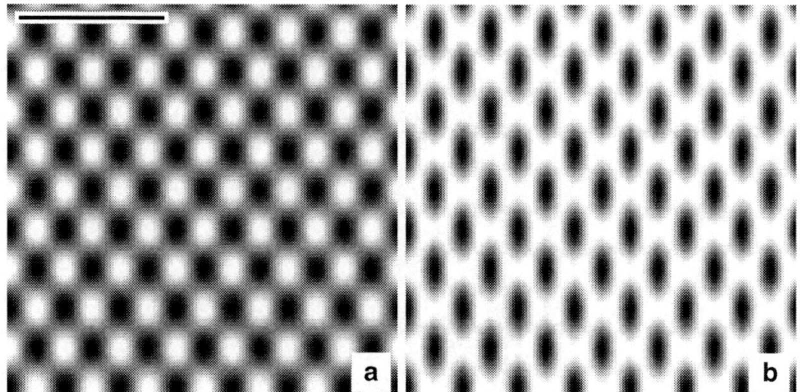

Fig. 5.17 Coherent bright field (BF) image of 110 silicon in the weak phase object approximation (128×128 pixels). (a) BF image at 200 keV, $\Delta f = 700$Å, $\alpha_{\max} = 10.37$ mrad. (b) BF image at 400 keV, $\Delta f = 566$Å, $\alpha_{\max} = 9.33$ mrad. Both have a spherical aberration of $C_s =$ 1.3 mm and Scherzer conditions. The scale bar in (a) is 10 Å. The image ranges are (a) 0.91–1.09, and (b) 0.87–1.07. White is a larger positive number and atoms should appear *black*

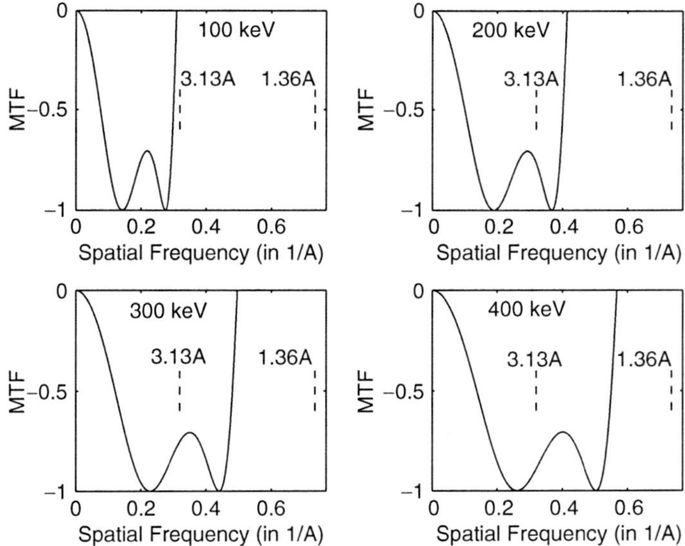

Fig. 5.18 The transfer function for a coherent bright field (BF) image in the weak phase approximation under Scherzer conditions. Spherical aberration is $C_s = 1.3$ mm and the electron energy varies from 100 kev to 400 keV. Two spacings relevant to the 110 projection of silicon are shown for comparison. About 1.36Å is the spacing between the dumbbells and 3.13Å is the spacing for the lowest order allowed reflection

at each point of reciprocal space (i.e., for each Fourier coefficient of the image). The use of the FFT as described earlier is much more efficient for coherent image formation.

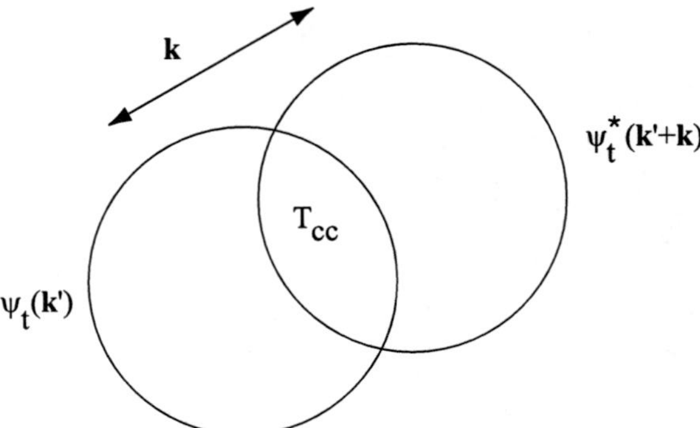

Fig. 5.19 The transmission cross coefficient in reciprocal space. Each *circle* represents the objective aperture. Only the overlap region labeled T_{cc} contributes to Fourier coefficient $\mathbf{k}$ in the image

The transmission cross coefficient however provides a mechanism for including the effects of partial coherence of strong phase objects as was discussed by O'Keefe [267], Ishizuka [173] and Pulvermacher [290]. (Bonevich and Marks [31] have also considered some higher order terms that are not discussed here.) If the specimen is thin enough to satisfy the weak phase object approximation then the form of the partial coherence derived in Sect. 3.2 is sufficient, however if the specimen contains heavy atoms or is many atoms thick then it may not be a weak phase object and a more detailed accounting of partial coherence is required. From the discussion on partial coherence in Sect. 3.2 the actual image will be formed with a small distribution of illumination angles and defocus values:

$$g(\mathbf{x}) = \int \left| \psi_t(\mathbf{x}) \otimes h_0(\mathbf{x}, \Delta f + \delta_{\mathrm{f}}, \mathbf{k}_\beta) \right|^2 p(\mathbf{k}_\beta) p(\delta_{\mathrm{f}}) \mathrm{d}\delta_{\mathrm{f}} \mathrm{d}^2\mathbf{k}_\beta, \tag{5.34}$$

where δ_{f} is the deviation in defocus, $\mathbf{k}_\beta$ is the deviation in illumination angle and $p(\delta_{\mathrm{f}})$ and $p(\mathbf{k}_\beta)$ are their distributions. If the specimen is thin enough so that its geometrical extent along the optic axis can be ignored (as has been assumed in this chapter) then the illumination angle can be included in either the specimen or the transfer function of the objective lens. When included as part of the objective lens the integration over δ_{f} and $\mathbf{k}_\beta$ can be completely contained within the transmission cross coefficient without modifying the transmitted wave function $\psi_t(\mathbf{x})$. The transmission cross coefficient with partial coherence becomes:

$$\begin{aligned} T_{cc}(\mathbf{k}', \mathbf{k}' + \mathbf{k}) = \\ \int \exp\left[-\mathrm{i}\chi(\mathbf{k}' + \mathbf{k}_\beta, \Delta f + \delta_{\mathrm{f}}) + \mathrm{i}\chi(\mathbf{k}' + \mathbf{k} + \mathbf{k}_\beta, \Delta f + \delta_{\mathrm{f}})\right] \\ A(\mathbf{k}' + \mathbf{k}_\beta) A(\mathbf{k}' + \mathbf{k} + \mathbf{k}_\beta) p(\delta_{\mathrm{f}}) p(\mathbf{k}_\beta) \mathrm{d}\delta_{\mathrm{f}} \mathrm{d}^2\mathbf{k}_\beta. \end{aligned} \tag{5.35}$$

If the deviations from the coherent mode are assumed to be small then this expression can be Taylor expanded to first order in δ_f and $\mathbf{k}_\beta$ as:

$$\begin{aligned}
&T_{\text{cc}}(\mathbf{k}',\mathbf{k}'+\mathbf{k}) \\
&= T_{\text{cc}}^{\text{coh}} \int \exp\{-\mathrm{i}[\mathbf{k}_\beta \cdot \frac{\partial}{\partial \mathbf{k}} + \delta_\text{f} \frac{\partial}{\partial \Delta f} + \delta_\text{f} \mathbf{k}_\beta \cdot \frac{\partial^2}{\partial \mathbf{k} \partial \Delta f} + \cdots] \\
&\qquad [\chi(\mathbf{k}') - \chi(\mathbf{k}'+\mathbf{k})]\} p(\delta_\text{f}) p(\mathbf{k}_\beta) \mathrm{d}\delta_\text{f} \mathrm{d}^2\mathbf{k}_\beta, \qquad (5.36)
\end{aligned}$$

where the small variation of the aperture function with $\mathbf{k}_\beta$ has been ignored (or equivalently the aperture diameter is much larger than the maximum spatial frequency of interest in the image). The indicated derivatives are:

$$\begin{aligned}
\mathbf{W}_{C1} &= \frac{\partial}{\partial \mathbf{k}} [\chi(\mathbf{k}') - \chi(\mathbf{k}'+\mathbf{k})] \\
&= 2\pi\lambda^3 C_s [|\mathbf{k}'|^2 \mathbf{k}' - |\mathbf{k}'+\mathbf{k}|^2 (\mathbf{k}'+\mathbf{k})] + 2\pi\lambda \Delta f \mathbf{k} \qquad (5.37) \\
W_{C2} &= \frac{\partial}{\partial \Delta f} [\chi(\mathbf{k}') - \chi(\mathbf{k}'+\mathbf{k})] \\
&= -\pi\lambda [|\mathbf{k}'|^2 - |\mathbf{k}'-\mathbf{k}|^2] \qquad (5.38) \\
\frac{\partial^2}{\partial \mathbf{k} \partial \Delta f} &[\chi(\mathbf{k}') - \chi(\mathbf{k}'+\mathbf{k})] = 2\pi\lambda \mathbf{k}, \qquad (5.39)
\end{aligned}$$

where the auxiliary symbols $\mathbf{W}_{C1}$ and W_{C2} will be used to simplify the derivation. Substituting for the derivatives yields:

$$\begin{aligned}
&T_{\text{cc}}(\mathbf{k}',\mathbf{k}'+\mathbf{k}) \\
&= T_{\text{cc}}^{\text{coh}} \int \exp\{-\mathrm{i}\mathbf{k}_\beta \cdot \mathbf{W}_{C1} - \mathrm{i}\delta_\text{f} W_{C2} - 2\pi \mathrm{i}\lambda \delta_\text{f} \mathbf{k}_\beta \cdot \mathbf{k}\} \\
&\qquad p(\delta_\text{f}) p(\mathbf{k}_\beta) \mathrm{d}\delta_\text{f} \mathrm{d}^2\mathbf{k}_\beta. \qquad (5.40)
\end{aligned}$$

Note that $\mathbf{W}_{C1}$ is a two dimensional vector quantity, as are all odd powers of $\mathbf{k}$. Assume that both the defocus and illumination angles have a Gaussian distribution:

$$p(\delta_\text{f}) = \frac{1}{\Delta_0 \sqrt{\pi}} \exp(-\delta_\text{f}^2/\Delta_0^2) \qquad (5.41)$$

$$p(\mathbf{k}_\beta) = \frac{1}{k_s^2 \pi} \exp(-\mathbf{k}_\beta^2/k_s^2), \qquad (5.42)$$

where δ_f and k_s are the $1/e$ widths of the two distributions. First perform the integration with respect to $\mathbf{k}_\beta$ giving:

$$\begin{aligned}
&T_{\text{cc}}(\mathbf{k}',\mathbf{k}'+\mathbf{k}) \\
&= T_{\text{cc}}^{\text{coh}} \int \exp\left[-\mathrm{i}\delta_\text{f} W_{C2} - k_s^2 |\mathbf{W}_{C1} + 2\pi\lambda \delta_\text{f} \mathbf{k}|^2/4\right] p(\delta_\text{f}) \mathrm{d}\delta_\text{f}. \qquad (5.43)
\end{aligned}$$

Note that this is equivalent to Fourier transforming the distribution function $p(\mathbf{k}_\beta)$. Next perform the integration with respect to δ_f, and use $\beta = \lambda k_s$ (i.e., the condenser angle in radians) to give:

$$T_{cc}(\mathbf{k}',\mathbf{k}'+\mathbf{k}) = T_{cc}^{coh}\frac{1}{\sqrt{1+\pi^2\beta^2\Delta_0^2k^2}}$$

$$\times \exp\left[-\frac{\beta^2}{4\lambda^2}W_{C1}^2 + \frac{\Delta_0^2}{4}\frac{(\pi\beta^2\mathbf{k}\cdot\mathbf{W}_{C1}/\lambda + iW_{C2})^2}{1+\pi^2\beta^2\Delta_0^2k^2}\right]. \tag{5.44}$$

The total transmission cross coefficient is a complicated damping factor whose net effect is similar to that for the weak phase approximation with linear imaging. However, the total transmission cross coefficient is much more complicated to calculate (note the vector nature of several components in the exponential factor). It is not separable into factors that depend only on $\mathbf{k}'$ and factors that depends only on $\mathbf{k}'+\mathbf{k}$. The image calculation with this form of partial coherence cannot be done using FFT's but must be done by an explicit weighted convolution in two dimensions (at each point in reciprocal space). This adds considerably to the required computer time but it is valid for a strongly scattering specimen with a negligible geometrical thickness such as several atomic layers of heavy atoms. If one of the deviations (defocus or angle) is small it may be more efficient to perform one integration analytically and the other numerically (possibly using FFTs) than to calculate the two dimensional convolution.

5.5 ADF STEM Images of Thin Specimens

In the STEM the objective lens acts on the electron beam before the beam interacts with the specimen in opposite order from the CTEM. The electrons that pass through the specimen and get scattered at high angles form the annular dark field or ADF signal. The wave function $\psi_p(\mathbf{x})$ of the focused probe incident upon the specimen at position $\mathbf{x}_p$ is formed by integrating the aberration wave function $\exp[-i\chi(\mathbf{k})]$ over the objective aperture as:

$$\psi_p(\mathbf{x},\mathbf{x}_p) = A_p\int_0^{k_{max}} \exp[-i\chi(\mathbf{k}) - 2\pi i\mathbf{k}\cdot(\mathbf{x}-\mathbf{x}_p)]d^2\mathbf{k}, \tag{5.45}$$

where $\lambda k_{max} = \alpha_{max}$ is the maximum angle in the objective aperture and A_p is a normalization constant chosen to yield;

$$\int |\psi_p(\mathbf{x},\mathbf{x}_p)|^2 d^2\mathbf{x} = 1. \tag{5.46}$$

In a practical sense the probe wave functions is easiest to calculate in Fourier space and then apply an inverse FFT.

$$\psi_p(\mathbf{x}, \mathbf{x}_p) = A_p FT^{-1} \left\{ \exp[-i\chi(\mathbf{k}) + 2\pi i \mathbf{k} \cdot \mathbf{x}_p] A(\mathbf{k}) \right\}, \tag{5.47}$$

where $A(\mathbf{k})$ is the aperture function (5.28). A graph of the ADF-STEM probe wave functions with Scherzer conditions is shown in Fig. 5.20 and 5.21. This is a complex valued function and the relative weighting of the real and imaginary parts can change dramatically with defocus etc.

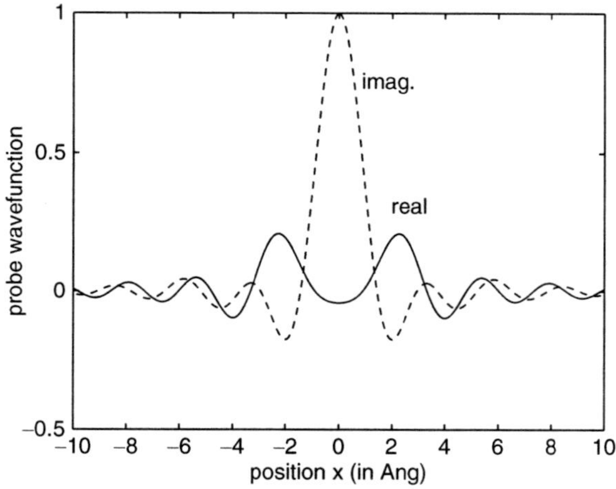

Fig. 5.20 Profile of the ADF-STEM probe wave functions (real and imaginary parts) for an electron energy of 200 keV, spherical aberration $C_s = 1.3$ mm., defocus $\Delta f = 700$Å, and an objective aperture of $\alpha_{\max} = 10.37$ mrad (Scherzer conditions). The effects of a finite source size have been ignored

The probe wave function is incident on the specimen and is modulated by the specimen transmission function $t(\mathbf{x})$ as it passes through the specimen. The wave function transmitted through the specimen is:

$$\begin{aligned} \psi_t(\mathbf{x}, \mathbf{x}_p) &= \psi_p(\mathbf{x}, \mathbf{x}_p) t(\mathbf{x}) \\ &= \psi_p(\mathbf{x}, \mathbf{x}_p) \exp[i\sigma v_z(\mathbf{x})]. \end{aligned} \tag{5.48}$$

In the STEM mode the transmitted wave function already has the effects of the objective lens aberration in it, unlike the CTEM mode where the objective lens effects enter after the wave function passes through the specimen. After passing through the specimen the transmitted wave function is then diffracted into the far field (i.e., the diffraction plane) and hits the ADF detector. Diffraction into the far field is equivalent to a Fourier transform of the transmitted wave function.

$$\Psi_t(\mathbf{k}) = \mathrm{FT}[\psi_t(\mathbf{x})]. \tag{5.49}$$

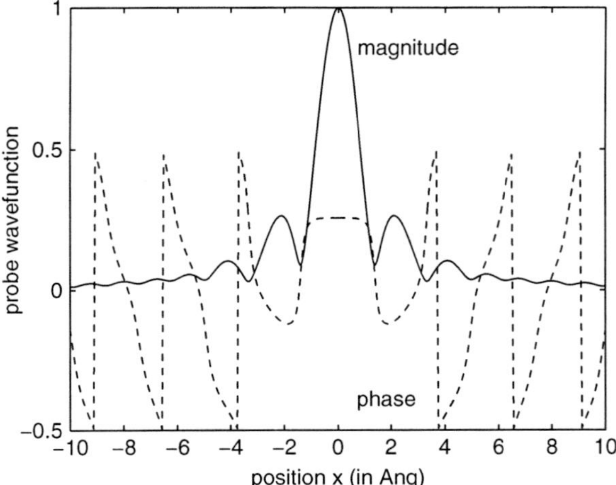

Fig. 5.21 Figure 5.20 replotted as a magnitude and phase. The phase is in units of 2π radians

The detector integrates the square modulus of the wave function in the diffraction plane to form the ADF-STEM signal at this point in the image (corresponding to the probe position $\mathbf{x}_p$):

$$g(\mathbf{x}_p) = \int D(\mathbf{k})|\Psi_t(\mathbf{k},\mathbf{x}_p)|^2 \mathrm{d}^2\mathbf{k}, \tag{5.50}$$

where $D(\mathbf{k})$ is the detector function:

$$\begin{aligned} D(\mathbf{k}) &= 1 \quad \text{on the detector} \\ &= 0 \quad \text{otherwise.} \end{aligned} \tag{5.51}$$

This process is repeated for each new position of the probe as it is scanned across the specimen (usually in a raster fashion). If $D(\mathbf{k})$ is a small point on the axis then the image is a bright field image and the discussion of Sect. 5.4 also applies via the reciprocity theorem. If the detector is a large annulus covering only high angle scattering then the image is an ADF (or annular dark field) image which is the focus of this section. This procedure is restated in algorithmic form in Table 5.3. The incoherent image model (3.66) captures many of the features of this calculation but is much faster (approximately as fast as the phase grating calculation) and may be a better approach in many cases.

5.5.1 Single Atom Images

The ADF detector in the STEM typically covers very large angles which will requires some changes to the sampling of the potential. First consider an image

Table 5.3 Steps in the calculation of STEM images of thin specimens

Step 1	Calculate the projected atomic potential $v_z(\mathbf{x})$ from (5.19) or (5.21).
Step 2	Calculate the transmission function $t(\mathbf{x}) = \exp[i\sigma v_z(\mathbf{x})]$ (5.25) and symmetrically bandwidth limit it.
Step 3	Calculate the probe wave function $\psi_p(\mathbf{x}, \mathbf{x}_p)$ at position $\mathbf{x}_p$ (5.45,5.47)
Step 4	Multiply the probe wave function by the specimen transmission function $t(\mathbf{x}) = \exp[i\sigma v_z(\mathbf{x})]$ to get the transmitted wave function $\psi_t(\mathbf{x})$.
Step 5	Fourier transform the transmitted wave function to get the wave function in the far field (diffraction plane).
Step 6	Integrate the intensity (square modulus) of the wave function in the diffraction plane including only those portions that fall on the annular detector (5.50). This is the signal for one point or pixel in the image.
Step 7	Repeat step 3 through step 6 for each position of the incident probe $\mathbf{x}_p$.

calculation of the five single atoms used in Fig. 5.13 for BF-CTEM phase contrast. The image is sampled in a 50Å by 50Å area. To get a scattering angle of 200 mrad at 200 keV requires that $0.5\lambda N_x/a \geq 200$ mrad. Therefore, the potential and wave functions should be sampled with 1024 pixels in each direction to get the large angles required for the ADF detector.

Figure 5.22 shows a line scan through the five single atoms used in Fig. 5.13. The transmission function (Fig. 5.11) is the same for both CTEM and STEM. The vertical axis is the portion of the incident probe intensity that falls on the ADF detector. The ADF signal (Fig. 5.22) is normalized slightly differently from the BF signal (Fig. 5.11) because of the different way in which they are generated. The ADF signal is relative to the total incident beam current but the BF signal is relative to the incident beam current density (the incident beam has a uniform intensity of one at all positions).

Even though the ADF signal and the BF signal are normalized differently it is apparent that the ADF signal is much weaker than the BF signal. However, the ADF signal shows a much stronger contrast between heavy and light atoms. It is even possible to image single heavy atoms on thin carbon supports many atoms thick because of the large increase in signal with atomic number Z (see for example Crewe et al. [68], Isaacson et al. [172] and Langmore et al. [222]). The peak single atom signal in ADF-STEM for all atoms is shown in Fig. 5.23 as a function of atomic number Z for three different electron beam energies. The ADF-STEM signal varies as approximately $Z^{1.5}$ to $Z^{1.7}$. Spherical aberration was fixed at $C_s = 1.3$ mm and Scherzer conditions were used for the defocus and the objective aperture. The inner angle of the ADF detector was set to four times the objective aperture and the outer angle was twenty times the objective aperture (this should integrate everything scattered beyond the inner angle). It is interesting to note that the single atom signal decreases with increasing electron beam energy contrary to BF phase contrast

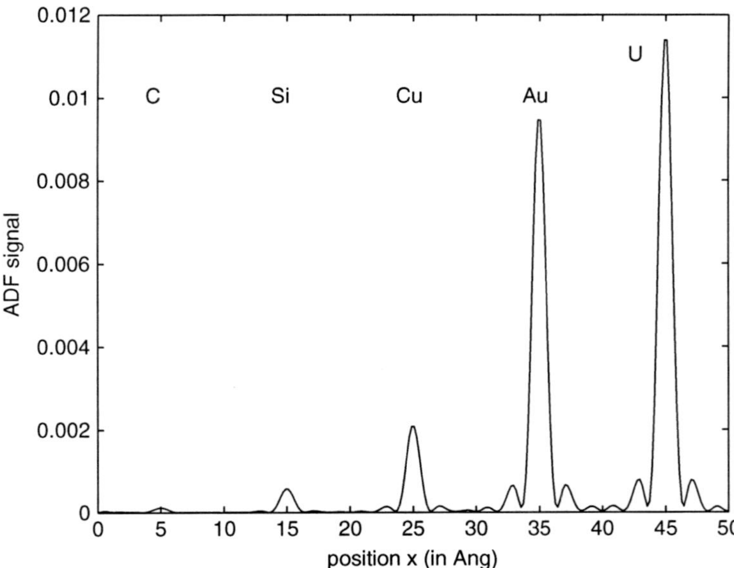

Fig. 5.22 Line scan of the ADF-STEM image intensity through the center of five isolated single atoms and an incident electron beam energy of 200 keV. The electron optical parameters are spherical aberration $C_s = 1.3$ mm, defocus $\Delta f = 700$ Å and an objective aperture of 10.37 mrad (Scherzer conditions). The ADF detector covers 40–200 mrad. This was calculated with a sampled image of 1024×1024 pixels (atomic number Z = 6,14,29,79,92)

signal shown in Fig. 5.14. As energy increases the resolution also increases meaning that the objective aperture also increases. The inner angle of the detector should be held at about 3–4 times the objective angle to remove any phase contrast effects. This means that the inner angle of the detector is also getting bigger so that there is less signal at high beam voltage (and higher resolution). The apparent signal decreases with increasing resolution in ADF-STEM. Also note that the weak phase object approximation may have small errors at 100 keV for very heavy atoms (Z near 100).

5.5.2 Thin Specimen Images

To simulate an image of (110) silicon also requires more sampling points than you might think at first. The image sequence given in Fig. 5.16 and 5.17 has a supercell size of $a \times b = 26.89 \times 27.15$ Å. The ADF-STEM detector collects large angles, and to get $\theta_d \sim 200$ mrad on the detector requires that the number of pixels in each direction be increased to about $N_x > 2a\theta_d/\lambda \sim 429$ pixels. The nearest power of 2 (for the FFT's) is 512×512 pixels.

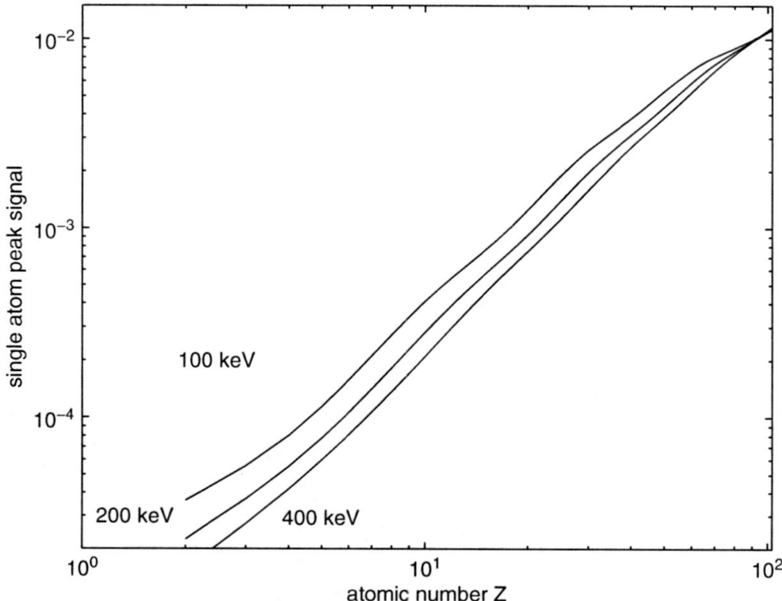

Fig. 5.23 Single atom peak signal in ADF-STEM vs. atomic number Z for 100 keV, 200 keV, and 400 keV. Spherical aberration was fixed at $C_s = 1.3$ mm and Scherzer conditions were used for defocus and objective aperture

Figure 5.24 shows the ADF-STEM image in the weak phase object approximation for three different energies (compare to Fig. 5.17). The specimen was four atoms thick and the signal is of the order of 10^{-3} to 10^{-4} of the incident beam intensity. The ADF detector spanned an angular range of 40–200 mrad for the 200 and 400 keV images and 45–200 mrad for the 100 keV image. The ADF signal with a large detector is an incoherent image mechanism and gives a slightly better resolution, however with a smaller signal. The characteristic dumb bell structure of silicon atom pairs (in the 110 projection) should be resolved at these beam energies and spherical aberration values. This ADF image does not rely on complicated scattering mechanisms like thermal diffuse scattering or high order Laue zones, but does qualitatively reproduce the features of an ADF-STEM image.

The simulated images in Fig. 5.24 are consistent with the incoherent annular dark field transfer functions as shown in Fig. 5.25. Both of the indicated spacings for the 110 projection of silicon can be resolved at the higher beam energies (compare to Fig. 5.25).

An experimental image of the 110 projection of silicon taken at 100 keV with an aberration corrected instrument is shown in Fig. 5.26. The dumbell spacing of 1.36Å is just barely resolved in this image. If only the geometrical lens aberration were considered for this image the probe size should be 1.0Å or less (and 1.36Å should be better resolved), however this image is most likely source limited.

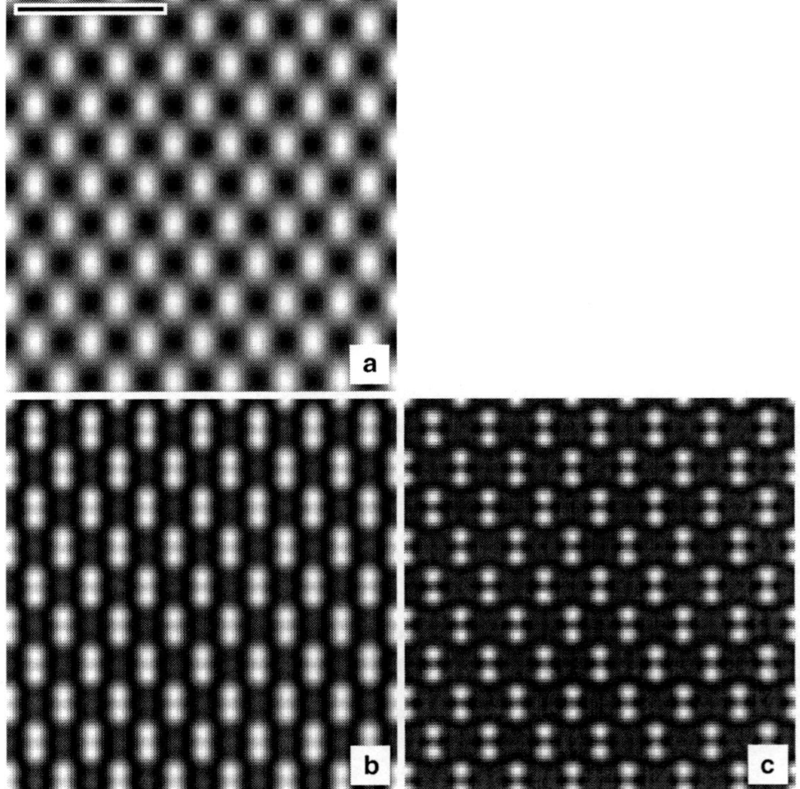

Fig. 5.24 Simulated ADF-STEM image of 110 silicon (4 atoms thick) in the weak phase object approximation with spherical aberration of $C_s = 1.3$ mm and three different beam energies. (a) 100 keV $\Delta f = 850$Å, $\alpha_{max} = 11.4$ mrad. (b) 200 keV, $\Delta f = 700$Å, $\alpha_{max} = 10.37$ mrad. (c) 400 keV, $\Delta f = 566$Å, $\alpha_{max} = 9.33$ mrad. All are Scherzer conditions. The wave function was sampled with 512×512 pixels but the final image is calculated for 128×128 pixels. The scale bar in (a) is 10 Å. The image ranges are (a) 0.00033–0.00133, (b) 0.00018–0.00072 and (c) 0.00003–0.00039. White is a larger positive number. Atoms should appear white in both images

The tip was nearing the end of its life and probably produces a larger than normal source size (lower brightness). A sequence of images calculated in the simple incoherent image model (3.66) with different source sizes (3.74) is shown in Fig. 5.27. There is a reasonable match to image d) with a 1Å source size. This produces more beam current (good) but reduces the obtainable resolution (usually bad). Part of this apparent source size blurring is probably due to a small stage vibration of about 0.5Å which appears not to have a preferred direction. The higher the resolution the more severe are the requirements on instrumental stability of all types.

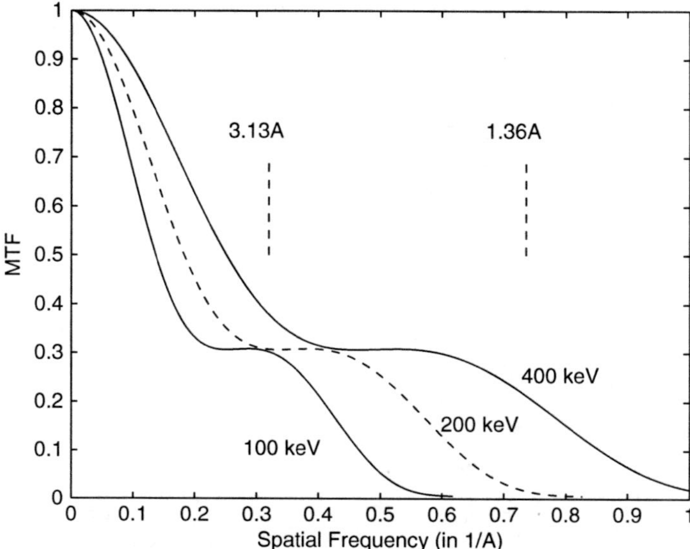

Fig. 5.25 The transfer function for an incoherent annular dark field STEM image in under Scherzer conditions. Spherical aberration is $C_s = 1.3$ mm and the electron energy varies from 100 kev to 400 keV. Two spacings relevant to the 110 projection of silicon are shown for comparison. About 1.36Å is the spacing between the dumbbells and 3.13Å is the spacing of the lowest order allowed reflection in the projected unit cell

Fig. 5.26 Experimental ADF-STEM image of 110 silicon recorded on a NION Ultra-STEM at 100 keV (aberration corrected to third order with reduced fifth order) with 512×512 pixels and a 35 mrad objective aperture. The scale bar is 5 Å. Atoms should appear *white*. This image has been averaged over four frames and low pass filtered to a little above the instrumental resolution

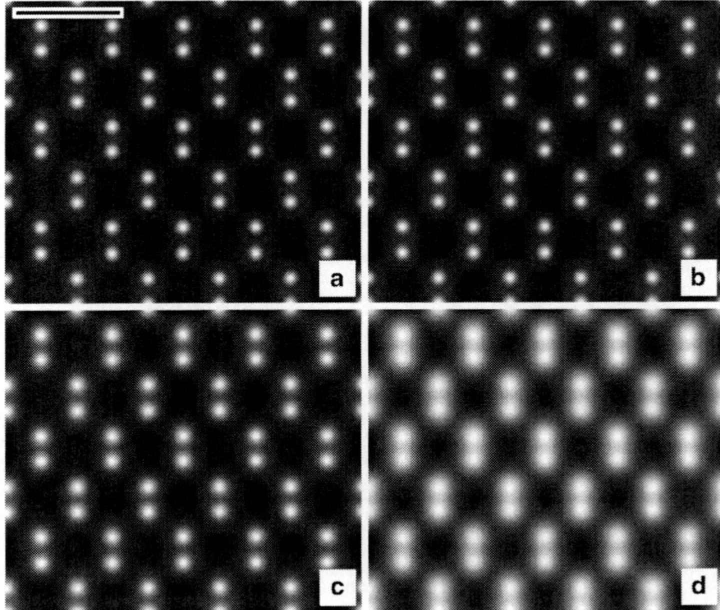

Fig. 5.27 Simple incoherent ADF-STEM images at 100 keV (C_{s3} =0 and C_{s5} = 5mm, 0 defocus and a 35 mrad obj. apert.). The scale bar is 5 Å. The source size is (a) 0.1Å, (b) 0.2Å, (c) 0.5Å, and (d) 1.0Å

5.6 Summary of Sampling Suggestions

The previous sections discussed sampling requirements for some specific cases in an anecdotal manner. There is really no easy way to calculate the required pixel size (sampling requirements) for every case in general. Usually you have to make an estimate of the sampling requirements, calculate an image with this estimate and compare to another calculation with better sampling (more and smaller pixels, etc.). If there is no significant change between the calculated images at two different pixels sizes (and number of pixels per image) then this is an indication that the image is correct. The deviation between two different calculations with different pixels sizes and number of pixels can also be used to estimate the error of the simulation itself. Table 5.4 gives a list of suggestions for the initial sampling sizes.

Table 5.4 Initial sampling suggestions for calculating images of thin specimens

1. The transmission function $t(\mathbf{x})$ (5.25) should be symmetrically bandwidth limited.
2. The real space pixel size Δx and Δy should be less than about $d_0/3$ to $d_0/4$ where d_0 is the resolution of the instrument.
3. The reciprocal space pixel size Δk_x and Δk_y should be less than about $k_{obj}/10$ where λk_{obj} is the maximum semiangle in the objective aperture.
4. The maximum angle in reciprocal space λk_{x-max} and λk_{y-max} should be about twice the maximum angle in the objective aperture (for CTEM) or slightly bigger than the maximum angle on the ADF-STEM detector (for STEM).

Chapter 6
Theory of Calculation of Images of Thick Specimens

Abstract This chapter describes the theory of calculating transmission electron microscope image of thick specimens (more than a few atoms thick), including the effects of multiple (or plural) scattering. Two popular methods are presented; Bloch wave methods and multislice methods. These approximations are typically good for specimens up to a few thousand Angstroms thick.

This chapter discusses how to calculate images of thick specimens including the effects of multiple or plural scattering in the specimen and the geometrical extent of the specimen along the optic axis of the electron microscope (the z direction). The electron interacts strongly with the specimen and can scatter more than once as it passes through specimens as thin as 10–50Å. When the electron can scatter more than once as it passes through the specimen the scattering is said to be dynamical. If the electron can only scatter once when passing through the specimen the scattering is said to be kinematical. The electron interaction in Chap. 5 is kinematical and the scattering processes discussed in this chapter are dynamical. Dynamical scattering also exists in X-ray diffraction (Batterman and Cole [21]) of thick specimens.

The instrumental aspects of the electron microscope and the passage of the electrons through the microscope are identical to what has already been described in previous chapters. This chapter will focus almost entirely on the relatively short portion of the electron trajectory as it passes through the specimen. Although this is a very short part of the electron's trajectory it is the most difficult to fully calculate because the electron interacts strongly with the specimen and can be scattered many times while passing through the specimen. In many ways this is one of the most interesting portions of the electron's path through the microscope because of the information about the specimen that it reveals.

The theory of dynamical electron diffraction has been studied by many authors over a large portion of this century. Bethe [24] first discussed dynamical scattering in 1928 in the context of electron diffraction (i.e., electron microscopes had not yet been invented). Bethe started with the Schrödinger equation and Fourier expanded the crystal potential and the electron wave function with components that

E.J. Kirkland, *Advanced Computing in Electron Microscopy*,
DOI 10.1007/978-1-4419-6533-2_6, © Springer Science+Business Media, LLC 2010

match the underlying periodicity of the crystal lattice. The Fourier components of the wave function have since become known as Bloch waves in analogy with Bloch's Theorem in solid state physics (see for example Ashcroft and Mermin [17] or Kittel [211]). Bethe solved for the three dimensional eigenvalues of the electron wave function in a crystalline specimen with the appropriate boundary conditions on the entrance and exit face of the crystal. Niehrs and Wagner [264], Fujimoto [118], and Sturkey [337] later organized the Bloch wave solution into a scattering matrix solution and Tournaire [346] developed a related reciprocal space matrix solution. Fertig and Rose [100] and Pennycook and Jesson [282–284] and Nellist and Pennycook [261] have extended the Bloch wave analysis to scanning transmission (STEM) microscopy. Howie and Whelan [163] used a different starting point but ended up with a set of coupled first order differential equations similar to the Bloch wave solution. Van Dyck [352] and Jap and Glaeser [186] have independently developed a path integral formulation of dynamical scattering. The history of the development of the theory of dynamical diffraction of electrons has been given by Cowley [59], Self et al. [315], Van Dyck [357], and Watanabe [372].

Cowley and Moodie [63] considered the dynamical scattering problem by starting from a physical optics point of view and derived a method that has become known as the multislice method. In this method the specimen is divided into thin two-dimensional slices along the electron beam direction (like a loaf of sliced bread). The electron beam alternately gets transmitted through a slice and propagates to the next slice. Each slice is thin enough to be a simple phase object and the propagation between slices is determined using Fresnel diffraction. Goodman and Moodie [127] later expanded the multislice theory into an accessible form appropriate for numerical implementation on a computer and showed how various methods of dynamical scattering calculations were related. Allpress et al. [10] and Lynch and O'Keefe [231] first implemented the multislice method on a computer to confirm the interpretation of high-resolution CTEM images of niobium oxides. Comparison of simulated images to images of known structures confirmed that image simulation with the multislice method could reliably simulate the observed image structure (Allpress and Sanders [11]). The availability of simulation programs such as SHRLI (O'Keefe et al. [270] and O'Keefe and Buseck [269]) lead to wide spread use of simulation in high-resolution image interpretation in CTEM. Ishizuka and Uyeda [179] produced a more rigorous quantum mechanical derivation of the multislice method and coincident with Bursill and Wilson [42] introduced the use of the fast Fourier transform or FFT which greatly reduced the computer time required for an image simulation. Cowley and Spence [66] have extended the multislice method to the calculation of convergent beam electron diffraction patterns (CBED) and Kirkland et al. [207] also extended it to include ADF-STEM image calculations, and Ishizuka [176] has presented some alternative suggestions to include the thermal diffuse scattering. Van Dyck [355, 356, 358], Coene and Van Dyck [50, 51] and Kilaas and Gronsky [196] recently proposed the so-called real space method which is related to the multislice method but performs the convolution and transmission in real space. The real space method may have advantages in some situations. O'Keefe [268] and Ishizuka [177] have recently reviewed the history of the multislice method.

If N is the number of Fourier components (also referred to as beams or Bloch waves) then any direct matrix solution will require the storage of N^2 elements in computer memory. The computer time required for a matrix multiplication scales as N^2 and the time for the direct solution of a matrix equation or Eigenvalue problem scales as N^3 (see for example Press et al. [288]). When there are a large number of beams (more than about 10 or 20) a direct matrix (Bloch wave) solution becomes very inefficient and the multislice solution using the FFT will prove to be much more efficient in computer time and memory requirements. Bloch wave matrix solutions can be obtained by hand with pencil and paper if there are only a small number of Bloch waves involved (two or three). This typically means that the specimen has to be a perfect crystal with a small unit cell. Analytical Bloch wave solutions can provide valuable insight into the imaging process, however small unit cell bulk crystal structures are already known (usually from X-ray diffraction) and there is no more than an academic interest in examining them again in the electron microscope. Most (nonbiological) specimens of interest contain interfaces or defects or are entirely amorphous. These specimens require relatively large numbers (many thousands) of beams or Bloch waves making a Bloch wave matrix solution impractical. The multislice method is generally much more efficient and easier to implement numerically on the computer and it is flexible enough to simulate defects and interfaces. The storage requirements for the multislice method scale as N and with the addition of the fast Fourier transform the computer time scales as approximately $N \log_2 N$. The multislice method is usually much more efficient for calculating dynamical electron diffraction patterns and images (Goodman and Moodie [127] and Self et al. [315]).

The multislice method solves the problem of propagation of a quantum mechanical wave packet through a potential. This is a rather general problem and it is not surprising that similar methods have evolved in fields other than electron microscopy. For example, the propagation of radio waves through an atmosphere can be treated in a method similar to the multislice method (Cordes et al. [54], French and Lovelace [114]) and Pidwerbetsky and Lovelace [285]). Molecular dynamics problems in chemistry also can be studied using at a method similar to the multislice method (Kosloff [217]). The split-step Fourier method (for example, Feit and Fleck [98, 107] or Agrawal [2] Sect. 2.4) is a numerical method used to calculate the propagation of light through a nonlinear media that is essentially the same as the multislice method.

As discussed in Sect. 2.3 the electrons in the microscope have enough energy that they should be treated with the relativistic Dirac wave equation with spin for a precise calculation. However, the simpler approach of ignoring the electron spin and using the nonrelativistic Schrödinger equation with the relativistic mass and wavelength will again be used because it is much easier to work with. This approximation is reasonably accurate at 100 keV but may introduce small errors at energies of order 1,000 keV or higher.

The specimen is almost always placed near the peak magnetic field of the objective lens (of order 1 Tesla or 10 kGauss). The electrons travel through the specimen while in a large magnetic field. Solving the problem with both the electrostatic interaction between the imaging electrons and the specimen and the imaging electrons and the magnetic field can be very difficult. The electron path deviation due to the magnetic field of the objective lens is on the scale of the focal length of the objective lens (about 1 mm). The electron path deviation due to the electrostatic interaction with the atoms in the specimen is on the scale of the specimen thickness (a few hundred Angstroms). Therefore, it is a reasonable approximation to neglect the action of the objective lens magnetic field on the scale of the specimen thickness. The electron trajectories while passing through the specimen will be calculated as if there were no external magnetic field present. This makes the problem somewhat easier to deal with.

Calculating the image given the atomic structure of the specimen is difficult enough (as will be seen). The inverse problem of calculating the atomic structure given the recorded image is even more difficult. The idea of an inverse multislice calculation has been around for a long time but few have had the courage to try it. Beeching and Spargo [22, 23]) have proposed approaches for an inverse multislice calculation. Allen et al. [5, 7, 8] and Spence et al. [329, 331] have considered the inverse problem using Bloch waves.

6.1 Bloch Wave Eigenvalue Solution

There is a long history of Bloch wave eigenvalue calculations starting with Bethe [24], with many worthwhile treatments in the literature, too numerous to mention all of them. Some recent reviews have been given by (for example) Humphreys [164], Spence [330], Spence and Zuo [334], and deGraf [129]. Bloch wave solutions of ADF-STEM have been considered by Fertig and Rose [100], Pennycook and Jesson [282], Nellist and Pennycook [260], Watanabe et al. [375], and Allen et al. [6] Bloch wave soultions for scanning confocal have been considered by Mitsuishi et al. [244]

6.1.1 Bloch Waves

The electron wave function can be expressed as a linear combination of any complete basis set. However, there is an advantage to using a basis set that also satisfies the Schrödinger equation in the specimen (with the same periodicity as the crystal) which are called Bloch waves. Expand the electron wave function in Bloch waves $b(\mathbf{k}_j, \mathbf{r})$ and their associated parameters $\mathbf{k}_j$ (scattering wave vectors on the Ewald sphere):

$$\psi(x,y,z) = \psi(\mathbf{r}) = \sum_j \alpha_j b_j(\mathbf{k}_j, \mathbf{r}). \tag{6.1}$$

With these Bloch waves any set of coefficients α_j are allowed inside the (crystal) specimen but only one set will also match the the incident wave function. There is an implicit assumption that the specimen is crystalline (periodic) so this approach may not work well for amorphous specimens.

Conceptually, the specimen can be thought of as a converter or filter that converts the incident electron (plane wave for BF CTEM, and probe for STEM) into a superposition of Bloch waves inside the (crystalline) specimen as in Fig. 6.1. The wave function and its first derivative must be continuous at the (top) entrance surface, which determines which Bloch waves will be initiated in the specimen. After entering the specimen the electrons propagate as Bloch waves and finally leave through the exit surface (bottom) of the specimen and are imaged by the objective lens. The characteristics of these Bloch waves determine how the electrons travel through the specimen. Hopefully, the strongest Bloch waves are representative of the actual structure in the specimen.

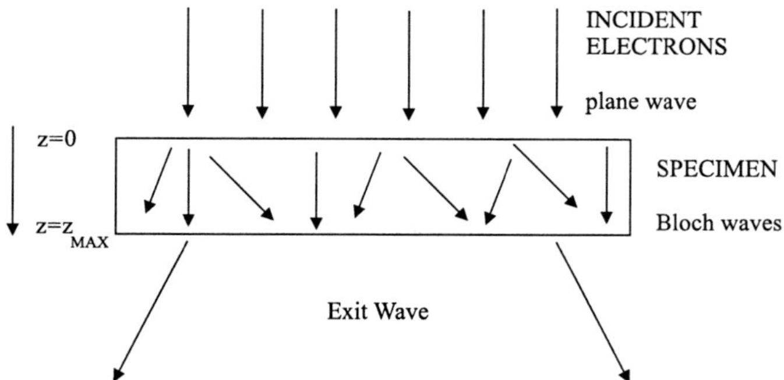

Fig. 6.1 The Bloch wave picture. The incident electron wave (plane wave for CTEM) becomes a superposition of Bloch waves inside the specimen. The properties of the Bloch wave determine the form of the wave that exits the specimen

Each Bloch wave must satisfy the Schrödinger equation in the (usually crystal) specimen:

$$\left[-\frac{\hbar^2}{2m}\nabla^2 - eV(x,y,z)\right] b_j(\mathbf{k}_j,\mathbf{r}) = E b_j(\mathbf{k}_j,\mathbf{r}), \tag{6.2}$$

where $\hbar = h/2\pi$ is Planck's constant divided by 2π, $m = \gamma m_0$ is the relativistic mass of the electron, $e = |e|$ is the magnitude of the charge of the electron, E is the kinetic energy of the electron and $-eV$ is the potential energy of the electron. V is the potential of the atoms in the specimen. The energy E will remain constant because elastic scattering is assumed, except for a slight increase due to the average inner potential of the specimen (with a related decrease in wavelength). The wavevector k_0 and energy of the incident wave (inside the specimen) are:

$$k_0 = \frac{1}{\lambda} \tag{6.3}$$

$$E = \frac{h^2 k_0^2}{2m} = \frac{h^2}{2m\lambda^2}, \tag{6.4}$$

where λ is the relativistically corrected electron wavelength. The Schrödinger equation becomes:

$$\left[-\frac{\hbar^2}{2m}\nabla^2 - eV(x,y,z)\right] b_j(\mathbf{k}_j,\mathbf{r}) = \frac{h^2k_0^2}{2m} b_j(\mathbf{k}_j,\mathbf{r}) \tag{6.5}$$

$$\begin{aligned} \left[\nabla^2 + 4\pi^2 k_0^2\right] b_j(\mathbf{k}_j,\mathbf{r}) &= -4\pi^2\left(\frac{2me}{h^2}\right) V(\mathbf{r}) b_j(\mathbf{k}_j,\mathbf{r}) \\ &= -4\pi^2 U(\mathbf{r}) b_j(\mathbf{k}_j,\mathbf{r}) \end{aligned} \tag{6.6}$$

$$U(\mathbf{r}) = \frac{2me}{h^2} V(\mathbf{r}) = \frac{\sigma}{\pi\lambda} V(\mathbf{r}), \tag{6.7}$$

where σ is the interaction parameter.

Each Bloch wave is basically a plane wave that is forced to have the periodicity of the (crystalline) specimen by multiplying by a linear combination of plane waves in three dimensions (a Fourier series):

$$\begin{aligned} b_j(\mathbf{k}_j,\mathbf{r}) &= \exp[2\pi i \mathbf{k}_j \cdot \mathbf{r}] \sum_{\mathbf{G}} C_{\mathbf{G}j} \exp[2\pi i \mathbf{G}\cdot\mathbf{r}] \\ &= \sum_{\mathbf{G}} C_{\mathbf{G}j} \exp[2\pi i (\mathbf{k}_j + \mathbf{G})\cdot\mathbf{r}]. \end{aligned} \tag{6.8}$$

The set of vectors $\mathbf{G} = (G_x, G_y, G_z) = (h/a, k/b, l/c)$ are typically the reciprocal lattice vectors of the specimen. The unit cell size of the specimen is (a,b,c) and (h,k,l) are integer indexes. In principle there are an infinite number of $\mathbf{G}$ vectors but in practice only a small number will typically be used. There is a different set of $C_{\mathbf{G}j}$ coefficients for each Bloch wave j.

6.1.2 Periodic Potential

Expand the specimen potential in a three-dimensional Fourier series as:

$$V(x,y,z) = V(\mathbf{r}) = \sum_{\mathbf{G}} V_{\mathbf{G}} \exp[2\pi i \mathbf{G}\cdot\mathbf{r}] \tag{6.9}$$

$$\begin{aligned} V_{\mathbf{G}} &= \frac{h^2}{2\pi m_0 e}\frac{1}{\Omega} \sum_j f_{ej}(|\mathbf{G}|) \exp(-2\pi i \mathbf{G}\cdot\mathbf{r}_j) \\ &= \frac{2\pi e a_0}{\Omega} \sum_j f_{ej}(|\mathbf{G}|) \exp(-2\pi i \mathbf{G}\cdot\mathbf{r}_j) \\ &= \frac{47.86}{\Omega} \sum_j f_{ej}(|\mathbf{G}|) \exp(-2\pi i \mathbf{G}\cdot\mathbf{r}_j), \end{aligned} \tag{6.10}$$

where $f_{ej}(q)$ is the electron scattering factor (in Angstroms) in the first Born approximation of the jth atom, e and m are the charge and mass of the electron, a_0 is the Bohr radius, and Ω is the unit cell volume (in cubic Angstroms) and $V_{\mathbf{G}}$ is in volts. The summation over j is over all atoms in the unit cell. The $\mathbf{G} = 0$ term in (6.9) is a uniform potential inside the specimen. There is a small computational advantage in absorbing this term into the electron wavelength (as already done in (6.3) so this term will be excluded in later equations.

If the specimen (crystal) has a center of symmetry (centro-symmetric) then for every atom at position $\mathbf{r}$ there is an identical atom at position $-\mathbf{r}$ and the terms in (6.10) appear as pairs of complex conjugates making $V_{\mathbf{G}}$ real, otherwise it may be complex. One possible strategy for finding the important $\mathbf{G}$ values is to calculate all $V_{\mathbf{G}}$ up to some maximum magnitude $|\mathbf{G}|$ (scan through integers (h,k,l)) and keep all with $|V_{\mathbf{G}}| > \varepsilon |V_{\mathbf{G}=0}|$, where ε is a small positive number chosen by the user (perhaps of order 10^{-5}).

6.1.3 Matrix Equation

Insert (6.9) for the potential [also using (6.7)] and (6.8) for the Bloch waves into the Schrödinger equation (6.6). The summation indexes $\mathbf{H}$ and $\mathbf{G}$ will both be over the full range of allowed reciprocal lattice vectors $\mathbf{G}$ but different letters are used to separate two different sums. The mean inner potential of the specimen $V_{\mathbf{G}=0}$ will be included in the wavelength so is dropped from the sum for the potential and the incident k_0 includes a small change of potential inside the specimen. Also dropping the common factor of $4\pi^2$ yields:

$$\sum_{\mathbf{G}} \left(k_0^2 - |\mathbf{k}_j + \mathbf{G}|^2\right) C_{\mathbf{G}j} \exp[2\pi i(\mathbf{k}_j + \mathbf{G}) \cdot \mathbf{r}]$$

$$= -\left[\sum_{\mathbf{H}\neq 0} U_{\mathbf{H}} \exp[2\pi i \mathbf{H} \cdot \mathbf{r}]\right] \left[\sum_{\mathbf{G}} C_{\mathbf{G}j} \exp[2\pi i(\mathbf{k}_j + \mathbf{G}) \cdot \mathbf{r}]\right] \tag{6.11}$$

$$= -\sum_{\mathbf{H}\neq 0} U_{\mathbf{H}} \left[\sum_{\mathbf{G}} C_{\mathbf{G}j} \exp[2\pi i(\mathbf{k}_j + \mathbf{G} + \mathbf{H}) \cdot \mathbf{r}]\right] \tag{6.12}$$

$$= -\sum_{\mathbf{H}\neq 0} U_{\mathbf{H}} \left[\sum_{\mathbf{X}} C_{(\mathbf{X}-\mathbf{H})j} \exp[2\pi i(\mathbf{k}_j + \mathbf{X}) \cdot \mathbf{r}]\right] \tag{6.13}$$

$$= -\sum_{\mathbf{H}\neq 0} U_{\mathbf{H}} \left[\sum_{\mathbf{G}} C_{(\mathbf{G}-\mathbf{H})j} \exp[2\pi i(\mathbf{k}_j + \mathbf{G}) \cdot \mathbf{r}]\right] \tag{6.14}$$

$$= -\sum_{\mathbf{G}} \left[\sum_{\mathbf{H}\neq 0} U_{\mathbf{H}} C_{(\mathbf{G}-\mathbf{H})j}\right] \exp[2\pi i(\mathbf{k}_j + \mathbf{G}) \cdot \mathbf{r}] \tag{6.15}$$

$$= -\sum_{\mathbf{G}} \left[\sum_{\mathbf{H}\neq \mathbf{G}} U_{\mathbf{G}-\mathbf{H}} C_{\mathbf{H}j}\right] \exp[2\pi i(\mathbf{k}_j + \mathbf{G}) \cdot \mathbf{r}], \tag{6.16}$$

where the substitution $\mathbf{X} = \mathbf{G} + \mathbf{H}$ was used in the second line and returned to $\mathbf{G}$ at the end. Some of these steps are not strictly valid unless the range of summation is infinite, which may not be true in practice (a small approximation). The $\mathbf{H} = \mathbf{G}$ term is excluded from the summation because the $\mathbf{G} = 0$ term (inner potential) has been transferred into the electron wavelength (via k_0). Equating coefficients of $\exp[2\pi\mathrm{i}(\mathbf{k}_j + \mathbf{G}) \cdot \mathbf{r}]$ yields:

$$\left(k_0^2 - |\mathbf{k}_j + \mathbf{G}|^2\right) C_{\mathbf{G}j} + \sum_{\mathbf{H} \neq \mathbf{G}} U_{\mathbf{G}-\mathbf{H}} C_{\mathbf{H}j} = 0. \tag{6.17}$$

This is a matrix equation where the first term (on the left hand side) is the diagonal elements and the second term is the off-diagonal elements. So far no serious approximations have been made and this expression is reasonably exact. However, both the $\mathbf{k}_j$ vectors and the $C_\mathbf{G}$ coefficients are unknown, so there is no obvious way to solve this equation. Next, several approximation will be made to get this equation into a linear form that can be solved. In the process half of the solutions will be lost, which turn out to be the backscattered electrons (deGraf [129]). These approximations are similar to those used in the multislice method.

A typical specimen is a thin slab of material perpendicular to the beam direction, and the incident electron beam is a very high energy traveling in a predominantly z direction. The incident beam is of order 100 keV or higher and the inner potential V_0 of the specimen is or order 10 eV so the kinetic energy does not change dramatically in the specimen.

$$k_0 \sim |\mathbf{k}_j| >> |\mathbf{G}| \tag{6.18}$$

This is sometimes referred to as the high-energy approximation. Also within this approximation all backscattered electrons will be ignored. In general some small percentage of electron will reverse direction (backscattered) but these will be ignored to simplify the mathematics.

Both the electron wave and its first derivative must be continuous across the entrance surface of the specimen. The inner potential of the specimen (V_0) may change the kinetic energy in the beam direction ($k_{0,z}$) but the transverse wave vector must be continuous at the entrance of the specimen. Approximate this requirement as applying to each individual Bloch wave rather than the wave function as a whole (weighted sum of Bloch wave vectors). Therefore, approximate the Bloch wave vector as the incident wave vector $\mathbf{k}_0$ plus a small term along the beam direction $\hat{z}$:

$$\mathbf{k}_j \sim \mathbf{k}_0 + \gamma_j \hat{z}, \tag{6.19}$$

where γ_j is a small quantity. Inserting this expression yields:

$$\begin{aligned} k_0^2 - |\mathbf{k}_j + \mathbf{G}|^2 &= k_0^2 - |\mathbf{k}_0 + \gamma_j \hat{z} + \mathbf{G}|^2 \\ &= k_0^2 - |\mathbf{k}_0 + \mathbf{G}|^2 - 2\gamma_j (\mathbf{k}_0 + \mathbf{G}) \cdot \hat{z} - \gamma_j^2 \end{aligned} \tag{6.20}$$

The γ_j^2 term may be ignored because γ_j is a small quantity. Also expanding terms leaves:

$$\begin{aligned} k_0^2 - |\mathbf{k}_j + \mathbf{G}|^2 &= 2k_0 s_{\mathbf{G}} - 2\gamma_j(\mathbf{k}_0 + \mathbf{G}) \cdot \hat{z} \\ &= 2k_0 s_{\mathbf{G}} - 2\gamma_j(k_{0,z} + G_z) \end{aligned} \tag{6.21}$$

$$2k_0 s_{\mathbf{G}} = k_0^2 - |\mathbf{k}_0 + \mathbf{G}|^2 = -2k_{0,z}G_z - |\mathbf{G}|^2, \tag{6.22}$$

where $s_{\mathbf{G}}$ is a general excitation error. Insert this expression back into (6.17) to obtain:

$$[2k_0 s_G - 2\gamma_j(k_{0,z} + G_z)]C_{\mathbf{G}j} + \sum_{\mathbf{H} \neq \mathbf{G}} U_{\mathbf{G}-\mathbf{H}} C_{\mathbf{H}j} = 0 \tag{6.23}$$

The high-energy approximation says that:

$$|\mathbf{G}| << |k_{0,z}| \tag{6.24}$$

so the G_z term is usually dropped. However, at 100 kV $k_0 \sim 27\text{Å}^{-1}$ and $|G_{\text{max}}|$ may be 4 or 5 Å^{-1} for high resolution, in which case G_z is important so this approximation may not be that good at high resolution (or for the higher order Laue zones). Rearranging terms and making the high-energy approximation leaves:

$$2k_0 s_{\mathbf{G}} C_{\mathbf{G}j} + \sum_{\mathbf{H} \neq \mathbf{G}} U_{\mathbf{G}-\mathbf{H}} C_{\mathbf{H}j} = 2\gamma_j k_{0,z} C_{\mathbf{G}j}. \tag{6.25}$$

This equation is repeated for each **G**. This is an eigenvalue matrix equation for the eigenvalues $2\gamma_j k_{0,z}$ and the eigenvectors (set of $C_{\mathbf{G}j}$). If the specimen is a centrosymmetric crystal then the matrix is real and symmetric, otherwise it may be complex and Hermitian. The eigenvectors are orthogonal and can be normalized (most subroutines automatically normalize their results), which will be assumed later. The eigenvalues will be real.

With N Bloch waves then this is said to be an N-beam calculation. There will also be N vectors **G**, N eigenvalues and N eigenvectors (N^2 coefficients $C_{\mathbf{G}j}$). The memory storage requirements scale as $\mathcal{O}(N^2)$. Solving the eigenvalue equation is a computationally difficult problem. The CPU time scales as $\mathcal{O}(N^3)$ so this approach is not competitive with the methods that will be discussed later for large N. However, small unit cells with a lot of symmetry may work well (small N). Finding eigenvalues and eigenvectors is also a difficult process. There are several existing free software libraries with well developed subroutine for this purpose. Lapack [13] (www.netlib.org) is probably one of the better packages and in general it is probably best to use this or similar packages. Lapack is mainly in Fortran but there are translations in several different programming languages. The GNU Scientific Library or GSL (www.gnu.org/software/gsl/) is written in C and also has some useful eigenvalue subroutines.

The structure of (6.25) is easier to see when a specific example is written out. In practice there will be many hundreds or thousands of beams, but for simplicity use only a few for this example. The $\mathbf{G} = 0$ term is almost always included to match the incident beam (typically a plane wave) in CTEM. If there are only four beams $\mathbf{G} = 0$, $\mathbf{D}$, $\mathbf{E}$, and $\mathbf{F}$, that are important then (6.25) becomes:

$$\begin{bmatrix} 0 & U_{-D} & U_{-E} & U_{-F} \\ U_D & 2k_0s_D & U_{D-E} & U_{D-F} \\ U_E & U_{E-D} & 2k_0s_E & U_{E-F} \\ U_F & U_{F-D} & U_{F-E} & 2k_0s_F \end{bmatrix} \begin{bmatrix} C_0 \\ C_D \\ C_E \\ C_F \end{bmatrix} = 2\gamma k_{0,z} \begin{bmatrix} C_0 \\ C_D \\ C_E \\ C_F \end{bmatrix}. \tag{6.26}$$

The off diagonal element are essentially the convolution of U_G with itself which may require the calculation of many more values than in the original small set. There will be four different eigenvalues $2\gamma k_{0,z}$ (with associated eigenvectors) in this example. In condensed matrix notation:

$$\mathsf{AC} = 2\gamma k_{0,z}\mathbf{C}. \tag{6.27}$$

If the crystal (specimen) is centro-symmetric then $U_G = U_{-G}^*$ and the matrix is real and symmetric otherwise it is Hermetian.

6.1.4 Initial Conditions and the Exit Wave

If the allowed Bloch wave have been determined (eignevalues and eigenvectors as earlier) then the wave function anywhere in the (crystalline) specimen is known if the weighting coefficients, α_j are known. The weighting coefficients can be determined by matching the wave function at the entrance surface of the specimen to the given incident wave function. First write the Fourier transform of the wave function with an explicit z dependence and a propagation in the beam direction $\mathbf{k}_0$:

$$\psi(\mathbf{r}) = \sum_{\mathbf{G}} \psi_{\mathbf{G}}(z) \exp[2\pi \mathrm{i}(\mathbf{k}_0 + \mathbf{G}) \cdot \mathbf{r}]. \tag{6.28}$$

Now equate this equation to the Bloch wave expansion, (6.1), (6.8), and (6.19):

$$\sum_{\mathbf{G}} \psi_{\mathbf{G}}(z) \exp[2\pi \mathrm{i}(\mathbf{k}_0 + \mathbf{G}) \cdot \mathbf{r}] = \sum_j \alpha_j b_j(\mathbf{k}_j, \mathbf{r}) \tag{6.29}$$

$$= \sum_{\mathbf{G}} \sum_j \alpha_j \, C_{\mathbf{G}j} \exp[2\pi \mathrm{i}(\mathbf{k}_j + \mathbf{G}) \cdot \mathbf{r}] \tag{6.30}$$

$$= \sum_{\mathbf{G}} \left[\sum_j \alpha_j \, C_{\mathbf{G}j} \exp[2\pi \mathrm{i}\gamma_j z] \right] \exp[2\pi \mathrm{i}(\mathbf{k}_0 + \mathbf{G}) \cdot \mathbf{r}]. \tag{6.31}$$

Now equate coefficients of $\exp[2\pi \mathrm{i}(\mathbf{k}_0 + \mathbf{G}) \cdot \mathbf{r}]$ to obtain:

$$\psi_{\mathbf{G}}(z) = \sum_j \alpha_j \, C_{\mathbf{G}j} \exp[2\pi \mathrm{i}\gamma_j z] \tag{6.32}$$

which illustrates how each Fourier coefficient of the wave functions propagates with depth z in the specimen. This equation is also in the form of a matrix equations. Using the previous example of four **G** values 0,D,E,F and adding four eigenvalues 0,1,2,3 an example of this equation becomes:

$$\begin{bmatrix} \psi_0(z) \\ \psi_D(z) \\ \psi_E(z) \\ \psi_F(z) \end{bmatrix} = \begin{bmatrix} C_{00} & C_{01} & C_{02} & C_{03} \\ C_{D0} & C_{D1} & C_{D2} & C_{D3} \\ C_{E0} & C_{E1} & C_{E2} & C_{E3} \\ C_{F0} & C_{F1} & C_{F2} & C_{F3} \end{bmatrix}$$
$$\times \begin{bmatrix} e^{2\pi i\gamma_0 z} & 0 & 0 & 0 \\ 0 & e^{2\pi i\gamma_1 z} & 0 & 0 \\ 0 & 0 & e^{2\pi i\gamma_2 z} & 0 \\ 0 & 0 & 0 & e^{2\pi i\gamma_3 z} \end{bmatrix} \begin{bmatrix} \alpha_0 \\ \alpha_1 \\ \alpha_2 \\ \alpha_3 \end{bmatrix} \tag{6.33}$$

In condensed matrix notation:

$$\psi(z) = \mathsf{C}[\exp(2\pi i\gamma_j z)]\alpha. \tag{6.34}$$

Where $\psi(z)$ is a column vector of $\psi_{\mathbf{G}}$. Each column of the C matrix containing the $C_{\mathbf{G}j}$ coefficients is an eigenvector of (6.25). The vector $\psi(z)$ contains the $\psi_{\mathbf{G}}(z)$ terms and $[\exp(2\pi i\gamma_j z)]$ denotes a diagonal matrix containing the eigenvalue dependence.

The form of the C matrix has the very useful property that its inverse is equal to its Hermetian adjoint (transpose plus complex conjugation) $\mathsf{C}^{-1} = \mathsf{C}^{\dagger}$ because the eigenvectors are orthogonal and normalized. Multiply each side by the inverse matrix for the special case of $z = 0$ (which is the known incident wavefunction) yields:

$$\mathsf{C}^{-1}\psi(z=0) = \mathsf{C}^{-1}\mathsf{C}\alpha = \alpha \tag{6.35}$$

$$\alpha = \mathsf{C}^{-1}\psi(z=0) = \mathsf{C}^{\dagger}\psi(z=0) \tag{6.36}$$

which determines the weighting coefficients α_j. Then the Fourier coefficients of the wave function $\psi_{\mathbf{G}}(z)$ may be calculated at any depth in the specimen from the eigenvalues, eigenvectors and weighting coefficients. This result can be elegantly summarized as a scattering matrix S (Sturkey [337]):

$$\psi(z) = \mathsf{C}[\exp(2\pi i\gamma_j)]\mathsf{C}^{-1}\psi(z=0) = \mathsf{S}\psi(z=0) \tag{6.37}$$

$$\mathsf{S} = \mathsf{C}[\exp(2\pi i\gamma_j)]\mathsf{C}^{-1}. \tag{6.38}$$

The wave function at the exit surface of the specimen is then calculated by inserting these weighting coefficients α_j (6.36) into (6.32) to obtain the set of $\psi_{\mathbf{G}}$ values at depth z. The slowly varying portion of the exit wave (similar to the multislice result), dropping the rapidly oscillating portion $\exp(2\pi i \mathbf{k}_0 \cdot \mathbf{r})$, is then given by an inverse Fourier transform (a single FFT) in two dimensions on (6.28);

$$\psi(x,y,z) = \mathrm{FT}_{xy}^{-1}\left[\sum_{\mathbf{G}} \psi_{\mathbf{G}} \exp(2\pi i G_z z)\right] \tag{6.39}$$

where FT_{xy}^{-1} is a 2D inverse (fast) Fourier transform over (G_x, G_y).

6.1.5 Bloch Wave Eigenvalue Summary

The steps in a Bloch wave eigenvalue calculation are summarized in algorithmic form in Table 6.1.

Table 6.1 Steps in the Bloch wave eigenvalue calculation of the wave function at the exit surface of the specimen

Step 1	Calculate the Fourier coefficients of the atomic potential $V_{\mathbf{G}}$ from (6.9) up to some maximum $|\mathbf{G}|$.
Step 2	Solve for the eigenvalues (proportional to γ_j) and eigenvectors ($C_{\mathbf{G}}$), both in (6.25).
Step 3	Find the weighting coefficients α_j to match the incident wave function at $z = 0$ (typically a plane wave for CTEM, or a focused probe for STEM) as in (6.36) .
Step 4	Calculate the electron wave function at the exit surface of the specimen (6.34) and (6.39).

The remaining steps to add the effects of the objective lens in CTEM (imaging) or STEM (probe forming) are similar to other approaches (not given here)

The scaling of computer (CPU) time of the Bloch wave method are compared to the multislice method in Fig. 6.2 for the 100 projection of aluminum. This specimen is centro-symmetric and the Bloch wave calculation took advantage of the real (not complex) matrix to be more efficient (about a factor of two) whereas the multislice calculation was a general complex valued calculation. Both calculations were carried out on the same inexpensive laptop computer (using a single CPU or thread) using the same operating system and compiler with similar compiler options. The Bloch wave calculation used a Lapack subroutine (in C using the CLAPACK package) to perform the eigenvalue calculation and the multislice calculation used an FFT. Both programs could be improved a little in terms of efficiency so the absolute time is not so important but it is reasonable to compare the relative performance. The Bloch wave method is the same time for all thickness but the multislice method scales linearly with thickness so two different specimen thickness are shown. Having all thickness available from a single Bloch wave calculation can be an advantage or a disadvantage (all thickness are calculated whether needed or not). The number of beams for the multislice method is taken to be the number of Fourier components in a single two-dimensional slice after removing aliasing. There is an argument for counting each slice as a different beam (not done here) in which case the two curves for the multislice method would likely come together and move to a larger N.

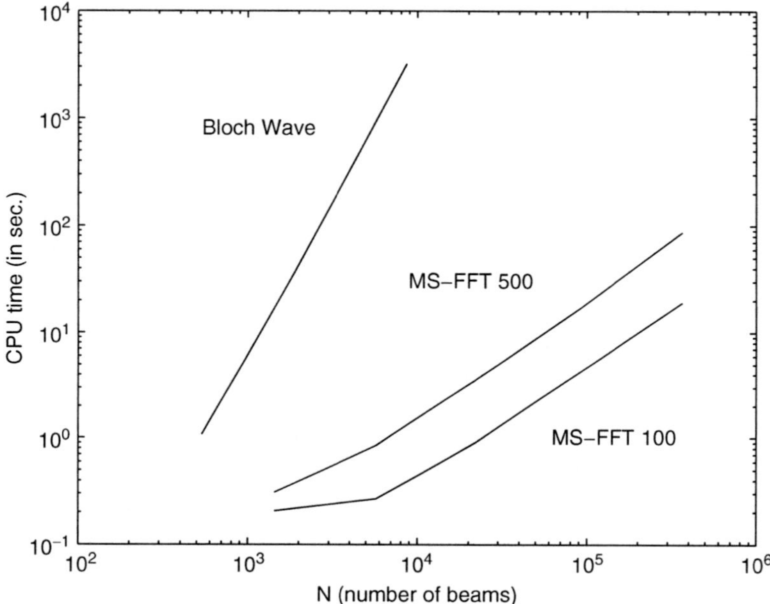

Fig. 6.2 Scaling of CPU time in the Bloch wave eigenvalue method with the number of beams N compared to the FFT multislice method for the 100 projection of aluminum. Two different thickness are shown for the multislice times (100Åand 500Å). The measured CPU time has an error that is some small fraction of a second

The Bloch wave CPU time grows very fast as the number of beams is increased which is a significant problem. Generally speaking, both methods require a similar number of beams and high resolution of nontrial specimens requires many beams, so the multislice method has a clear advantage except for some simple cases requiring only a small number of beams.

6.2 The Wave Equation for Fast Electrons

In principle the full time dependent Schrödinger equation could be used to solve for the time dependent electron wave function at all positions in the specimen or microscope column (in three dimensions) at each point in time. The initial wave function is a Gaussian wave packet that would propagate through the crystal. The time evolution could in principle be traced using something like the Crank-Nicholson numerical method (for example see Press et al. [288]). However, this approach is prohibitively expensive in both computer memory and computer time.

The wavelength of a 200 keV electron is $\lambda = 0.025$Å. A straightforward numerical sampling of this wavelength would require of order 10 points per wavelength. To sample a small specimen in a cube of 100 Å per side would require about $(100/0.0025)^3$ or 6.3×10^{13} points. Even in single precision (four bytes per value) this would require 2.4×10^5 Gbytes of memory, which is clearly not possible in the near future. Clearly some other approach must be found. The Bloch wave method solves this problem by expanding in a small basis set of plane waves so real space sampling is not really used. However, there is still a finite sample in reciprocal space.

An alternative approach described later will use the time independent Schrödinger equation because the image is in some sense stationary. This section will further approximate the Schrödinger for fast electrons and later sections will discuss numerical solutions with the goal of reducing the computer memory and time requirements.

The Schrödinger equation for the full wave function $\psi_f(x,y,z)$ as a function of three spatial coordinates (x,y,z) in an electrostatic potential $V(x,y,z)$ of the specimen is:

$$\left[-\frac{\hbar^2}{2m}\nabla^2 - eV(x,y,z)\right]\psi_f(x,y,z) = E\psi_f(x,y,z), \tag{6.40}$$

where $\hbar = h/2\pi$ is Planck's constant divided by 2π, $m = \gamma m_0$ is the relativistic mass of the electron, $e = |e|$ is the magnitude of the charge of the electron, E is the kinetic energy of the electron and $-eV$ is the potential energy of the electron. In an electron microscope the energy of the incident electrons (100–1,000 keV) is much greater than the additional energy they gains (or lose) inside the specimen $eV(x,y,z)$. The electron motion will be predominately in the forward z direction (i.e., along the optic axis of the microscope) and the specimen will be a relatively minor perturbation on the electron's motion. It is useful to separate the large velocity in the z direction from other small effects due to the specimen. First write the full wave function $\psi_f(x,y,z)$ as a product of two factors one of which is a plane wave traveling in the z direction and the other factor $\psi(x,y,z)$ is the portion of the wave function that varies slowly with position z:

$$\psi_f(x,y,z) = \psi(x,y,z)\exp(2\pi \mathrm{i}z/\lambda), \tag{6.41}$$

where λ is the electron wavelength. It is tempting to identify the phase in the exponent as proportional to the electron wave number in the z direction (k_z). However, the squared magnitude of the total wave vector is $k_x^2 + k_y^2 + k_z^2 = 1/\lambda^2$ so this is not strictly true once the electron scatters out of a pure plane wave and k_x and k_y are nonzero. Only elastic processes will be considered so the total kinetic energy of the electron is:

$$E = \frac{h^2}{2m\lambda^2}. \tag{6.42}$$

To use (6.41) in (6.40) requires the calculation of the following derivatives:

$$\nabla^2\psi_f(x,y,z) = \left[\frac{\partial^2}{\partial x^2} + \frac{\partial^2}{\partial y^2} + \frac{\partial^2}{\partial z^2}\right]\psi_f(x,y,z)$$

$$= \left[\nabla_{xy}^2 + \frac{\partial^2}{\partial z^2} \right] \psi_f(x,y,z)$$

$$= \exp(2\pi i z/\lambda) \nabla_{xy}^2 \psi(x,y,z) + \frac{\partial^2}{\partial z^2} [\psi(x,y,z) \exp(2\pi i z/\lambda)], \tag{6.43}$$

where ∇_{xy}^2 is the sum of second derivatives with respect to x and y only. Now concentrate on the derivative with respect to z. The first derivative is:

$$\frac{\partial}{\partial z} [\psi \exp(2\pi i z/\lambda)] = \exp(2\pi i z/\lambda) \left[\frac{\partial \psi}{\partial z} + \frac{2\pi i}{\lambda} \psi \right] \tag{6.44}$$

and the second derivative is:

$$\frac{\partial^2}{\partial z^2} [\psi \exp(2\pi i z/\lambda)] = \exp(2\pi i z/\lambda) \left[\frac{\partial^2 \psi}{\partial z^2} + \frac{4\pi i}{\lambda} \frac{\partial \psi}{\partial z} + \left(\frac{2\pi i}{\lambda} \right)^2 \psi \right]$$

$$= \exp(2\pi i z/\lambda) \left[\frac{\partial^2 \psi}{\partial z^2} + \frac{4\pi i}{\lambda} \frac{\partial \psi}{\partial z} \right] - \frac{4\pi^2}{\lambda^2} \psi_f. \tag{6.45}$$

Now substitute (6.45) into (6.40). The last term on the right hand side of (6.45) cancels the right hand side of (6.40) given the value for E in (6.42). Dropping the common factor of $\exp(2\pi i z/\lambda)$ leaves:

$$-\frac{\hbar^2}{2m} \left[\nabla_{xy}^2 + \frac{\partial^2}{\partial z^2} + \frac{4\pi i}{\lambda} \frac{\partial}{\partial z} + \frac{2meV(x,y,z)}{\hbar^2} \right] \psi(x,y,z) = 0. \tag{6.46}$$

The motion of the high energy electrons is predominately in the forward z direction meaning that ψ changes slowly with z and λ is very small. Therefore:

$$\left| \frac{\partial^2 \psi}{\partial z^2} \right| << \left| \frac{1}{\lambda} \frac{\partial \psi}{\partial z} \right| \tag{6.47}$$

so (6.46) may be approximated as:

$$\left[\nabla_{xy}^2 + \frac{4\pi i}{\lambda} \frac{\partial}{\partial z} + \frac{2meV(x,y,z)}{\hbar^2} \right] \psi(x,y,z) = 0. \tag{6.48}$$

Ignoring the term containing the second derivative with respect to z is sometimes referred to as ignoring the backscattered electrons which is appropriate for high energy electrons (Howie and Basinski [162], Lynch and Moodie [230]). However, it is probably more accurate to refer to (6.48) as the paraxial approximation to the Schrödinger equation. However including the second derivative term does not automatically include backscattered electrons because some initial conditions may further prohibit the backscattered electrons. Lewis et al. [225] have further considered the effect of neglecting the second order derivative term. Bird [29] has concluded that dropping the second order derivative (with respect to z) produces an error of about one percent in the position of the FOLZ (First Order Laue Zone) ring (the error decreases with increasing beam voltage).

The Schrödinger equation (6.40) for fast electrons traveling in the z direction may be written as a first order differential equation in z as:

$$\begin{aligned}\frac{\partial \psi(x,y,z)}{\partial z} &= \left[\frac{\mathrm{i}\lambda}{4\pi}\nabla_{xy}^2 + \frac{2me\mathrm{i}\lambda}{4\pi\hbar^2}V(x,y,z)\right]\psi(x,y,z)\\ &= \left[\frac{\mathrm{i}\lambda}{4\pi}\nabla_{xy}^2 + i\sigma V(x,y,z)\right]\psi(x,y,z),\end{aligned} \quad (6.49)$$

where $\psi(x,y,z)$ is defined in (6.41), λ is the wavelength of the incident electrons and $\sigma = 2\pi me\lambda/h^2$ is the interaction parameter (5.6). An identical equation would result if the exponent in (6.41) were replaced with $(2\pi z/\lambda - iEt/\hbar)$ (where t is time) and the time dependent Schrödinger equation were used.

6.3 A Bloch Wave Differential Equation Solution

If the specimen is periodic (i.e., crystalline) then the specimen potential can be expanded in a three dimensional Fourier series as:

$$V(x,y,z) = V(\mathbf{r}) = \sum_{\mathbf{G}} V_{\mathbf{G}} \exp[2\pi\mathrm{i}\mathbf{G}\cdot\mathbf{r}], \quad (6.50)$$

where the set of vectors $\mathbf{G} = (G_x, G_y, G_z)$ are typically the reciprocal lattice vectors of the specimen. Next expand the electron wave function in Bloch waves with the same crystal periodicity:

$$\psi(x,y,z) = \psi(\mathbf{r}) = \sum_{\mathbf{G}} \phi_{\mathbf{G}}(z) \exp[2\pi\mathrm{i}\mathbf{G}\cdot\mathbf{r}], \quad (6.51)$$

where the coefficients ϕ_G vary weakly with depth in the crystal z. Substituting (6.50) and (6.51) into the Schrödinger wave equation for fast electrons (6.49) yields:

$$\begin{aligned}&\sum_{\mathbf{G}}\left(\frac{\partial\phi_{\mathbf{G}}}{\partial z} + 2\pi\mathrm{i}G_z\phi_{\mathbf{G}}\right)\exp[2\pi\mathrm{i}\mathbf{G}\cdot\mathbf{r}]\\ &= \frac{\mathrm{i}\lambda}{4\pi}\sum_{\mathbf{G}}(-4\pi^2G_x^2 - 4\pi^2G_y^2)\phi_{\mathbf{G}}\exp[2\pi\mathrm{i}\mathbf{G}\cdot\mathbf{r}]\\ &+ \mathrm{i}\sigma\sum_{\mathbf{G}}\left[\sum_{\mathbf{G}'}V_{\mathbf{G}-\mathbf{G}'}\phi_{\mathbf{G}'}\right]\exp[2\pi\mathrm{i}\mathbf{G}\cdot\mathbf{r}]\end{aligned} \quad (6.52)$$

Equating coefficients of $\exp[2\pi i\mathbf{G}\cdot\mathbf{r}]$ yields:

$$\frac{\partial \phi_{\mathbf{G}}(z)}{\partial z} = -\pi i(2G_z + \lambda G_x^2 + \lambda G_y^2)\phi_{\mathbf{G}}(z) + i\sigma \sum_{\mathbf{G}'} V_{\mathbf{G}-\mathbf{G}'}\phi_{\mathbf{G}'}(z)$$
$$= 2\pi i s_{\mathbf{G}}\phi_{\mathbf{G}}(z) + i\sigma \sum_{\mathbf{G}'} V_{\mathbf{G}-\mathbf{G}'}\phi_{\mathbf{G}'}(z). \quad (6.53)$$

This set of first order differential equation is known as the Howie-Whelan [163] equations. If there are N Fourier coefficients in (6.50) and (6.51) then there are N coupled first order equations. The excitation error, $s_{\mathbf{G}} = G_z + 0.5\lambda(G_x^2 + G_y^2)$ for the reflection G is illustrated in Fig. 6.3 for the case where G_z=0. The incident wave vector $\mathbf{k}_z$ is along the positive z direction. Only elastic scattering is considered here so the scattered wave vector $\mathbf{k}_s$ must be the same length as the incident wave vector. This means that the end of the scattered wave vector must lie on the Ewald sphere centered at the beginning of the incident wave vector. The distance between the end of the scattered wave vector and the nearest reciprocal lattice point is labeled as $\mathbf{s}_G$.

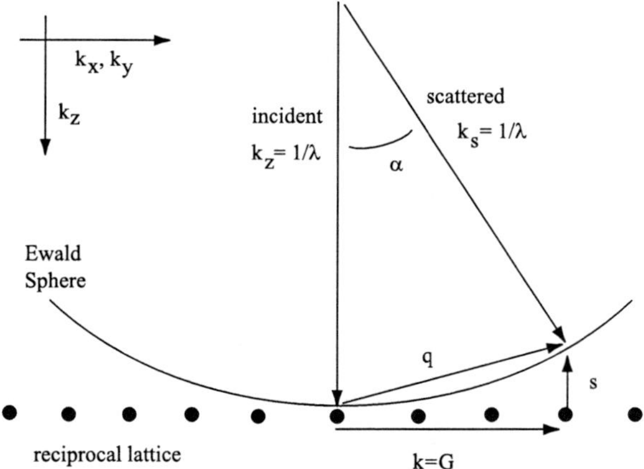

Fig. 6.3 Scattering geometry in reciprocal space. The incident electron is traveling in the positive z direction and is scattered through an angle α. The reciprocal lattice points of the specimen are shown as *solid dots*. Elastic scattering requires that the incident wave vector $\mathbf{k}_z$ and the scattered wave vector $\mathbf{k}_s$ both lie on the Ewald sphere. $\mathbf{s}$ is the extinction error for this scattering angle. In the small angle approximation $|\mathbf{q}| \sim \alpha/\lambda$ and $|\mathbf{s}| = |\mathbf{q}|\sin(\alpha/2) \sim 0.5\alpha^2/\lambda$

The Howie-Whelan equations (6.53) can also be written as a matrix equation for a column vector of N components $\phi_{\mathbf{G}}$ and a matrix of N^2 components $V_{\mathbf{G}-\mathbf{G}'}$. A general matrix multiplication on the right hand side would scale as N^2 for computer time. Calculating $V_{\mathbf{G}}$ in three dimensions with a thousand or more beams is itself a formidable task. In the CTEM all of the initial $\phi_{\mathbf{G}}$ components (at $z = 0$) would be zero except for a single plane wave in the z direction. This set of equations can be solved using standard numerical differential equations methods such as Runge-Kutta methods (see for example Press et al. [288]) to advance the vector of $\phi_{\mathbf{G}}$ components for each small step Δz. There are several different Bloch wave formulations of the dynamical scattering problem. This is only one specific example but does illustrate some of the general features of a Bloch wave or reciprocal space solution. The Howie-Whelan equations are conceptually very similar to the multislice equation taken up in the next section. The first term on the right hand side of (6.53) is similar to the propagator function and the second term is similar to the transmission function in the multislice method. The main difference is that the Howie-Whelan equations are stated completely in reciprocal space. Other, general analytical Bloch wave solutions are discussed at length in Hirsch et al. [159] and Reimer [295].

Watanabe et al. [373, 374] have recently proposed an alternate approach to the solution of the Howie-Whelan equations. They find that a direct integration of the equations (with the appropriate boundary conditions at the entrance and exit surface of the crystal) yields a method that has about the same accuracy as the multislice solution but is in between Bethe's eigenvalue solution and the multislice solution in efficiency (i.e., computer time).

6.4 The Multislice Solution

6.4.1 A Formal Operator Solution

The wave equation for fast electrons (6.49) can be written in operator form as:

$$\begin{aligned} \frac{\partial \psi(x,y,z)}{\partial z} &= [A+B]\psi(x,y,z) \\ A &= \frac{\mathrm{i}\lambda}{4\pi}\nabla^2_{xy} \\ B &= i\sigma V(x,y,z), \end{aligned} \tag{6.54}$$

where A and B are noncommuting operators. This equation has a formal operator solution of:

$$\psi(x,y,z) = \exp\left[\int_0^z [A(z') + B(z')]\mathrm{d}z'\right] \psi(x,y,0) \tag{6.55}$$

This can be verified by formal differentiation. Offsetting the initial value to z and integrating from z to $z+\Delta z$ yields:

$$\psi(x,y,z+\Delta z) = \exp\left[\int_z^{z+\Delta z} \left(\frac{\mathrm{i}\lambda}{4\pi}\nabla^2_{xy} + i\sigma V(x,y,z')\right)\mathrm{d}z'\right] \psi(x,y,z) \tag{6.56}$$

Δz will become a small slice of the specimen and this solution may be further simplified as:

$$\psi(x,y,z+\Delta z) = \exp\left[\frac{\mathrm{i}\lambda}{4\pi}\Delta z\nabla^2_{xy} + i\sigma v_{\Delta z}(x,y,z)\right] \psi(x,y,z) \tag{6.57}$$

where $v_{\Delta z}(x,y,z)$ is the projected potential of the specimen between z and $z+\Delta z$.

$$v_{\Delta z}(x,y,z) = \int_{z}^{z+\Delta z} V(x,y,z')\mathrm{d}z' \tag{6.58}$$

The appearance of the operator ∇^2_{xy} in the exponent complicates the solution somewhat because the $\exp(\cdots)$ factor must also be regarded as an operator. If A and B are noncommuting operators or matrices and ε is a small real number then:

$$\begin{aligned} \exp(A\varepsilon + B\varepsilon) &= 1 + (A+B)\varepsilon + \frac{1}{2!}(A+B)^2\varepsilon^2 + \cdots \\ &= 1 + (A+B)\varepsilon + \frac{1}{2!}(A^2 + AB + BA + B^2)\varepsilon^2 + \cdots \end{aligned} \tag{6.59}$$

If A and B were simple scalar variables then this expression could be easily factored, however the most common factorizations do not yield the expected results if A and B do not commute.

$$\begin{aligned} \exp(A\varepsilon)\exp(B\varepsilon) &= \left[1 + A\varepsilon + \frac{1}{2!}A^2\varepsilon^2 + \cdots\right]\left[1 + B\varepsilon + \frac{1}{2!}B^2\varepsilon^2 + \cdots\right] \\ &= 1 + (A+B)\varepsilon + \frac{1}{2!}(A^2 + 2AB + B^2)\varepsilon^2 + \cdots \end{aligned} \tag{6.60}$$

and:

$$\begin{aligned} \exp(B\varepsilon)\exp(A\varepsilon) &= \left[1 + B\varepsilon + \frac{1}{2!}B^2\varepsilon^2 + \cdots\right]\left[1 + A\varepsilon + \frac{1}{2!}A^2\varepsilon^2 + \cdots\right] \\ &= 1 + (B+A)\varepsilon + \frac{1}{2!}(B^2 + 2BA + A^2)\varepsilon^2 + \cdots \end{aligned} \tag{6.61}$$

By comparison (6.59) may be factored to lowest order in either of two ways:

$$\exp(A\varepsilon + B\varepsilon) = \exp(A\varepsilon)\exp(B\varepsilon) + \frac{1}{2}[B,A]\varepsilon^2 + \mathcal{O}(\varepsilon^3) \tag{6.62}$$

or

$$\exp(A\varepsilon + B\varepsilon) = \exp(B\varepsilon)\exp(A\varepsilon) + \frac{1}{2}[A,B]\varepsilon^2 + \mathcal{O}(\varepsilon^3) \tag{6.63}$$

where $[B,A] = BA - AB$ is the commutator of operators (or matrices) B and A. [A more accurate answer can be obtained simply by averaging (6.62) and (6.63).] This result is referred to as the Zassenhaus theorem (Goodman and Moodie [127]). Feynman [104], Weiss and Maradudin [377], and Wilcox [379] have given a more detailed discussion of exponentiation of operators.

The traditional form of the multislice solution uses one of (6.62) and (6.63) in (6.57).

$$\begin{aligned} \psi(x,y,z+\Delta z) &= \exp\left(\frac{\mathrm{i}\lambda\Delta z}{4\pi}\nabla^2_{xy}\right)\exp\left[i\sigma v_{\Delta z}(x,y,z)\right]\psi(x,y,z) + \mathcal{O}(\Delta z^2) \\ &= \exp\left(\frac{\mathrm{i}\lambda\Delta z}{4\pi}\nabla^2_{xy}\right)t(x,y,z)\psi(x,y,z) + \mathcal{O}(\Delta z^2) \end{aligned} \tag{6.64}$$

where $t(x,y,z)$ is the transmission function for the portion of the specimen between z and $z+\Delta z$ [compare to (5.7)].

$$t(x,y,z) = \exp\left[i\sigma \int_{z}^{z+\Delta z} V(x,y,z')\mathrm{d}z'\right] \tag{6.65}$$

The remaining factor of exp($\cdots$) is a little more difficult to interpret. First form the two-dimensional Fourier transform of the right hand side of (6.64):

$$\mathrm{FT}\left[\exp\left(\frac{\mathrm{i}\lambda\Delta z}{4\pi}\nabla_{xy}^2\right)(t\psi)\right] = \int\left[\exp\left(\frac{\mathrm{i}\lambda\Delta z}{4\pi}\nabla_{xy}^2\right)(t\psi)\right]\exp[2\pi\mathrm{i}(k_x x + k_y y)]\mathrm{d}x dy \tag{6.66}$$

The derivatives with respect to x and y commute so the exponential operator may be split into two factors (one with x and one with y) without further error.

$$\mathrm{FT}\left[\exp\left(\frac{\mathrm{i}\lambda\Delta z}{4\pi}\nabla_{xy}^2\right)(t\psi)\right] = \int\left[\exp\left(\frac{\mathrm{i}\lambda\Delta z}{4\pi}\frac{\partial^2}{\partial x^2}\right)\exp\left(\frac{\mathrm{i}\lambda\Delta z}{4\pi}\frac{\partial^2}{\partial y^2}\right)(t\psi)\right] \times\exp[2\pi\mathrm{i}(k_x x + k_y y)]\mathrm{d}x\mathrm{d}y \tag{6.67}$$

Each exponential operator may be expanded in a power series as:

$$\begin{aligned}\exp\left(\frac{\mathrm{i}\lambda\Delta z}{4\pi}\frac{\partial^2}{\partial x^2}\right) &= \sum_{n=0}^{\infty}\frac{1}{n!}\left(\frac{\mathrm{i}\lambda\Delta z}{4\pi}\frac{\partial^2}{\partial x^2}\right)^n \\ \exp\left(\frac{\mathrm{i}\lambda\Delta z}{4\pi}\frac{\partial^2}{\partial y^2}\right) &= \sum_{n=0}^{\infty}\frac{1}{n!}\left(\frac{\mathrm{i}\lambda\Delta z}{4\pi}\frac{\partial^2}{\partial y^2}\right)^n\end{aligned} \tag{6.68}$$

Inserting the power series expansions for the exponentials in (6.67) and repeatedly integrating each term by parts (with the assumption that $t\psi$ vanishes at infinity or obeys periodic boundary conditions) yields:

$$\begin{aligned}\mathrm{FT}\left[\exp\left(\frac{\mathrm{i}\lambda\Delta z}{4\pi}\nabla_{xy}^2\right)(t\psi)\right] &= \sum_{n=0}^{\infty}\frac{1}{n!}\left(-\mathrm{i}\pi\lambda\Delta z k_x^2\right)^n\sum_{n=0}^{\infty}\frac{1}{n!}\left(-\mathrm{i}\pi\lambda\Delta z k_y^2\right)^n\mathrm{FT}[(t\psi)] \\ &= \exp\left[-i\pi\lambda\Delta z(k_x^2+k_y^2)\right]\mathrm{FT}\left[(t\psi)\right] \\ &= P(k,\Delta z)\mathrm{FT}\left[t\psi\right],\end{aligned} \tag{6.69}$$

where $k^2 = (k_x^2 + k_y^2)$ and $P(k, \Delta z)$ is the propagator function. A multiplication in Fourier space converts to a convolution in real space so the operator may be interpreted as:

$$\exp\left(\frac{i\lambda\Delta z}{4\pi}\nabla_{xy}^2\right) = p(x, y, \Delta z)\otimes, \tag{6.70}$$

where $\otimes$ is a two-dimensional convolution (in x and y) and $p(x, y, \Delta z)$ is the propagator function in real space for a distance Δz:

$$\begin{aligned} P(k, \Delta z) &= \exp\left(-i\pi\lambda k^2 \Delta z\right) \\ p(x, y, \Delta z) &= \mathrm{FT}^{-1}\left[P(k, \Delta z)\right] = \frac{1}{i\lambda\Delta z}\exp\left[\frac{i\pi}{\lambda\Delta z}(x^2 + y^2)\right] \end{aligned} \tag{6.71}$$

Combining (6.69) and (6.64) yields:

$$\psi(x, y, z + \Delta z) = p(x, y, \Delta z) \otimes \left[t(x, y, z)\,\psi(x, y, z)\right] + \mathcal{O}(\Delta z^2) \tag{6.72}$$

If the slices in the specimen are labeled $n = 0, 1, 2, \ldots$ then the depth in the specimen is z_n ($z_n \sim n\Delta z$ if all slices were the same thickness). The wave function at the top of each slice is labeled $\psi_n(x, y)$ and the propagator and transmission functions for each slice are labeled as $p_n(x, y, \Delta z_n)$ and $t_n(x, y)$, respectively. The multislice equation (6.72) can be written in compact form as:

$$\psi_{n+1}(x, y) = p_n(x, y, \Delta z_n) \otimes \left[t_n(x, y)\,\psi_n(x, y)\right] + \mathcal{O}(\Delta z^2) \tag{6.73}$$

If the other identity (6.62) and (6.63) is used then another equally accurate expression results.

$$\psi_{n+1}(x, y) = t_n(x, y)\left[p_n(x, y, \Delta z_n) \otimes \psi_n(x, y)\right] + \mathcal{O}(\Delta z^2) \tag{6.74}$$

The initial wave function $\psi_0(x, y)$ is a plane wave in the CTEM and the probe wave function in the STEM.

The last term in (6.73) and (6.74) indicates the order of magnitude of the error caused by the approximation used to get the remaining terms in the equations and is called the error term. The error term should not be used in the actual calculation but is for informational purposes only. The error associated with one step of the multislice (6.72), (6.73), and (6.74) is of order Δz^2. This error is referred to as the local error. If the specimen is divided into N_s slices then typically $N_s \propto 1/\Delta z$. If the multislice equation is applied N_s times to advance the wave function all of the way through the specimen then the error term of the final result is reduced by approximately one order. The error of the final result is therefore of order Δz. The final error is referred to as the global error.

6.4.2 A Finite Difference Solution

The traditional numerical analysis approach to solving a differential equation given an initial value is to Taylor expand the dependent variable (ψ in this case) in powers of Δz and approximate the derivatives by their finite difference approximations. Expanding $\psi(x,y,z)$ about the point z (and temporarily dropping explicit reference to x and y in ψ for simplicity) yields:

$$\psi(z+\Delta z) = \psi(z) + \Delta z \frac{\partial \psi(z)}{\partial z} + \frac{1}{2!}\Delta z^2 \frac{\partial^2 \psi(z)}{\partial z^2} + \cdots \qquad (6.75)$$

Substituting the first derivative from (6.49) yields:

$$\begin{aligned} \psi(z+\Delta z) &= \psi(z) + \Delta z \left[\frac{\mathrm{i}\lambda}{4\pi}\nabla^2_{xy} + i\sigma V(x,y,z)\right]\psi(z) + \mathcal{O}(\Delta z^2) \\ &= \left\{1 + \Delta z\left[\frac{\mathrm{i}\lambda}{4\pi}\nabla^2_{xy} + i\sigma V(x,y,z)\right]\right\}\psi(z) + \mathcal{O}(\Delta z^2). \end{aligned} \qquad (6.76)$$

The Taylor series expansion for $\exp(x)$ with small x is:

$$\exp(x) = 1 + x + \frac{1}{2!}x^2 + \frac{1}{3!}x^3 + \cdots \qquad (6.77)$$

By comparison to (6.77), (6.76) may be written to lowest order (with the x, y dependence reinstated) as:

$$\psi(x,y,z+\Delta z) = \exp\left[\Delta z \frac{\mathrm{i}\lambda}{4\pi}\nabla^2_{xy} + i\sigma\Delta z V(x,y,z)\right]\psi(x,y,z) + \mathcal{O}(\Delta z^2). \qquad (6.78)$$

Keeping the leading factor as $\exp(\cdots)$ where the exponent is predominately imaginary tends to keep the total integrated intensity constant.

$$\int |\psi(x,y,z+\Delta z)|^2 \mathrm{d}x\mathrm{d}y = \int |\psi(x,y,z)|^2 \mathrm{d}x\mathrm{d}y = \text{constant}. \qquad (6.79)$$

This is physically relevant because elastic scattering should be unitary. Without this constraint the result is essentially Euler's method (for example see Press et al. [288]) which is known to require an excessively small step size (Δz)to be stable. Furthermore, to this level of accuracy the term containing the specimen potential may be written as:

$$\begin{aligned} V(x,y,z)\Delta z &= \int_z^{z+\Delta z} V(x,y,z')\mathrm{d}z' + \mathcal{O}(\Delta z^2) \\ &= v_{\Delta z}(x,y,z) + \mathcal{O}(\Delta z^2), \end{aligned} \qquad (6.80)$$

where $v_{\Delta z}(x,y,z)$ is the projected potential of the specimen between z and $z+\Delta z$. Therefore (6.78) may be written as:

$$\psi(x,y,z+\Delta z) = \exp\left[\Delta z \frac{\mathrm{i}\lambda}{4\pi}\nabla^2_{xy} + i\sigma v_{\Delta z}(x,y,z)\right] \psi(x,y,z) + \mathcal{O}(\Delta z^2). \quad (6.81)$$

This is just the previous result (6.57) and the rest of the derivation follows that of Sect. 6.4.1.

6.4.3 Free Space Propagation

In the special limiting case in which the specimen potential vanishes the formal multislice solution (6.57) takes the simpler form of:

$$\psi(x,y,z+\Delta z) = \exp\left[\frac{\mathrm{i}\lambda}{4\pi}\Delta z \nabla^2_{xy}\right] \psi(x,y,z) \quad (6.82)$$

which can be written in terms of the propagator function without introducing any further approximations (or errors) as:

$$\psi(x,y,z+\Delta z) = p(x,y,\Delta z) \otimes \psi(x,y,z) \quad (6.83)$$

Closer inspection of the error term in the full multislice solution (6.73), or (6.74) reveals that it should be written as $\mathcal{O}(\Delta z^2 v_{\Delta z})$ where $v_{\Delta z}$ is the projected atomic potential of the specimen and Δz is the slice thickness. This means that for a given slice thickness the accuracy of the multislice solution increases as the specimen potential gets smaller. The multislice solution should be more accurate for light atoms with low atomic numbers and less accurate for heavy atoms with high atomic numbers.

It is interesting to observe that the two components of the multislice solution (the transmission function and the propagation function) are nearly exact when used separately. The phase grating approximation (Sect. 5.1) of the transmission function and the propagation function are individually more accurate than when they are combined into the multislice solution (van Dyck [357]). The main error in the multislice solution is due to how these operations are combined.

6.5 Multislice Interpretation

If the initial value of the wave function $\psi(x,y,z=0)$ is given in an x,y plane at the entrance face of the specimen (i.e., unity for the CTEM and the probe wave function for the STEM) then the electron wave function can be calculated at any depth z inside the specimen given a description of the potential inside the specimen by repeated application of (6.72). The specimen is first divided into many thin slices as in Fig. 6.4. At each slice the electron wave function experiences a phase shift due to the projected atomic potential of all atoms in the slice and then is propagated along z for the thickness of the slice. In general each slice is independent of all other slices, so both the slice thickness Δz and transmission function $t(x,y,z)$ may vary from one slice to another.

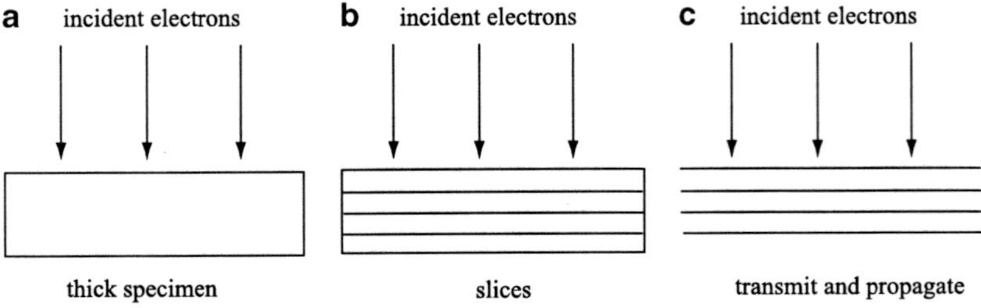

Fig. 6.4 Multislice decomposition of a thick specimen. (a) Original thick specimen, (b) the specimen divided into thin slices, (c) each slice is treated as a transmission step (*solid line*) followed by a propagator (vacuum between slices)

The original derivation of Cowley and Moodie [63] was done from a physical optics viewpoint. The transmission function $t(x,y,z)$ can be associated with the phase grating approximation of a thin specimen for the layer of the specimen between z and $z+\Delta z$ as in Sect. 5.1. The propagator function $p(x,y,\Delta z)$ can be associated with the Fresnel (near zone) diffraction over a distance Δz.

Huygens' principle (of classical optics) states that every point of a wave front gives rise to an outgoing spherical wave. These outgoing spherical waves propagate to the next position of the wavefront and interfere with one another. The wave function in an x,y plane at $z+\Delta z$ is the interference of all of these spherically outgoing waves that originated in an x,y plane at z as illustrated in Fig. 6.5

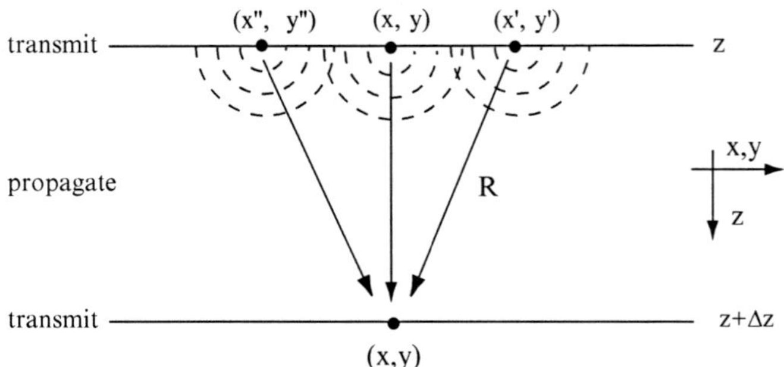

Fig. 6.5 Physical optics interpretation of the multislice propagator. The wave function in an x,y plane at z propagates to the x,y plane at $z+\Delta z$. Each point in the wavefront at z emits a spherically outgoing wave. All of these waves combine at each point in the x,y plane at $z+\Delta z$

The propagation of a wavefront as described by the Huygens principle may be calculated using the Fresnel-Kirchhoff diffraction integral (for example see Born and Wolf [34] Sect. 8.3.2).

$$\psi(x,y,z+\Delta z) = \frac{1}{2i\lambda}\int \psi(x',y',z)\frac{\exp(2\pi \mathrm{i} R/\lambda)}{R}(1+\cos\theta)\mathrm{d}x'\mathrm{d}y', \tag{6.84}$$

where θ is the angle between the plane of the initial wave front and the direction of the outgoing spherical wave at point (x',y'). R is the total distance the outgoing spherical wave must travel to point (x,y) on the next wavefront.

$$R = \sqrt{(x-x')^2+(y-y')^2+\Delta z^2}. \tag{6.85}$$

The maximum scattering angle for high energy electrons is only about 100–200 mrad. The angle θ is very close to zero for high energy electrons so (6.84) may then be written as:

$$\psi(x,y,\Delta z) = \frac{1}{i\lambda}\int \psi(x',y',z)\frac{\exp(2\pi \mathrm{i} R/\lambda)}{R}\mathrm{d}x'\mathrm{d}y' \tag{6.86}$$

Small angle scattering also means that the lateral distances ($|x-x'|$ and $|y-y'|$) are also small compared to Δz. For example if the slice thickness is $\Delta z \sim 3$Å then $|x-x'| \sim 0.3$Å at a scattering angle of 100 mrad. The distance R may then be approximated as:

$$\begin{aligned} R &= \Delta z\sqrt{1+(x-x')^2/\Delta z^2+(y-y')^2/\Delta z^2} \\ &\sim \Delta z\left[1+0.5(x-x')^2/\Delta z^2+0.5(y-y')^2/\Delta z^2+\cdots\right]. \end{aligned} \tag{6.87}$$

Substituting (6.87) into (6.86) and keeping only lowest powers of $(x-x')$ and $(y-y')$ yields:

$$\begin{aligned} \psi(x,y,\Delta z) &= \frac{1}{i\lambda}\frac{\exp(2\pi \mathrm{i}\Delta z/\lambda)}{\Delta z}\int \psi(x',y',z) \\ &\exp\left\{\frac{i\pi}{\lambda\Delta z}\left[(x-x')^2+(y-y')^2\right]\right\}\mathrm{d}x'\mathrm{d}y' \end{aligned} \tag{6.88}$$

The right hand side of this expression is just the convolution of ψ with the propagator function (6.71).

$$\psi(x,y,\Delta z) = \exp(2\pi \mathrm{i}\Delta z/\lambda)\left[\psi(x,y,z)\otimes p(x,y,\Delta z)\right], \tag{6.89}$$

where the propagator function is:

$$p(x,y,\Delta z) = \frac{1}{\mathrm{i}\lambda\Delta z}\exp\left[\frac{i\pi}{\lambda\Delta z}(x^2+y^2)\right] \tag{6.90}$$

The leading exponential factor in (6.88) is just the forward propagation of the plane wave and is part of the full wave function [as in the Fresnel-Kirchhoff diffraction integral, (6.84)] but not part of the slowly varying part of the wave function [see (6.41)] as used in the multislice derivation. Expression 6.88 is then identical to the convolution with the propagator derived in the multislice method (6.72). The multislice propagator function can be interpreted simply as the Fresnel diffraction over a distance Δz. Also note that this is equivalent to a defocus of $\Delta f = \Delta z$.

6.6 The Multislice Method and FFT's

When implemented numerically in the computer the wave function $\psi(x,y)$ will be sampled as discrete points in x and y. The solution of (6.73) and (6.74) will require computer storage for only a small number of two-dimensional arrays of $\psi(x,y)$ at any one time because only the values at two positions of z are needed at the same time. Furthermore, the convolution can be efficiently calculated using the fast Fourier Transform as first discussed by Ishizuka and Uyeda [179]. The Fourier convolution theorem says that:

$$\begin{aligned} f(x,y) \otimes h(x,y) &= \mathrm{FT}^{-1}\left[F(k_x,k_y)H(k_x,k_y)\right] \\ F(k_x,k_y) &= \mathrm{FT}\left[F(k_x,k_y)\right] \\ H(k_x,k_y) &= \mathrm{FT}\left[H(k_x,k_y)\right], \end{aligned} \tag{6.91}$$

where FT[] is a two-dimensional Fourier transform. Equations (6.73) and (6.74) can be rewritten using Fourier transforms as:

$$\psi_{n+1}(x,y) = \mathrm{FT}^{-1}\left\{P_n(k_x,k_y,\Delta z_n)\mathrm{FT}\left[t_n(x,y)\psi_n(x,y)\right]\right\} + \mathcal{O}(\Delta z^2) \tag{6.92}$$

and

$$\psi_{n+1}(x,y) = t_n(x,y)\mathrm{FT}^{-1}\left\{P_n(k_x,k_y,\Delta z_n)\mathrm{FT}\left[\psi_n(x,y)\right]\right\} + \mathcal{O}(\Delta z^2) \tag{6.93}$$

Equation (6.92) will turn out to be the best form to eliminate aliasing (see Sect. 6.8). If the wave function is sampled with N_x points in the x direction and N_y points in the y directions then there are $N = N_x N_y$ Fourier coefficients all together. Using the fast Fourier transform (Chap. 4) means that the total computation time then scales roughly as $N \log_2 N$ instead of N^2 as in a direct matrix solution (or the direct convolution in (6.73) and (6.74)). In a matrix algebra description the transmission function is diagonal in real space and the propagator is diagonal in reciprocal space. The FFT is a convenient and fast way to convert back and forth between real and reciprocal space. The combined efficiency of small memory requirements and fast computation makes the multislice method preferable over a direct matrix or eigenvalue solution.

The multislice solution is formally equivalent to a solution of the Howie-Whelan equation (6.53) if the slice thickness Δz is small enough. Both are derived from the wave equation for fast electron (6.49) and both neglect the second derivative of ψ with respect to z (6.47). The principle difference is the relative efficiency of the two solutions. When the multislice solution is implemented using the fast Fourier transform (FFT) it can be dramatically faster (i.e., less computer time) than a Bloch wave solution.

6.7 Slicing the Specimen

Generating a description of the specimen in a form that can be used in a multislice program can be the most difficult part of simulating an image (often involving many long hours staring at the crystallographic data for the specimen at hand). Overall the specimen needs to be described as a sequence of layers (x,y planes) and the spacing between each layer (thickness of the layer along z, the optic axis of the electron microscope). Deciding on a strategy (and input format) for listing the specimen parameters is part of designing the program and part of actually using the program once it is written. The FFT (fast Fourier transform) has an enormous advantage in computational efficiency (i.e., computer time) and is almost always used in a multislice program. The FFT is factorable into independent x and y components (see Chap. 4). The easiest way to use the FFT is to describe the specimen as a rectangular unit cell. Obviously not all specimens have a convenient rectangular unit cell so this may require some work by the person using the multislice simulation program. If the specimen naturally has a rectangular unit cell it is easy to simulate it if viewed along one of the major crystal axis. If the specimen is not viewed along a major crystal axis or is not naturally rectangular, it is necessary to redefine a larger unit cell to get something that is rectangular. It is technically possible to define a multidimensional Fourier transform with nonorthogonal coordinates (for example, Dudgeon and Mersereau [83]) however this is beyond the scope of this book (and this author). It is generally easier to write one program and rearrange the specimen coordinates than to write a new program for every possible specimen with different symmetry. There is no general procedure for all specimens. Each specimen may require a different approach to generate its multislice description.

The multislice method requires that the specimen be divided into a sequence of thin rectangular slices. Each slice must be thin enough to be a weak phase object and is in a plane perpendicular to the optic axis of the electron microscope (along z) as in Fig. 6.6. All of the atoms within z to $z+\Delta z$ are compressed into a flat plane or slice at z. When viewed along the optic axis each slice must be aligned with the natural periodicity of the specimen. The edges of the slice (in the x,y plane) must obey periodic boundary conditions (in x and y) in the plane of the slice. This is a requirement for using the FFT and is identical to the requirements of the phase grating calculation (Chap. 5). If the slices do not obey periodic boundary conditions in the x and y direction serious artifacts may be generated in the image due to the so called wrap around error (see Chap. 4). The transmission function for each slice should also be symmetrically bandwidth limited as in the phase grating calculation (Fig. 5.10) even though the sample spacing may be different in each direction (x and y).

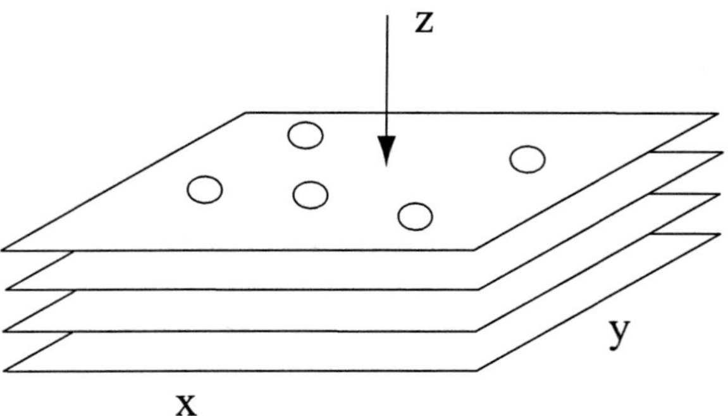

Fig. 6.6 The standard multislice description of the specimen as many *thin slices* perpendicular to the optic axis of the electron microscope (four slices shown). Each slice must be thin enough to be a weak phase object and is typically one atomic layer of the specimen. The optic axis of the electron microscope is in the positive z direction. Each slice should obey periodic boundary conditions in x and y

The slices do not necessarily have to be periodic along the optic axis (the z direction), although this is frequently the case. Many crystalline specimens of interest are organized into layers of atoms. If the specimen is aligned such that these layers are perpendicular to the electron beam direction then it is usually best to identify the slices with the atomic layers in the specimen. Many crystalline specimen can be described as a repetitive sequence of a small number of identical layers. For example the (111) projection of silicon has a stacking sequence of *abcabc*... with three repeating layers. Each layer can be used as one slice for the multislice method. In this case it is more efficient to calculate the transmission function for each of these three layers and store them in three arrays in computer memory. A multislice simulation program would then just reuse each layer over and over again without recalculation (to reduce the overall computer time required). The opposite extreme would be a completely disordered structure with no obvious repetitive structure along z. In this case there is no advantage to saving the transmission function of each slice. The transmission function for each slice can be calculated when needed and then discarded. This reduces the computer memory requirements significantly and has little or no effect on the computer time. The total computer time will increase with the number of slices that must be calculated but there is no advantage to saving the slices because they cannot be reused. These two strategies (precalculating the slices and reusing them vs. calculating each slices as needed) may produce two different types of computer programs.

Aligning the natural atomic layers of the specimen with the slices can have beneficial side effects in the accuracy of the multislice solution. Figure 6.7 illustrates how this can occur. The atomic potential is strongly peaked near the atomic nucleus and falls off quickly away from the nucleus (the potential is approximately a screened $1/r$ dependence where r is the distance from the nucleus). The effective range of the potential in the atoms in one layer can be smaller than the distance between layers (see the rectangular box labeled "atoms" in Fig. 6.7). The potential is identically zero in between the layers. The transmission in vacuum is nearly exact but the multislice error occurs only over the thickness of the layer Δz_a. The propagator may be cascaded as:

$$p(x,y,\Delta z) = p(x,y,\Delta z - \Delta z_a) \otimes p(x,y,\Delta z_a). \tag{6.94}$$

In effect the multislice equations are a transmission plus propagator over a distance Δz_a followed by a propagation over a distance $\Delta z - \Delta z_a$. The effective multislice error is of order Δz_a which can be significantly smaller than total slice thickness Δz.

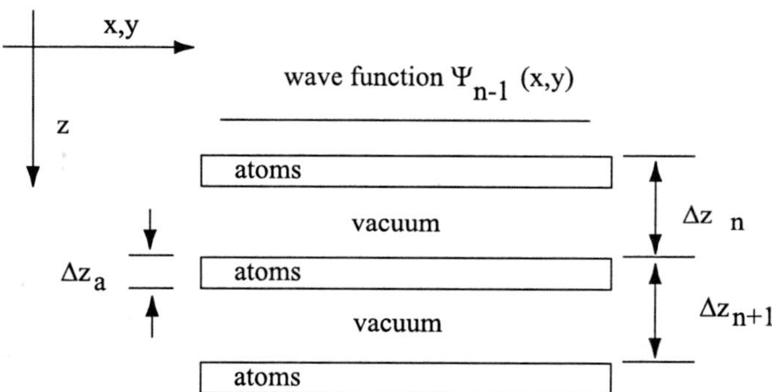

Fig. 6.7 The multislice error may be reduced by aligning atomic layers with the slice boundaries. The potential for each atomic layer is relatively thin (Δz_a) and the vacuum space between the slices is a free space propagation. The multislice slice thickness is Δz_n

Furthermore, if there are an integer number of slices in one repeat distance (along z) of the specimen then the multislice simulation can correctly reproduce the upper layer lines (higher order Laue zones) in the scattering process (Goodman and Moodie [127], Van Dyck [357]). If the specimen has a nonzero extent along the beam direction (the z axis) then its reciprocal lattice will have a definite three dimensional structure. The simple case of a cubic unit cell is shown in Fig. 6.8. The wave vector of the incident electron beam is shown as k_z. The multislice calculation considers only elastic scattering so the scattered electron wave vector k_s must lie on the Ewald sphere to keep the length (or energy) of the incident and scattered electrons equal. The reciprocal lattice of the specimen (dark spots in Fig. 6.8) represents the only allowed changes in electron wave vector that can occur via elastic scattering in the specimen. Only the electron scattering angles on the Ewald sphere that cross spots in the reciprocal lattice of the specimen are allowed to contribute to the final transmitted electron wave function. The reciprocal lattice sites with a nonzero offset along the k_z direction are referred to as the higher order Laue zones (HOLZ). Only the bottom layer is allowed in the thin specimen calculations in Chap. 5 and is referred to as the zero order Laue zone (ZOLZ). The first nonzero (along z) layer is referred to as the first order Laue zone (FOLZ), the second layer is the second order Laue zone (SOLZ), etc. The ZOLZ gives rise to a large region in the diffraction pattern near the origin. The HOLZ regions give rise to rings at successively

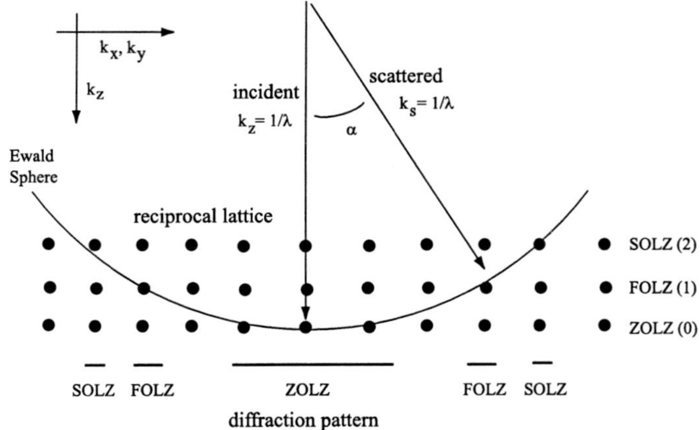

Fig. 6.8 The upper layer lines or higher order Laue zones (HOLZ). If the specimen is periodic in three dimensions (crystalline) then the reciprocal lattice will have additional diffraction conditions along the third dimension which can give rise to additional diffraction spots at high angle

higher angles. If the slice thickness in the multislice calculation matches the natural periodicity in the specimen (i.e., an integer number of slices in the repeat length of the specimen) then the multislice simulation will reproduce the higher order Laue zones. If the slice thickness does not match the specimen periodicity then beating can occur between the slice thickness and the specimen periodicity to produce artifacts in the image. The slice thickness can produce an artificial periodicity in the specimen if it does not match the natural periodicity of the specimen. If the multislice slice thickness is much larger than the natural periodicity of the specimen then false HOLZ lines can be created at:

$$k = \sqrt{\frac{2}{\Delta z \lambda}}, \tag{6.95}$$

where Δz is the multislice slice thickness, λ is the electron wavelength and $\alpha = \lambda k$ is the electron scattering angle. The slice thickness effectively takes the place of the normal crystal lattice spacing along z. An overly large slice thickness can produce serious artifacts in a simulated image and should be avoided. Kilaas et al. [197] have discussed methods of including the upper layers lines in more detail.

The standard multislice method [(6.92) and (6.93)] is only accurate to order Δz. It is tempting to just let Δz get smaller and smaller to obtain more accuracy. However, there is a competing effect that limits the minimum value of the slice thickness in most practical implementations. The standard multislice is frequently stated in terms of the total projected atomic potential $v_z(x,y)$ of the specimen, which is the integrated potential (along z) from minus infinity to plus infinity. Typically the total projected atomic potential is equated with the projected atomic potential for only the slice thickness:

$$v_{\Delta z}(x,y) = \int_z^{z+\Delta z} V(x,y,z)\mathrm{d}z \sim \int_{-\infty}^{+\infty} V(x,y,z)\mathrm{d}z = v_z(x,y). \tag{6.96}$$

As long as the slice thickness is large compared to the effective range of the atomic potential this approximation is valid. However, if the slice thickness Δz is made less than the range of the atomic potential (about 1Å) then this approximation is no longer valid. Therefore, if the total projected atomic potential is used (as is typical) the minimum slice thickness is about 1Å, which sets a limit on the maximum achievable accuracy of the multislice calculation. There is some question if 1Å is thin enough for heavy atoms, such as gold, at 100 keV, however higher voltage or lower atomic number should be all right (Watanabe [372]). The electron wavelength is typically of order 0.03Å, which is much smaller than the minimum slice thickness. If the full wave function were used then the slice thickness would have to be much smaller than the electron wavelength and the multislice method would not work at all. However, only the slowly varying part of the wave function is calculated so a slice thickness of several Angstroms is acceptable.

6.8 Aliasing and Bandwidth

Each slice in the multislice method produces two operations on the wave function (propagation and transmission). The first step is multiplication by the transmission function in real space and the second step is a convolution with the propagator function (a multiplication in reciprocal or Fourier space). Multiplication in real space is equivalent to a convolution in reciprocal space. Using a discretely sampled image for the wave function allows the use of the FFT with its computational efficiency but the discrete sampling also creates some subtle problems with aliasing (see Sect. 4.1) caused by multiplication by the transmission function in real space (or any other nonlinear operation).

In reciprocal space, multiplication of the wave function $\psi(x,y)$ by the specimen transmission function $t(x,y)$ is:

$$\begin{aligned} \mathrm{FT}[t(x,y)\psi(x,y)] &= T(k_x,k_y)\otimes\Psi(k_x,k_y) \\ &= \int T(k'_x,k'_y)\Psi(k_x-k'_x,k_y-k'_y)\mathrm{d}k'_x\mathrm{d}k'_y, \end{aligned} \tag{6.97}$$

where $T(k_x,k_y)$ is the Fourier transform of $t(x,y)$ and $\Psi(k_x,k_y)$ is the Fourier transform of $\psi(x,y)$. If each function is bandwidth limited to a maximum spatial frequency of $k_{\max}$ (required when discretely sampled) then each function is only nonzero within a circle whose radius is $k_{\max}$. As shown earlier in Fig. 5.10 it is best to symmetrically bandwidth limit each function as:

$$k_{\max} = \min\left[\frac{N_x}{2a},\frac{N_y}{2b}\right], \tag{6.98}$$

where the supercell size is $a\times b$ in real space with $N_x\times N_y$ pixels. A convolution is equivalent to sliding one circle across the other as in Fig. 6.9. The result of the convolution is a two-dimensional function whose value is the integration of the overlap between the two circles and the distance between the circles is the spatial frequency associated with this value. Multiplying two functions of bandwidth k_{max} in real space produces a function with twice the original bandwidth ($2k_{max}$).

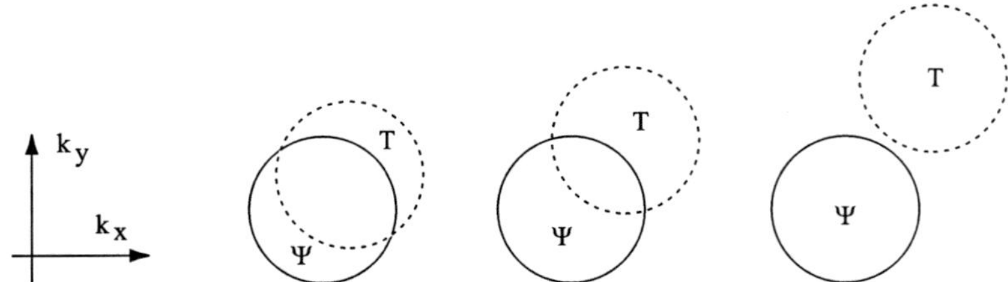

Fig. 6.9 The convolution of two continuous bandwidth limited function Ψ and T in reciprocal space (equivalent to multiplication in real space). When limited to a symmetrically maximum frequency (or bandwidth) of k_{max} each function is only nonzero within a *circle* of radius k_{max}. Each point in the convolution is a different offset of the two circles and the value is the integrated intensity in the overlap region

When the functions are discretely sampled the spectra of each function is periodically repeated in both direction as shown in Fig. 6.10. Both the wave function $\Psi(k_x, k_y)$ and the transmission function $T(k_x, k_y)$ are repeated periodically although only $\Psi(k_x, k_y)$ is shown as being repeated to make the drawing easier to understand. When a discretely sampled wave function and transmission function are convolved in reciprocal space, one function slides across the other and will overlap the adjacent periodically repeated functions. The other repeated functions improperly appear as very high spatial frequencies. It is equivalent to say that low spatial frequencies are aliased as high spatial frequencies. This can produce rather serious artifacts in the final simulation if not corrected.

The solution to this aliasing problem is to set the maximum spatial frequency (or bandwidth) of both functions to be 2/3 of the maximum sampling frequency (k_{max}) in the wave function (Self et al. [315]). This way one function no longer overlaps the other when its offset reaches the maximum allowed spatial frequency (Fig. 6.11). Frequencies greater than 2/3 still overlap more than one Ψ function so the final result must be explicitly bandwidth limited again to $(2/3)k_{max}$. If there are $N_x \times N_y$ pixels in the wave function and $N = N_x = N_y$ then there are a total of N^2 possible Fourier coefficients. Limiting the bandwidth to 2/3 of its maximum symmetrical value limits the number of Fourier coefficients to $\pi[(2/3)(N/2)]^2 = \pi N^2/9 = 0.35N^2$. Even though nearly two thirds of the Fourier coefficients must be set to zero the overall benefit of the FFT is still worth the effort. A convenient way of limiting both functions is to set both the specimen transmission function and the propagator function to zero outside of $(2/3)k_{max}$. When the electron wave function is multiplied by the transmission function (at each step or slice of the multislice method) its bandwidth doubles and is then reduced to the required maximum by convolution with the propagator function (multiplication in reciprocal space). This

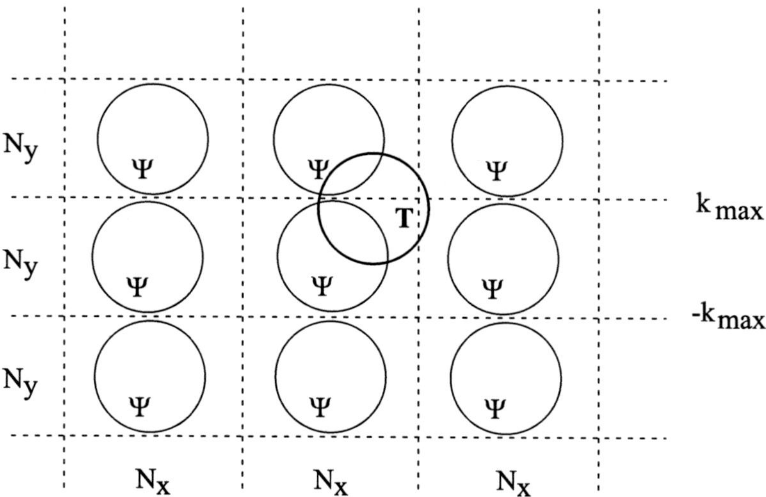

Fig. 6.10 The convolution of two discretely sampled bandwidth limited functions (the wave function Ψ and the transmission function T) in reciprocal space (equivalent to multiplication in real space). The bandwidth limit of each function appears as a *circle*

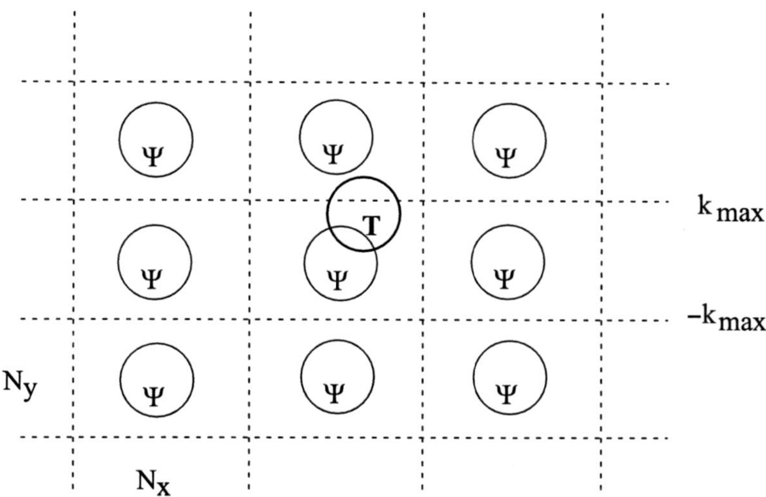

Fig. 6.11 The convolution of two discretely sampled bandwidth limited functions (the wave function Ψ and the transmission function T) in reciprocal space (equivalent to multiplication in real space). The bandwidth limit of each function appears as a *circle* and is limited to 2/3 of its maximum to eliminate aliasing

requires using the form of the multislice method in which convolution with the propagator function is performed after multiplication by the transmission function (6.73) and not (6.74).

Without the appropriate bandwidth limit the simulated image can have serious artifacts and dramatically differ from the correct result. Reducing the effective bandwidth can also reduce the convergence of the multislice method. Usually, the best recourse is to increase the number of pixels in one or both directions (N_x, N_y).

An alternate approach to eliminate aliasing was suggested by O'Keefe and Kilaas [271]. The wave function can be limited to $(1/2)k_{max}$ (before and after each slice) and the transmission function limited to k_{max} to get a similar elimination of aliasing.

6.9 Interfaces and Defects

Crystal defects and interfaces pose a problem for image simulation. The FFT is very efficient and produces a multislice calculation with an acceptable amount of computer time, but requires that the specimen potential and electron wave function be periodic in the xy plane (perpendicular to the optic axis of the microscope). The specimen does not need to be periodic along the z or beam direction although if it is not periodic along the beam it may require significantly more computer time.

An isolated point defect at a particular point in the specimen is obviously not periodic. If the defect is simply placed in the super cell of the specimen then the discrete nature of the sampled potential periodically reproduces an infinite number of point defects with the periodicity of the super cell as shown in Fig. 6.12. If the super cell dimensions $a \times b$ are too small then the image from one defect can interfere with the image from the periodically produced adjacent defects. The solution to this problem is to use an artificially large unit cell. If the super cell dimensions $a \times b$ are large then the point defects become essentially isolated and do not interfere. A good starting point is to keep the defects separated by about 25–30 Å. This method has been given the name periodic continuation and has been discussed by Grinton and Cowley [134], MacLagan et al. [233], Fields and Cowley [105], Anstis and Cockayne [16], Wilson and Spargo [381], and Matsuhata et al. [236].

A similar problem occurs for crystal interfaces as shown in Fig. 6.13. Consider an interface between material A on the left and material B on the right. The interface between the two materials in the center can be correctly modeled, however the so-called wrap around effect (or periodic continuation similar to Fig. 6.12) causes an additional interface to be produced when the right hand side of material B wraps around to touch material A on the right for a second time. In this problem the super cell should again be made large enough so that these two interfaces do not interfere with one another. Usually the edges of the image (with the second unintended interface) can just be ignored, and the interface in the center will contain the correct simulation (assuming that the super cell size is large enough). Alternately, the presence of two interfaces can be explicitly acknowledged by sandwiching a narrow region of material B in the center of two regions of material A on the left and right. Then two interfaces are simulated. Note that the two regions of material A must match in orientation when wrapped around.

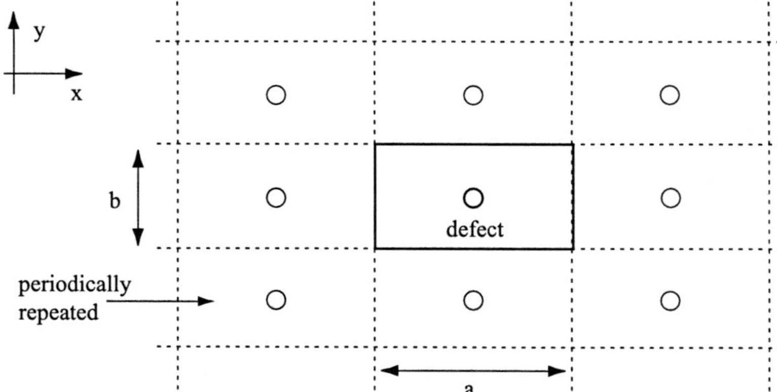

Fig. 6.12 Modeling a point defect (*the small circle*) in the center of the specimen of size $a \times b$. Periodic continuation of the image causes the defect to be repeated infinitely many times in both directions. The unit cell dimension $a \times b$ should be increased so that adjacent (periodically continued) defects do not interfere

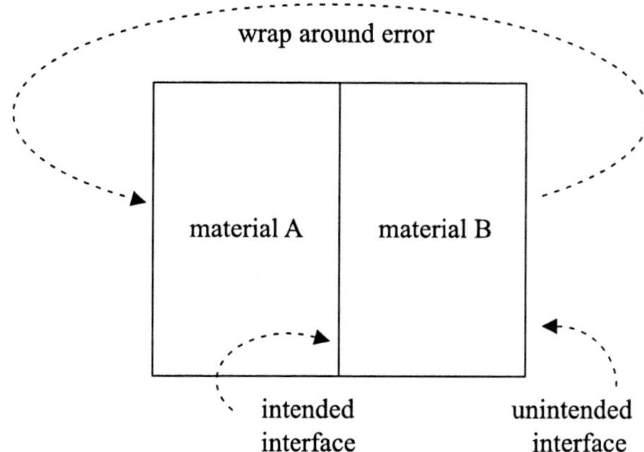

Fig. 6.13 Modeling an interface in the center of the sampled area. The wrap-around error causes an extra interface to be created due to the interaction of the *right hand side* and the *left hand side* of the image are

6.10 Multislice Implementation

The multislice method is particularly well suited for numerical implementation on a computer. There are several subtle problems and strategies associated with the multislice method, some of which will be discussed in this section. O'Keefe and Buseck [269] were the first to describe a specific computer implementation. Several other programs that have been described in the literature are listed in Table 6.2. Many other multislice program have likely been produced but have not been specifically described in the literature.

Table 6.2 Some image simulation software packages appearing in the literature or on-line

Program	Author	Year	Type	Comments
SHRLI	O'Keefe and Buseck [269]	1978,9	M	
TEMPAS	Kilaas [194]	1987	M	
EMS	Stadelmann [335]	1987	B	
NCEMSS	O'Keefe and Kilaas [271]	1988	M	
MacTEMPAS	Kilaas [195]	?	M	on-line
TEMSIM	Kirkland [205]	1998	M	CD, on-line
?	Ishizuka [178]	2001	B, M	online
?	deGraf [129]	2003	B	online
JEMS	Stadelmann [336]	2004	B, M	online
WebEMAPS	Zuo [334, 393]	2005	B	online
EDM	Marks et al [235]	2006	B,M	online
SimulaTEM	Gómez-Rodríguez et al. [123]	2010	M	online

Type M is multislice and type B is Bloch-Wave. Some of the listed programs may be commercial. Many other private programs likely exist

Table 6.3 Steps in the simulation of CTEM images of thick specimens

Step 1	Divide the specimen into thin slices.
Step 2	Calculate the projected atomic potential $v_{zn}(\mathbf{x})$ [(5.19) or (5.21)] for each slice and symmetrically bandwidth limit them.
Step 3	Calculate the transmission function $t_n(\mathbf{x}) = \exp[i\sigma v_{zn}(\mathbf{x})]$ (5.25) for each slice and symmetrically bandwidth limit each to 2/3 of it maximum to prevent aliasing.
Step 4	Initialize the incident wave function $\psi_0(x,y) = 1$.
Step 5	Recursively transmit and propagate the wave function through each slice $\psi_{n+1}(x,y) = p_n(x,y,\Delta z_n) \otimes [t_n(x,y)\psi_n(x,y)]$ using FFT's as in (6.92). Repeat until the wave function is all the way through the specimen
Step 6	Fourier transform the wave function at the exit surface of the specimen $\Psi_n(k_x,k_y) = \mathrm{FT}[\psi_n(x,y)]$.
Step 7	Multiply the transmitted wave function $\Psi_n(k_x,k_y)$ by the transfer function of the objective lens, $H_0(k)$ (5.27) to get the image wave function in the back focal plane $\Psi_i(\mathbf{k}) = H_0(k)\Psi_n(\mathbf{k})$.
Step 8	Inverse Fourier transform the image wave function $\psi_i(\mathbf{x}) = \mathrm{FT}^{-1}[\Psi_i(\mathbf{k})]$.
Step 9	Calculate the square modulus of the image wave function (in real space) to get the final image intensity $g(\mathbf{x}) = \|\psi_i(\mathbf{x})\|^2 = \|\psi_n(\mathbf{x}) \otimes h_o(\mathbf{x})\|^2$.

If there are a small number of distinct layers repeated several times then the transmission function for each can be calculated and stored otherwise they can be calculated as needed and discarded

The basic procedure for calculating CTEM and STEM images is similar to the procedure for simulating thin specimens given in Tables 5.1 and 5.3 except that the steps that propagate the electron wave function through the specimen are changed into the recursive relation (6.73) or (6.92). The portion of the electron path in the electron microscope that does not involve the specimen is identical for thin and thick specimens. Tables 6.3 and 6.4 summarize the multislice method for CTEM and STEM.

Table 6.4 Steps in the simulation of STEM images of thick specimens

Step 1	Divide the specimen into thin slices.
Step 2	Calculate the projected atomic potential $v_{zn}(\mathbf{x})$ [(5.19) or (5.21)] for each slice and symmetrically bandwidth limit them.
Step 3	Calculate the transmission function $t_n(\mathbf{x}) = \exp[i\sigma v_{zn}(\mathbf{x})]$ (5.25) for each slice and symmetrically bandwidth limit each to 2/3 of it maximum to prevent aliasing.
Step 4	Calculate the probe wave function $\psi_p(\mathbf{x}, \mathbf{x}_p)$ at position $\mathbf{x}_p$ (5.45,5.47)
Step 5	Recursively transmit and propagate the probe wave function through each slice $\psi_{n+1}(x,y) = p_n(x,y,\Delta z_n) \otimes [t_n(x,y)\psi_n(x,y)]$ using FFT's as in (6.92). Repeat until the wave function is all the way through the specimen
Step 6	Fourier transform the transmitted wave function to get the wave function in the far field (diffraction plane).
Step 7	Integrate the intensity (square modulus) of the wave function in the diffraction plane including only those portions that fall on the detector (5.50). This is the signal for one point or pixel in the image.
Step 8	Repeat step 4 through step 7 for each position of the incident probe $\mathbf{x}_p$.

There is a multislice simulation for each point in the final image. If there are a small number of distinct layers repeated several times then the transmission function for each can be calculated first and stored otherwise they can be calculated as needed and discarded. In a parallel computing environment, many probes can be propagated at the same time on different processors

6.10.1 The Propagator Function and Specimen Tilt

Small amounts of specimen tilt may be included with a small modification to the propagator function (6.71):

$$P(k, \Delta z, \theta) = \exp\left[-i\pi\lambda k^2 \Delta z + 2\pi \mathrm{i} \Delta z (k_x \tan\theta_x + k_y \tan\theta_y)\right] \qquad (6.99)$$

where θ_x, θ_y is the crystal tilt in the x, y directions, $k^2 = k_x^2 + k_y^2$ and Δz is the slice thickness. This is equivalent to shifting the wave function between slices and is only valid for small tilts of no more that about 1 degree (Cowley [59]). A specimen tilt is not the same as a beam tilt because the beam direction has a strong interaction with the electron optical aberrations of the objective lens. Ishizuka [174] and Chen et al. [48] have given a more detailed discussion of tilt in multislice simulations.

The propagator function requires a small but significant amount of computation because of the transcendental functions. If the specimen contains many layers with the same slice thickness it is advantageous to calculate the propagator once, in advance, and then reuse it for each slice. The whole two-dimensional propagator can require a significant amount of computer memory. However the propagator function may be factored into an x component and a y component as:

$$\begin{aligned} P(k, \Delta z, \theta) &= P_x(k_x, \Delta z, \theta_x) P_y(k_y, \Delta z, \theta_y) \\ P_x(k_x, \Delta z, \theta_x) &= \exp\left[-i\pi\lambda k_x^2 \Delta z + 2\pi \mathrm{i} \Delta z k_x \tan\theta_x\right] \\ P_y(k_y, \Delta z, \theta_y) &= \exp\left[-i\pi\lambda k_y^2 \Delta z + 2\pi \mathrm{i} \Delta z k_y \tan\theta_y\right]. \end{aligned} \qquad (6.100)$$

This factorization can be put to good use when programming the multislice simulation. $P_x(k_x, \Delta z, \theta_x)$ and $P_y(k_y, \Delta z, \theta_y)$ can be precalculated and stored in two one-dimensional arrays that require much less computer memory than a whole two-dimensional array and then multiplied together when the whole two-dimensional propagator function is needed. This produces about the same reduction in computer time as precalculating the whole two-dimensional propagator (and reusing it for each slice) but requires only a relatively small amount of additional computer memory.

6.10.2 Convergence Tests

The computer will merrily calculate the wave function with total disregard for the accuracy of the calculation. It is the responsibility of the human user to interpret the results and decide if they are correct or not. There are a variety of approximation that go into the derivation of the multislice method and it is not guaranteed to give the right answer for every possibly set of input parameters. The user must specify the atomic coordinates of the specimen, the slice thickness, the size and number of pixels in the wave function, and specimen potentials. In a practical sense there are a lot more ways to make the simulation fail than there are to make it succeed. When using a multislice simulation program (or any other simulation program) it is always a good idea to be a little skeptical and do some testing to try and verify that the result is correct or at least internally consistent.

The multislice algorithm has a built in parameter that can be used to verify that it is running correctly. The electron wave function only interacts with the specimen with elastic scattering as far as the simulation is concerned. This means that the total number of electrons should be conserved. One easy test is to watch the total integrated intensity of the electron wave function as it progresses through the specimen. If the wave function incident on the specimen (plane wave for CTEM and a focused probe for STEM) is normalized such that:

$$I_n = \int |\psi_n(x,y)|^2 \mathrm{d}x\mathrm{d}y = 1 \quad \text{for n} = 0, \tag{6.101}$$

where n is the slice index and $n = 0$ for the incident wave function, then I_n should remain constant. The integrated intensity can become less than one if the sampling is inadequate. When the specimen scatters electrons to high angle some of them may be scattered outside of the maximum allowed angle (the 2/3 maximum bandwidth limit required to eliminate aliasing). Once they are scattered outside of this maximum limit then they are effectively lost to the calculation and I_n decreases. The actual value of I_n has no particular physical significance other than indicating whether or not the simulation is working. The electrons scattered outside the maximum limit are just as likely to continue to higher angles or to be scattered back to low angles. An I_n less than one only indicates the relative precision of the simulation. In practice this number is not very sensitive to the accuracy of the calculation.

A value of $I_n \leq 0.90$ is probably wrong (although it may give a qualitatively valid image) and the sampling should be increased. This may mean a smaller pixel size in real space or reciprocal space, more pixels or some combination of these. Values of $0.95 \leq I_n \leq 1.00$ are typical for well behaved calculations. On the other extreme values of $I_n > 1.0$ can also occur. An integrated intensity greater than one is also a clear indication that the calculation is not correct (typically this means that the slice thickness is too large).

Even if the total integrated intensity is within bounds the simulation is not guaranteed to be correct either. One of the best tests is to compare two different simulations performed with slightly different sampling sizes (usually about a factor of two apart is good). The multislice simulation is only accurate in the limit of an infinitesimally small pixel size and slice thickness. If the sampling is adequate then reducing the pixel size or slice thickness by a factor of two should have no effect. Therefore, if two simulations with different sampling sizes produce the same result then the simulation is likely to have the proper sampling. The phrase "pixel size" refers to both the real space pixel size and the reciprocal space pixel size. If one simulation is performed with an image size of $a \times b$ and $N_x \times N_y$ pixels then increasing the number of pixels to $2N_x \times 2N_y$ only tests the pixel size in real space but not in reciprocal space. To test both pixel sizes the image size should be increase to about $\sqrt{2}a \times \sqrt{2}b$ and the number of pixels doubled in each direction to $2N_x \times 2N_y$. This approach tests the pixel size in both real space and reciprocal space at the same time.

All of the schemes for testing the accuracy of the simulation discussed earlier are only testing the internal consistency of the calculation. In the end no amount of internal consistency will guarantee that the simulation is correct. The only real test is to compare to experimentally observed images. Comparisons between real experimentally recorded image and theoretically simulated images have been performed and the multislice simulation is generally agreed to produce acceptable simulation of real images. Note however that most comparisons are done in a rather qualitative manner. The simulated images are subjectively judged to look like the experimentally observed image. This is due in part to the lack of quantitative experimental image data. Usually important parameters such as defocus or the incident beam intensity are simply not known, so a detailed quantitative comparison is difficult.

6.10.3 Partial Coherence in BF-CTEM

The electron microscope image is never perfectly coherent as assumed in Table 6.3. There is always a small spread in illumination angles from the condenser lens and a small spread in defocus values due to small instabilities in the high voltage and lens current supplies. When these effects are included the image is said to be partially coherent. Sections 3.2 discussed partial coherence in the linear image model and Sect. 5.4.3 introduced the transmission cross coefficient for partial coherence in nonlinear imaging of thin specimen.

If the specimen is not too thick and the amount of incoherence is small (small condenser angle) then the transmission cross coefficient (Sect. 5.4.3) can be applied to the wave function transmitted through the specimen via the multislice algorithm. However, if the spread in illumination angles is significant and the specimen is thick then the transmission cross coefficient is not quit right. Each illumination angle incident on the specimen may interact differently with the specimen. Different illumination angles may satisfy different diffraction conditions in the specimen. The best way to simulate a large spread in illumination angels in a thick specimen is to perform a multislice simulation for each angle in the condenser aperture and sum the results incoherently (OKeefe and Sanders [272]) assuming the source is incoherent. If $\mathbf{k}_\beta$ is one angle in the condenser aperture then the initial wave function is:

$$\psi_0(\mathbf{x}) = \exp(2\pi i \mathbf{k}_\beta \cdot \mathbf{x}). \tag{6.102}$$

This wave function is then recursively transmitted through the specimen using the multislice algorithm (6.92) yielding a transmitted wave function of $\psi_n(\mathbf{x}, \mathbf{k}_\beta)$. Each of these wave functions should then be convolved with the objective lens point spread function $h_0(\mathbf{x})$ and incoherently summed to give a final image intensity of:

$$g(\mathbf{x}) = \frac{1}{N_\beta} \sum_{\mathbf{k}_\beta} |\psi_n(\mathbf{x}, \mathbf{k}_\beta) \otimes h_0(\mathbf{x})|^2, \tag{6.103}$$

where N_β is the number of condenser illumination angles used. The illumination angles vary in two dimension and must match the existing periodic boundary conditions of the specimen (i.e., only integer coordinantes in the Fourier transform). This summation can require a significant amount of computer time but is not excessive on currently available computer hardware for reasonable sizes of condenser angles. If the condenser aperture is not uniformly illuminated then a suitable weighting factor can be added to (6.103). A defocus spread can also be included by simply integrating over a range of defocus values Δf as:

$$\begin{aligned} g(\mathbf{x}) &= \frac{1}{CN_\beta} \sum_{\mathbf{k}_\beta} \sum_{\Delta f} p(\Delta f) |\psi_n(\mathbf{x}, \mathbf{k}_\beta) \otimes h_0(\mathbf{x}, \Delta f)|^2 \\ C &= \sum_{\Delta f} p(\Delta f), \end{aligned} \tag{6.104}$$

where $p(\Delta f)$ is the probability distribution of defocus values. The integration over defocus is relatively quick compared to the integration over illumination angles because it does not require a multislice simulation for each defocus value.

6.10.4 Parallel Computing

The current trend in computer hardware is to couple many processors (or CPU's) into one unit. Improvements in the speed of individual processors appear to be nearing a physical speed limit, so the trend of multiprocessing may be here to stay. Even inexpensive computers intended for personal use can have much more than a single processor, so this trend is worth exploiting. Most of the algorithm development in the past was modeled on a single processor approach, so many algorithms require a complete reorganization to use more than one processor at a time and many new algorithms are being developed for this mode of computation. Fortunately, there are several relatively easy ways to use multiple CPUs in the multislice simulation method, which have been used by many authors (including this one) in recent years.

There are currently two different popular methods of connecting multiple processors together, that mainly differ in how the memory is accessed. One is the shared memory processing (SMP) approach in which many processors share the same large block of memory with a fast data path to and from this memory. The processors communicate by reading or writing data from or to this fast memory. This approach is usually limited to a small number (of order 5 or 10, but increasing every year) of processors because of the hardware difficulty of connecting multiple processors to the same memory, but has the advantage of being easier to program. The other approach uses distributed memory and many physically separate computers, each with their own memory, connected together on a very fast network switching hub (connection), collectively called a cluster. This has the advantage of using existing computers and allows a vast numbers (thousands) of processors to be coupled into one system. The disadvantage is that the communications between processors is relatively slow (compared to SMP) which is a little harder to program in some cases. A computing cluster can have both modes of operation. Robertson [303] and Grillo et al. [45, 133] have implemented ADF-STEM on an MPI cluster. There is a third evolving approach in which specialized graphics processors with hundreds of very inexpensive processors have been re-targeted for numerical processing. These are difficult to program but can be very inexpensive. Dwyer [84] has given some preliminary results. It is not clear how this will evolve at the time of this writing.

Multiprocessing has been around for a long time. Two recent developments have made it a lot more attractive in recent years. The SMP hardware has moved into the mainstream of inexpensive every day computers (not just expensive research computers) and the software tools have become easy to use. Two packages are worth mentioning; openMPI which is a message passing environment for a distributed memory cluster and openMP for a shared memory (SMP) environment (see, for example, Quinn [291]). Many compilers now support one or both of these in some manner. This software is relatively easy to use (compared to previous vendor specific approaches) and is nonproprietary. It is being implemented on a variety of different computing platforms, so is likely to last for a long time.

There are several ways to utilize multiple processors in parallel in a multislice calculation. It is not so obvious for Bloch wave calculation, although the ScaLapack subroutine package targets a many processor distributed cluster and may help with the eigenvalue calculation which is the main portion of the calculation. At a low level the multidimensional FFT (in this case 2D, used in the multislice method) can easily be split into many parallel paths. Each row can be done be a different processor in one direction and then each col. in the other direction. Various sums in the transmission function can also be split into separate processors. Both of these methods are a relatively fine grain parallelization and work best on an SMP machine. A distributed memory cluster may not yield an improvement this way due to the increased communications overhead. An ADF-STEM calculation can be extremely compute intensive. However, it is also very easy to expand into a multiprocessor mode. The transmission of each probe position in the image can be put on a separate processor and run in parallel. This has the advantage of a coarse grain parallelization and may work well in a distributed memory environment. The ADF-STEM code used here uses this approach in an SMP mode.

6.11 More Accurate Slice Methods

The standard multislice expression is only accurate to first order in Δz. There is some incentive to find more accurate solutions in which the error term is a higher power (or order) of the small quantity Δz. An obvious idea is to simply average (6.73) and (6.74) to increases the accuracy by one order in Δz. This approach is more accurate but unfortunately also doubles the computer time. Averaging would also have the benefit of insuring that reciprocity is obeyed.

6.11.1 Operator Solutions

One obvious weak spot in the multislice derivation is factoring the combined operator (6.62) and (6.63). An alternate approach to factorizing the combined operator is:

$$\begin{aligned}
&\exp(A\varepsilon/2)\exp(B\varepsilon)\exp(A\varepsilon/2)\\
&\quad=\left[1+A\frac{\varepsilon}{2}+\frac{1}{2!}A^2\left(\frac{\varepsilon}{2}\right)^2+\cdots\right]\left[1+B\varepsilon+\frac{1}{2!}B^2\varepsilon^2+\cdots\right]\\
&\qquad\qquad\times\left[1+A\frac{\varepsilon}{2}+\frac{1}{2!}A^2\left(\frac{\varepsilon}{2}\right)^2+\cdots\right]\\
&\quad=1+(A+B)\varepsilon+\frac{1}{2!}(A^2+AB+BA+B^2)\varepsilon^2+\mathcal{O}(\Delta z^3),
\end{aligned}\tag{6.105}$$

where A and B are noncommuting matrices or operators and ε is a small scalar quantity. By comparison to (6.59) the operator relevant to the multislice method can be written in symmetrical form as:

$$\exp(A\varepsilon+B\varepsilon)=\exp(A\varepsilon/2)\exp(B\varepsilon)\exp(A\varepsilon/2)+\mathcal{O}(\varepsilon^3).\tag{6.106}$$

The error term is proportional to $\mathcal{O}(\Delta z^3)$ which is one order better than the previous identities (6.62 and 6.63) used in the standard multislice derivation. This symmetrical operator approximation has been used to simulate the paraxial propagation of laser beams by Fleck et al. [107] with a method very similar to the multislice method.

van Dyck [353, 357] has pointed out that this new operator identity allows the standard multislice equation to be re-interpreted in a more accurate manner. Applying (6.106) to the multislice equation (6.57) yields:

$$\psi_{n+1}(x,y) = p_n(x,y,\Delta z/2) \otimes \{t_n(x,y)\,[p_n(x,y,\Delta z/2) \otimes \psi_n(x,y)]\} + \mathcal{O}(\Delta z^3) \tag{6.107}$$

where $\psi_n(x,y)$, $p_n(x,y,\Delta z/2)$, $t_n(x,y)$ are the wave function, propagator function and transmission functions of the nth layer. The propagator function of the nth layer is split into two component each with thickness $\Delta z/2$. When (6.107) is applied recursively the convolution with the propagator on the right combines with the convolution with the propagator of the left to yield:

$$p_{n+1}(x,y,\Delta z_{n+1}/2) \otimes p_n(x,y,\Delta z_n/2) \otimes = p(x,y,\Delta z) \otimes \tag{6.108}$$

if $\Delta z_n = \Delta z_{n+1} = \Delta z$. The final result is a succession of convolutions with the propagator and multiplication by the transmission function that is identical to the standard multislice formulation with only one additional propagation by $\Delta z/2$ at the end. This means that the standard multislice method ((6.73) and (6.74)) can be interpreted as being accurate to $\mathcal{O}(\Delta z^3)$ locally and $\mathcal{O}(\Delta z^2)$ globally if the result is offset by one half of a slice thickness. The defocus is rarely known with much accuracy so this offset is negligible. Van Dyck [354,356,357] and Chen [47,49] have proposed other possible higher order methods.

6.11.2 Finite Difference Solutions

Another approach to increasing the accuracy is to include more terms in the Taylor series expansion for ψ in Δz. Consider the following expansions for both positive and negative steps in z and temporarily drop explicit reference to the independent variables x and y in ψ.

$$\psi(z+\Delta z) = \psi(z) + \Delta z \frac{\partial \psi(z)}{\partial z} + \frac{1}{2!}\Delta z^2 \frac{\partial^2 \psi(z)}{\partial z^2} + \frac{1}{3!}\Delta z^3 \frac{\partial^3 \psi(z)}{\partial z^3} + \cdots \tag{6.109}$$

$$\psi(z-\Delta z) = \psi(z) - \Delta z \frac{\partial \psi(z)}{\partial z} + \frac{1}{2!}\Delta z^2 \frac{\partial^2 \psi(z)}{\partial z^2} - \frac{1}{3!}\Delta z^3 \frac{\partial^3 \psi(z)}{\partial z^3} + \cdots \tag{6.110}$$

Subtracting (6.109 and 6.110) yields:

$$\frac{\partial \psi(z)}{\partial z} = \frac{\psi(z+\Delta z) - \psi(z-\Delta z)}{2\Delta z} + \mathcal{O}(\Delta z^2). \tag{6.111}$$

All terms containing even derivatives vanish identically leaving only the higher order error term proportional to Δz^2. If instead (6.109 and 6.110) are added then:

$$\frac{\partial^2 \psi(z)}{\partial z^2} = \frac{\psi(z+\Delta z) - 2\psi(z) + \psi(z-\Delta z)}{\Delta z^2} + \mathcal{O}(\Delta z^2), \tag{6.112}$$

where all terms containing odd derivatives vanish identically leaving only the error term proportional to Δz^2. These are the finite difference approximations to both the first and second derivatives of $\psi(z)$ with respect to z. The Schrödinger wave equation (6.46) for the slowly varying portion of the wave function contains both first and second derivatives with respect to z. Rewriting it using the interaction parameter $\sigma = 2\pi m e\lambda/h^2$:

$$\left[\frac{\partial^2}{\partial z^2} + \frac{4\pi i}{\lambda}\frac{\partial}{\partial z} + \nabla^2_{xy} + \frac{4\pi\sigma}{\lambda}V(x,y,z)\right]\psi(x,y,z) = 0 \tag{6.113}$$

Next substitute the finite difference approximations for the first and second derivatives (6.111 and 6.112) without ignoring the second derivative with respect to z as was done for the Howie-Whelan equations and the traditional multislice solution.

$$\frac{\psi(z+\Delta z) - 2\psi(z) + \psi(z-\Delta z)}{\Delta z^2} + \frac{4\pi i}{\lambda}\frac{\psi(z+\Delta z) - \psi(z-\Delta z)}{2\Delta z} + \nabla^2_{xy}\psi(z) + \frac{4\pi\sigma}{\lambda}V(x,y,z)\psi(z) + \mathcal{O}(\Delta z^2) = 0 \tag{6.114}$$

Multiply by Δz^2, rearrange terms and restore the reference to the x,y dependence in ψ:

$$\psi(x,y,z+\Delta z) = \frac{1}{c_+}\left[2 - \Delta z^2\left(\nabla^2_{xy} + \frac{4\pi\sigma}{\lambda}V(x,y,z)\right)\right]\psi(x,y,z) - \left(\frac{c_-}{c_+}\right)\psi(x,y,z-\Delta z) + \mathcal{O}(\Delta z^4/c_+) \tag{6.115}$$

$$c_+ = 1 + 2\pi i\Delta z/\lambda$$
$$c_- = 1 - 2\pi i\Delta z/\lambda$$

$\Delta z/\lambda >> 1$ so that dividing by c_+ does not necessarily reduce the order of the error term. The values of $\psi(x,y)$ are calculates in an xy plane at $z+\Delta z$ from ψ in an xy plane at z and $z-\Delta z$. This is also a slice method but of higher accuracy because the error term is a higher power of the small quantity Δz and because the second derivative of ψ with respect to z has not been ignored. This method can be referred to as a three plane slice method because the values of $\psi(x,y)$ in three planes are related to one another in each iteration. The standard multislice method can be referred to as a two plane slice method. When using a discretely sampled wave function and potential this method should also be bandwidth limited to $2/3$ of its maximum (both $\psi_n(x,y)$ and $V(x,y)$) just as in the standard multislice method to avoid aliasing. The potential $V(x,y,z)$ is not a very well behaved function. It is strongly peaked near the center of each atom (with a logarithmic singularity) which may defeat the higher accuracy of this equation. For comparison repeating the earlier derivation without the second derivative term yields the result:

$$\psi(x,y,z+\Delta z) = 2\left[\Delta z \frac{i\lambda}{4\pi}\nabla_{xy}^2 + i\sigma\Delta z V(x,y,z)\right]\psi(x,y,z) + \psi(x,y,z-\Delta z) + \mathcal{O}(\Delta z^3) \quad (6.116)$$

The second derivative term $\nabla_{xy}^2\psi$ can be approximated with its finite difference equivalent:

$$\nabla_{xy}^2\psi(x,y) = \frac{\psi(x+\Delta x,y) - 2\psi(x,y) + \psi(x-\Delta x,y)}{\Delta x^2} + \frac{\psi(x,y+\Delta y) - 2\psi(x,y) + \psi(x,y-\Delta y)}{\Delta y^2} + \mathcal{O}(\Delta x^2) + \mathcal{O}(\Delta y^2) \quad (6.117)$$

or by using a pair of FFT's as:

$$\nabla_{xy}^2\psi(x,y) = \mathrm{FT}^{-1}\left[-4\pi^2k^2\mathrm{FT}[\psi(x,y,z)]\right] \quad (6.118)$$

where FT is the two-dimensional Fourier transform with respect to x and y and $k^2 = k_x^2 + k_y^2$ is the spatial frequency. Using the Fourier transform does not add a significant error if the sampling is adequate.

Equation (6.115) uses the atomic potential directly (without projection) and does not suffer from a minimum slice thickness limitation as in the standard multislice method although it has an opposite problem. If the slice thickness is too thick (greater than about 1 Å) then (6.115) can possibly lose some atoms in between slices.

This method (6.115) is accurate to about one order better than the standard multislice method and is more complete because it includes the second derivative (with respect to z) term. The local (or per step) error is $\mathcal{O}(\Delta z^3)$ and the total or global error is $\mathcal{O}(\Delta z^2)$ if each slice is symmetrical about its center. This solution requires two FFT's to calculate $\nabla_{xy}^2\psi(x,y)$ plus several pairs of FFT's to bandwidth limit the appropriate functions (to avoid aliasing) for each iteration. This amount of computation is similar to the amount of computation in the standard multislice method. However, it requires the storage of an additional set of wave function values $\psi(x,y,z-\Delta z)$ (which is only a mild inconvenience with the current low cost of memory) and it requires that each slice thickness be identical (which can be a significant problem with some specimens). Also, the initial values ψ are required at two values of z. If there are $N = N_x N_y$ Fourier coefficients then the computer memory requirements scale as N and the computer time scales as $N\log_2 N$, similarly to the standard multislice method. This method (6.115) is similar to that tested by Kosloff and Kosloff [216] for solution of the time dependent Schrödinger wave equation in molecular dynamics calculations and may eventually have some value in electron microscope image simulation.

Although more accurate this method is not stable for an arbitrary slice thickness. At each step the current solution $\psi_n(x,y,z)$ will include some small error either from the finite precision arithmetic of the computer or from the errors arising from truncating the Taylor series at only the first few terms. It is important to consider how the recursive calculation in (6.115) treats this error. If a small error increases with each step then the method is unstable and if it decreases with each step then the method is stable. First consider a small error with a particular two-dimensional spatial frequency $\mathbf{k}_e = (k_{xe}, k_{ye})$ and amplitude ε_n in the wave function $\psi_n(x,y)$. This error will give rise to an associated error in the wave function in the next plane $\psi_{n+1}(x,y)$.

$$\begin{aligned} \psi_n(x,y) &\rightarrow \psi_n(x,y) + \varepsilon_n \exp(2\pi i \mathbf{k}_e \cdot \mathbf{x}) \\ \psi_{n+1}(x,y) &\rightarrow \psi_{n+1}(x,y) + \varepsilon_{n+1} \exp(2\pi i \mathbf{k}_e \cdot \mathbf{x}). \end{aligned} \tag{6.119}$$

Substituting these into (6.115) yields an expression relating the amplitude of the error at each step (neglecting a possible error in $\psi_{n-1}(x,y)$).

$$\varepsilon_{n+1} \exp(2\pi i \mathbf{k}_e \cdot \mathbf{x}) = \frac{1}{c_+}\left[2 - \Delta z^2\left(\nabla_{xy}^2 + 4\pi\sigma V/\lambda\right)\right]\varepsilon_n \exp(2\pi i \mathbf{k}_e \cdot \mathbf{x}) \tag{6.120}$$

The rate of growth g of the error must be less than or equal to one to produce a stable recursive solution.

$$g = \left|\frac{\varepsilon_{n+1}}{\varepsilon_n}\right| = \left|\frac{1}{c_+}\left[2 - \Delta z^2\left(-4\pi^2 k_e^2 + 4\pi\sigma V/\lambda\right)\right]\right| \leq 1 \tag{6.121}$$

This is the von Neuman stability analysis (for example Press et al. [288]). Further rearranging (6.121) yields:

$$k_e^2 < \frac{1}{4\pi^2\Delta z^2}\left[\sqrt{1 + 4\pi^2\Delta z^2/\lambda^2} - 2\right] + \frac{\sigma V}{\lambda\pi} \tag{6.122}$$

Note that the last term is always positive. This expression means the bandwidth of the calculation (the maximum of $|k_e|$) is coupled to the slice thickness. Either the bandwidth or the slice thickness should be reduced to produce a stable solution. Unfortunately, the slice thickness must typically be reduced to about the same size as the pixel (of order 0.1Å) so this method is not competitive with the standard multislice method (with a slice thickness of order two to three Å). This argument does however explain why the standard multislice method is stable and well behaved. Each multislice step is unitary meaning that ψ_n is multiplied by a factor whose magnitude is identically one (i.e., the exponentially of a purely imaginary number). This means that each step of the standard multislice is unconditionally stable (there is a distinction between stability and accuracy however).

In the long run any higher order solution may be defeated by the ill behaved nature of the specimen potential $V(x,y,z)$ (i.e., it has a sharp cusp or singularity near the atomic nucleus). The error term is proportional to Δz^3, and the other factors of the the error term involve various derivatives of the wave function which indirectly involve various derivatives of the potential. If there are singularities in the potential then the higher order derivatives in the Taylor series can grow abnormally large making convergence slow at best.

Chapter 7
Multislice Applications and Examples

Abstract The multislice method of calculating electron microscope images of thicker specimen is applied to several different specimen. First, some simple examples are worked through to help understand how to use the method and then some more complicated specimens are investigated to illustrate the method.

The multislice method simulates electron transmission in a thick specimen including dynamical scattering. Chapter 6 presented the theory of the multislice method and discussed how to use it in general terms. This chapter will gives some specific examples of performing a multislice simulation. The examples serve to illustrate some typical multislice results and also provide a more detailed description of using the multislice method.

A prerequisite for image simulation is a detailed description of the coordinates of the atoms in the specimen. The two volume set of Wyckoff [385] is a good starting point for many common materials with a crystalline structure. The books by Megaw [241] and Vainshtein et al. [350,351] give a thorough discussion of crystal structure. There are also numerous journal articles (see for example the journal Acta Cryst.) with crystal structure information.

7.1 Gallium Arsenide

Gallium Arsenide (GaAs) is a relatively simple structure similar to that of silicon (the diamond structure) except that adjacent atoms alternate between gallium and arsenic in the zinc sulfide (or zincblende) structure. The cubic lattice constant is slightly bigger (5.65Å) than that of silicon (5.43Å) so that it may be slightly easier to image in the electron microscope. However, the atoms in GaAs are significantly heavier (atomic number $Z = 31$ and 33 for Ga and As, respectively) and scatter

E.J. Kirkland, *Advanced Computing in Electron Microscopy*,
DOI 10.1007/978-1-4419-6533-2_7, © Springer Science+Business Media, LLC 2010

more strongly than silicon (Z = 14). GaAs may be expected to have more dynamical scattering than silicon, so a simple phase grating calculation (as in Chap. 5) may not be sufficient. This is a good place to start testing the multislice method because this specimen may produce significant dynamical scattering but is relatively simple to describe.

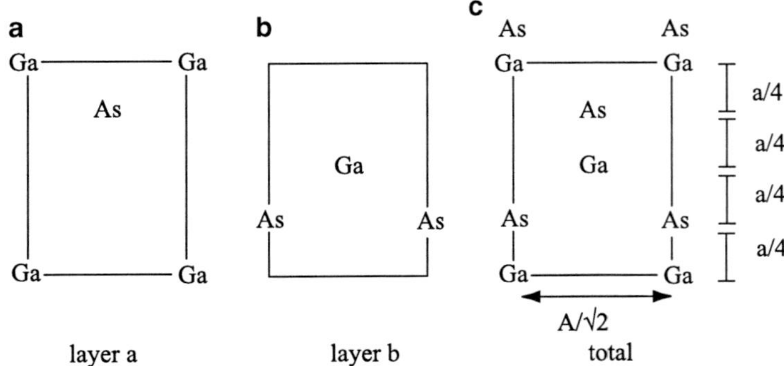

Fig. 7.1 The structure of gallium arsenide (GaAs) projected along the 110 direction. GaAs has a zincblende structure with a cubic lattice constant of a_{GA} = 5.65Å. (a) and (b) are two different atomic layers suitable for use in a multislice calculation and (c) is the projection of whole specimen. The layers are stacked in the sequence *ababa*.... Each layer is 1.998Å thick (along z or the optic axis) and 3.995 × 5.65Å in x and y (in the plane of the paper)

In the 110 projection GaAs has a projected structure as shown in Fig. 7.1. The main difference between silicon (Fig. 5.15) and GaAs is that there are two different types of atoms (Ga and As). In the 110 projection, GaAs naturally divides into two different layers as (labeled a and b in Fig. 7.1). The layers are stacked in the sequence *ababa*... with a spacing along the z axis (optic axis) of $\sqrt{2}a_{GA}/4 = 1.9976$Å. Silicon can be divided in the same manner if the Ga and As atoms are replaced with Si and the lattice constant is changed. It is best to use these two layers as the slices in the multislice method. The thickness is small enough to get an accurate answer (which is usually the case for the natural layers in a small unit cell crystal) and using the natural layers of the specimen will ensure that the multislice simulation correctly reproduces the HOLZ (or upper layer lines) portion of the diffraction patterns. Although the HOLZ lines may not contribute directly to a BF CTEM phase image of a thin specimen, getting the HOLZ lines correct is frequently required for an accurate simulation of thick specimens in BF (to get the phase of the low order reflections correct) and is usually required for ADF-STEM images because the ADF detector collects large scattering angles (where the HOLZ lines are). The actual atomic coordinates for each layer are shown in Table 7.1.

Table 7.1 The normalized coordinates for two layers in one projected unit cell of the (110) projection of GaAs

Atom	Layer	x/a_0	y/b_0
Ga	a	0	0
As	a	0.5	0.75
As	b	0	0.25
Ga	b	0.5	0.5

The two dimensional unit cell dimensions are $a_0 \times b_0$ where $a_0 = a_{GA}/\sqrt{2} = 3.995$Å and $b_0 = a_{GA} = 5.65$Å. a_{GA} is the cubic lattice constant of GaAs

7.1.1 BF-CTEM Simulation

Once the structure of the specimen has been determined and a convenient way of layering or slicing the specimen has been identified, the next step is to find the correct sampling for the simulation. The general guidelines for simulating thin specimens (Table 5.4) are a good starting point. Dynamical scattering in a thick specimen will also introduce further constraints on the sampling. The easiest way to discover these sampling requirements is to simply try a range of super cell sizes with different numbers of pixels. A sequence of trial multislice runs (using an incident plane wave) is shown in Table 7.2 for two different thickness of 110 GaAs. The super cell sizes were chosen to yield a reciprocal space sampling size of about one mrad. at 200 keV to get adequate sampling inside the objective aperture. The columns labeled "Intensity" refer to the total integrated intensity in the final electron wave function at the exit surface of the specimen. As discussed in Sect. 6.10.2 the integrated intensity should remain constant at unity if there is adequate sampling. Typically this value will decrease with thickness but should not go below about 0.90 for a reasonable simulation, and a value of 0.95 or higher is typical for a good simulation. Table 7.2 shows that this simulation probably needs about 512×512 pixels (or more) to simulate a thickness of 200 Å.

The magnitude $|\psi(x,y)|$ of the electron wave function $\psi(x,y)$ after passing through a thickness of 10(ab) and 50(ab) layers (40 and 200 Å) is shown in Fig. 7.2. An ideal perfect microscope with amplitude contrast would produce an image similar to that in Fig. 7.2. A phase grating calculation (as in Chap. 5) would yield identically $|\psi(x,y)| = 1$ across the whole area because it approximates the transmission process as a pure phase shift. Figure 7.2 illustrates that the specimen is not a pure phase object. The electron intensity accumulates near the atom sites (a white spot in Fig. 7.2). The imaging electrons see a large positive charge at the atom nucleus that is screened by the bound electrons of the atom. Far away from the atom the imaging electrons (at 200 keV in this example) do not see any net charge (for neutral atoms the positive charge on the nucleus equals the number of negatively charged bound electrons), but near the nucleus the imaging electrons see the large positive charge on the nucleus and are attracted to the center of the atoms in the specimen. The

Table 7.2 The effects of sampling on a multislice calculation of 110 GaAs

Unit cells	Size (in Å)	Number of pixels	Max angle (mrad)	Intensity at 10(ab)	Intensity at 50(ab)
$5a_0 \times 3b_0$	19.98×16.95	128×128	53.6	0.969	0.821
"	"	256×256	107.1	0.987	0.906
"	"	512×512	214.3	0.997	0.978
$6a_0 \times 4b_0$	23.97×22.2	128×128	44.6	0.964	0.802
"	"	256×256	89.3	0.980	0.871
"	"	512×512	178.6	0.995	0.967
$7a_0 \times 5b_0$	27.97×28.25	128×128	37.9	0.967	0.812
"	"	256×256	75.8	0.975	0.849
"	"	512×512	151.5	0.993	0.951
$8a_0 \times 6b_0$	31.96×33.9	128×128	31.6	0.969	0.807
"	"	256×256	63.1	0.972	0.834
"	"	512×512	126.3	0.991	0.938
"	"	1024×1024	252.5	0.998	0.982

The two-dimensional unit cell dimensions are $a_0 = a_{GA}/\sqrt{2} = 3.995$Å and $b_0 = a_{GA} = 5.65$Å. a_{GA} is the cubic lattice constant for GaAs. 10(ab) means that there are 10 repeats of both the a and b layers (about 40Å thick). 50(ab) results in a thickness of about 200Å

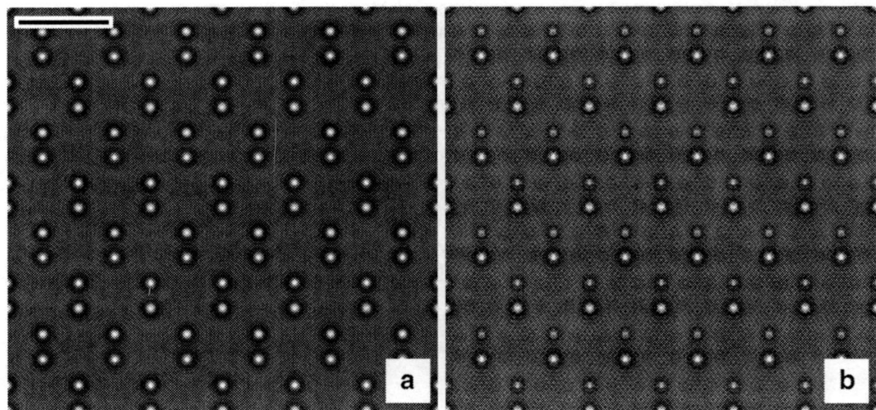

Fig. 7.2 The magnitude of the electron wave function $|\psi(x,y)|$ (in real space) after passing through (a) 40Å and (b) 200Å of 110 GaAs at an electron energy of 200 keV. The super cell size is $6a_0 \times 4b_0$ with 512×512 pixels. The scale bar in (a) is 5 Å. The numerical range of each image is (a) 0.18–3.70 and (b) 0.01–3.30 (*white is a larger positive number*)

imaging electrons effectively get channeled into the atomic columns in the specimen. The black ring surrounding each atom site is a depletion of electron intensity.

Channeling is not a static process either. As the imaging electrons progress further through the specimen they may be scattered out of the atomic column as well. Figure 7.2b shows this effect. The As atoms have a slightly larger positive charge on the nucleus. The white spot at the As sites is less bright than the white spot at the Ga sites indicating that the imaging electrons are diminishing in intensity at the As sites. This is a dynamical scattering effect. The relative intensities of the channeling peaks on the Ga and As sites will likely oscillate with thickness. There is a faint structure in the magnitude of the electron wave function that looks like standing waves between the atom sites. This occurs when the HOLZ lines appear and it is not always clear whether this is the real space manifestation of the HOLZ lines or simply a sampling error.

Figure 7.3 show the intensity and phase of the $k_x = k_y = 0$ Fourier coefficient (or zero order beam) of the wave function of the electron as it is passing through the specimen. Both the intensity and the phase oscillate with depth, clearly indicating the dynamical nature of the scattering process. All of the nonzero Fourier coefficients or beams will oscillate with depth, although with a different period. The period of oscillation is referred to as the extinction depth for the particular beam (or Fourier coefficient). The dynamical nature of the scattering process is completely lacking in a phase grating calculation (as in Chap. 5) and the multislice calculation (or equivalently a Bloch wave calculation) is required to correctly simulate strongly scattering specimen (more than about 10–20 Angstroms in the case of 110 GaAs).

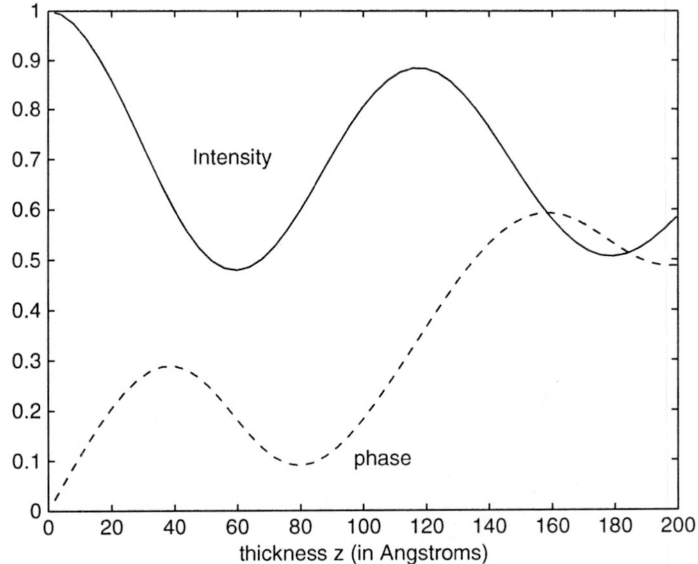

Fig. 7.3 The intensity and phase of the $k_x = k_y = 0$ Fourier coefficient of the 200 keV electron wave function as it passes through the 110 projection of GaAs. The super cell size is $6a_0 \times 4b_0$ with 512×512 pixels

The simulated bright field (BF) phase contrast image is shown in Fig. 7.4 for various thickness of the specimen. Scherzer conditions were used for the defocus and partial coherence was approximately included using the transmission cross coefficient (Sect. 5.4.3). Figure 7.4a is for two layers of the specimen and is approximately the same as a phase grating calculation. Each pair of atoms (Ga and As) in the dumbbell appears a a black ellipse. Figure 7.4b is for a slightly thicker crystal. The relative phases of the Fourier coefficients have already changed significantly. Some features that should be white are black and vice versa. The image has a lot of artifacts and the overall periodicity is double of what it should be. These artifacts can vary dramatically with small changes in defocus and the size of the objective aperture. Figure 7.4c is a thick crystal. Although the overall periodicity is again correct there has been a contrast reversal. This illustrates why image simulation is necessary to interpret a high-resolution phase contrast image.

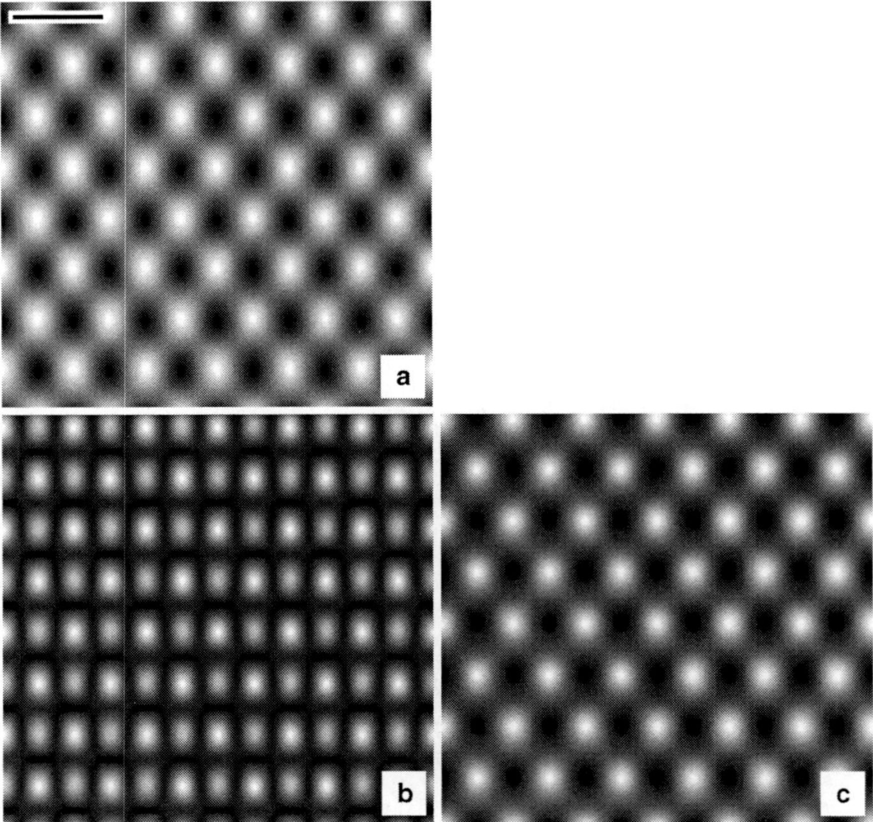

Fig. 7.4 Simulated bright field phase contrast images for different thickness of 110 GaAs at 200 keV. $C_s = 1.3$ mm, $\Delta f = 700$Å, obj. apert. 12 mrad, with partial coherence (1 mrad spread in illumination and 100 Å defocus spread). (a) 4 Å (2 layers), (b) 40 Å (20 layers), and (c) 200 Å (100 layers). The super cell size is $6a_0 \times 4b_0$ with 512×512 pixels. The range of each image is (a) 0.91–1.09, (b) 0.57–1.07, (c) 0.247–1.48 (*white is a larger positive number*)

7.1.2 ADF-STEM Simulation

The annular dark field (ADF) STEM image is a small signal (compared to phase contrast BF) and directly involves high angle scattering. Both of these effects are more difficult to calculate and an ADF-STEM image simulation requires more attention to the accuracy and tolerance of the calculation (i.e., sampling requirements). Finding the super cell size and number of pixels to achieve a total integrated intensity of nearly unity (as in Table 7.2) is the first requirement. It is also a good idea to test the accuracy more directly. This is conveniently done by comparing two different simulations with slightly different sampling. If the sampling is adequate then there should be no difference between using $N_x \times N_y$ pixels and $2N_x \times 2N_y$ pixels with a similar real space super cell size. There are two different sampling requirements that should be tested. The pixel size (and number of pixels) in both real space and reciprocal space is important. The number of pixels is restricted to powers of two (for the FFT) so the two simulations must differ by a factor of two in the number of pixels. If the real space super cell size were left the same then the sampling in real space (Δx and Δy) would double however the pixel size in Fourier or reciprocal space ($\Delta k_x = 1/a$ and $\Delta k_y = 1/b$) would remain the same. To vary both at the same time the super cell size should be increased by about a factor of $\sqrt{2}$ (within the constraints of the specimen periodicity) to get approximately the same reduction in the pixel size in both real space and reciprocal space.

Figure 7.5 shows a comparison between two different simulations with different sampling sizes on the 110 projection of GaAs. This is only a test of the sampling, so a full image simulation is not necessary. An ADF-STEM multislice simulation requires a full multislice calculation for each position of the scanned probe which can require a lot of computer time. A simple one dimensional line scan through an appropriate feature of the specimen is usually sufficient to test the sampling, but requires substantially less computer time. Figure 7.5 shows a scan through adjacent Ga and As atoms along the direction of there closest position (vertical in Fig. 7.1). The low resolution simulation had a super cell size of $6a_0 \times 4b_0$ with 512×512 pixels and the high-resolution simulation had a super cell size of $8a_0 \times 6b_0$ with 1024×1024 pixels. There is a good agreement between the two curves so this sampling is probably adequate for performing the simulation. The difference between the two curves can also serve as an estimate of the sampling error in the simulation. The As atom is slightly heavier ($Z = 33$) than the Ga atom ($Z = 31$) so the peak on the As position is slightly higher (stronger scattering at the As position). Also, there is a slight peak in between the main dumbbells (at about $y = 3.5$Å in Fig. 7.5). This is caused by the tails of the probe (compare to Fig. 3.11).

Figure 7.6 shows the magnitude of the electron wave function $|\psi(x,y)|$ of the focused probe as it is passing through the specimen. The probe was positioned at an offset of (8Å,10Å) with (0,0) being in the lower left corner of the image. Although the initial probe is smooth and round the atomic columns of the specimen again attract the imaging electron which get channelled into the atomic columns. After a thickness of 200Å (Fig. 7.6b) the electron distribution in the probe can get fairly distorted by the specimen.

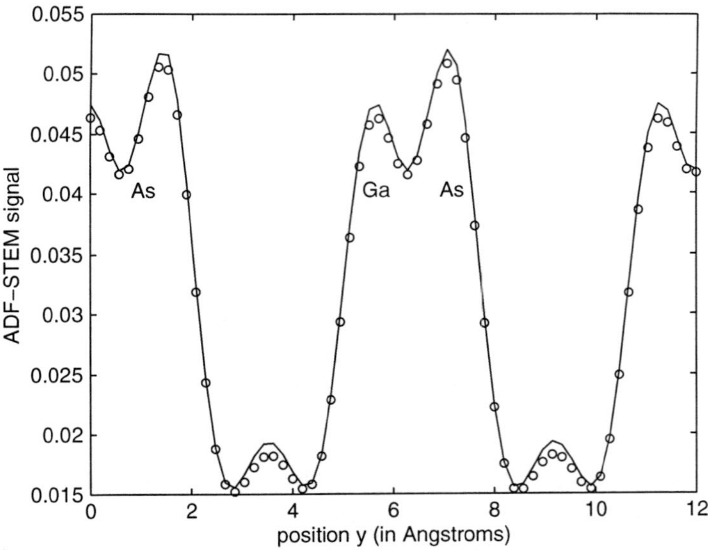

Fig. 7.5 Sampling test for ADF-STEM of 110 GaAs (40Å thick) at an electron energy of 200 keV. The curve shows the ADF signal in a *vertical line* through pairs of Ga and As atoms. The optical parameters were $C_s = 1.3$ mm, $\Delta f = 700$Å, with an objective aperture of 10.37 mrad. (Scherzer conditions). The ADF detector covered 40–175 mrad. The *solid curve* is for a super cell size of $8a_0 \times 6b_0$ with 1024×1024 pixels and the *circles* are for a super cell size of $6a_0 \times 4b_0$ with 512×512 pixels. The agreement between the two curves indicates that the sampling is sufficient

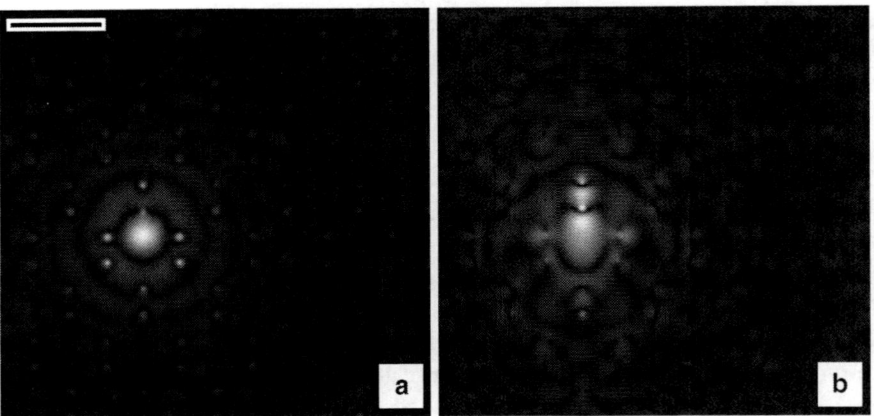

Fig. 7.6 The intensity distribution in the electron probe as it passes through 110 GaAs at an electron energy of 200 keV. (a) 40Å thick and (b) 200Å thick. The optical parameters were $C_s = 1.3$ mm, $\Delta f = 700$Å, with an objective aperture of 10.37 mrad. (Scherzer conditions). The super cell size had dimensions of $6a_0 \times 4b_0$ with 512×512 pixels. The range of each image is (a) 0.0–13.5, (b) 0.0–9.92 (*white is a larger positive number*). The scale bar in (a) is 5Å

The actual (simulated) ADF-STEM images calculated using the multislice method are shown in Fig. 7.7. These images require a multislice calculation for each point (or pixel) in the image. To save computer time only 32×32 pixels in one unit cell were calculated, and the unit cell was duplicated to fill the same area as the BF-CTEM simulation (Fig. 7.4). The specimen potential and electron wave function were each sampled with 512 × 512 pixels. The effective extinction distance for scattering to high angles (as on the ADF detector) is much larger than that for low angle scattering, which makes the ADF-STEM image much less sensitive to specimen thickness in the way that BF phase contrast is (compare to Fig. 7.4). The atomic columns are imaged as white in both the 40Å (Fig. 7.7a) and 200Å (Fig. 7.7b) images. The dumbbells are nearly resolved in these images. The magnitude of the ADF signal increases with thickness but there is not a contrast reversal as there was in BF phase contrast. The price paid for this improvement in image interpretation is a much smaller overall signal.

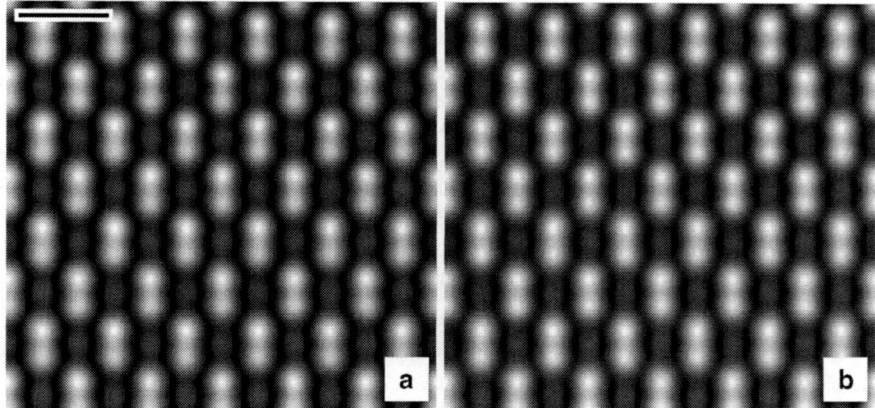

Fig. 7.7 Simulated ADF-STEM images of 110 GaAs at an electron energy of 200 keV for a thickness of (a) 40Å and (b) 200Å. The optical parameters were $C_s = 1.3$ mm, $\Delta f = 700$Å, with an objective aperture of 10.37 mrad. (Scherzer conditions). The specimen super cell size had dimensions of $6a_0 \times 4b_0$ with 512×512 pixels. The image was calculated as 32×32 pixels (in one unit cell) and periodically replicated to 96×64 pixels for display. *Black* is (a) 0.011, (b) 0.036 and *white* is (a) 0.051, (b) 0.144 where the total incident beam current is one. The scale bar in (a) is 5Å

7.1.3 Channeling

When the specimen is aligned along a major zone axis columns of atoms line up in a row along the optic axis. This is usually desirable because the atom columns may be imaged in their corresponding atomic location (in the 2D projection). As the electrons travel through the specimen they tend to channel along these atomic columns in a process called channeling as is already evident in Figs. 7.2 and 7.6. This name can also apply to ion scattering in which the ions travel through the hollow space between atomic columns. Electrons are negatively charged and are attracted to the positive charge on the nucleus, so they tend to get pulled into the atomic columns and channel down a narrow channel that can be very small. The channeling width can be smaller than the outer electron shell in the atom and have a significant effect on low loss Electron Energy Loss Spectroscopy (EELS) signals (Kirkland [206]).

One advantage of a simulation is the ability to display signals that are not normally accessible in a real experiment. Figure 7.8a shows the electron intensity along a row of atom pairs in 110 GaAs vs. depth in the specimen (vertical direction) in an y,z plane (along a vertical line in Fig. 7.7) for an incident plane wave. An incident uniform plane wave is incident at the bottom of the image and exists the specimen at the top of the image (the electron wave is traveling up in this particular image). The channeling peak intensity increases significantly about half way through this specimen and then spreads out again as it travels further through the specimen. Channeling will oscillate with specimen thickness. The channeling peak will form (or oscillate) quicker with heavier atoms (and atom density along the beam) and lower electron energy. This particular example was chosen to reasonably fit on the page (low energy so it oscillates in a short distance). Channeling occurs in both fixed beam (CTEM) and scanned probe (STEM).

Figure 7.8b shows the channeling for an incident aberration correct probe placed in between atom pairs. The probe slowly gets pulled onto the atom cloumns as it passes through the specimen. Figure 7.8c is the same as (b) but with a saturated greyscale to bring out the low intensity portion of the image.

Channeling is sometimes an annoyance, but can be used to advantage. If the specimen (substrate) thickness is chosen so that the channeling peak is sharp and intense at the exist surface any atoms deposited on the exit surface will be strongly imaged only if they are on atomic columns but not in between, which may help distinguish where these atoms are.

In the Bloch wave picture, the electron eigenfunctions (or eigenvectors) have a high density near the atomic columns. In a two-dimensional plane perpendicular to the beam direction the electrons appear as states loosely bound to the atomic columns much like the lower energy atomic electrons are bound to the nucleus (but in this case the high energy beam electrons are bound to the screened nucleus). These states can be identified using atomic quantum numbers 1s, 1p, etc. (Kambe et al. [191] and Buxton et al. [44]) with similar symmetry. These eigenfunctions indicate a preference for propagation along the atomic columns (channeling). Pennycook and Jesson [283] were able to develop an intuitive understanding of ADF-STEM imaging based on the s-state eigenstates for well separated atomic columns. Anstits [14, 15] have shown that more states are needed for high resolution if the atomic columns are close. An atomic sized probe (with aberration corrector) placed on one atomic columns with another atomic column close by, may oscillate between columns as it propagated through the specimen.

Fig. 7.8 Calculated electron intensity in 110 GaAs at an electron energy of 100 keV. (a) Plane wave incident. (b) Aberration corrected probe in between atom pairs ($C_{S3} = C_{S5} = \Delta f = 0$, objective aperture of 25 mrad). (c) same as (b) but with smaller *grey scale* range to bring out the low intensity portion of the image (*at top*). The specimen super cell size had dimensions of $6a_0 \times 4b_0$ with 512×512 pixels. The total thickness (*bottom to top*) was 100Å and the scale bar in (a) is 5Å

7.2 Silicon Nitride

Silicon nitride (specifically the β phase, β-Si_3N_4) has a hexagonal unit cell (Wyckoff [385]) as shown in Fig. 7.9. Each side of the unit cell is 7.606Å and there are two layer with a total repeat length of 2.909Å (perpendicular to the plane of the paper in Fig. 7.9). The hexagonal unit cell contains 14 atoms (6 silicon atoms and 8 nitrogen atoms).

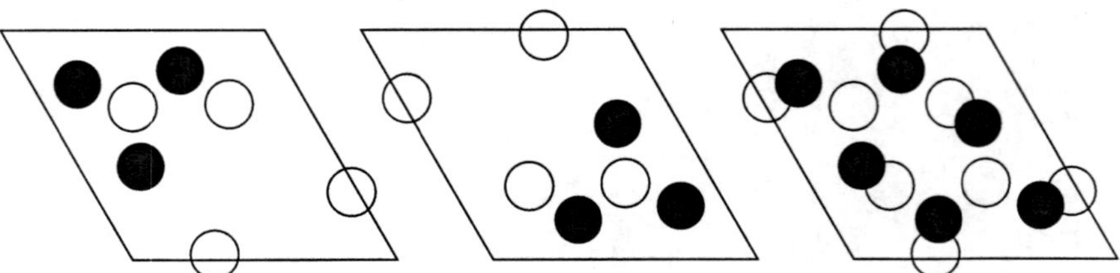

Fig. 7.9 The 001 projection of the hexagonal primitive unit cell of silicon nitride (β-Si_3N_4). The *open circles* are the positions of the nitrogen atoms and the *filled circles* are the positions of the silicon atoms. The a layer is on the *left*, the b layer is in the *middle* and the total projection is on the *right*. The *solid line* indicates the unit cell boundaries. The side of the unit cell is 7.606Å

Silicon nitride is an example of a more complicated unit cell that does not have rectangular symmetry. The multislice method of image simulation is most efficient when the FFT is used and the FFT is separable in x and y. This means that an FFT version of the multislice method works best with a rectangular unit cell. To simulate this specimen requires finding a larger unit cell with rectangular symmetry. One possible choice is shown in Fig. 7.10. The rectangular unit cell contains 28 atoms (12 silicon atoms and 16 nitrogen atoms). This particular specimen is easy to redefine a larger rectangular unit cell but an arbitrary specimen may be more difficult to describe this way. However, this step is required when using a multislice implementation using the FFT (required for an efficient calculation). Each specimen may require a different strategy to generate an equivalent rectangular unit cell.

Figure 7.11 shows the electron wave function after passing through about 50Å of the specimen. The magnitude in a) again shows that the electrons are attracted to the positively charged atomic nuclei and the specimen is not a pure phase object. The exit wave function, shown in b) and c) has both a strong real part and strong imaginary part. The complex transfer function of the objective lens can mix these two components in rather nonintuitive ways.

Figure 7.12 shows a simulated defocus series of silicon nitride with a thickness of 49.5Å (stacking sequence 17(ab)). The specimen was modeled as two layers with a stacking sequence of *ababa*.... The wave function and potentials were sampled with 256×256 pixels with a size of 38.03×39.52Å or $5a_0\times3b_0$ using the rectangular unit

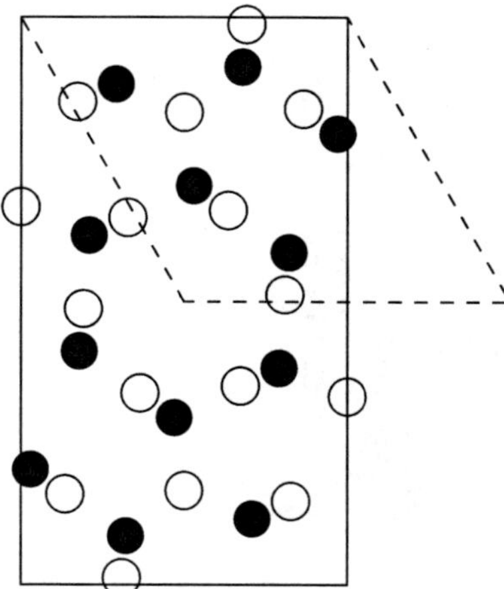

Fig. 7.10 The unit cell of silicon nitride (β-Si_3N_4) expanded to fill a rectangular area suitable for simulating using the multislice method with rectangular FFT's. The *open circles* are the positions of the nitrogen atoms and the *filled circles* are the positions of the silicon atoms. The position of the primitive hexagonal unit cell is shown as a *dashed line*. The unit cell has dimensions of $a_0 = 7.606$Å, and $b_0 = 13.174$Å, There are two layers each with a thickness of 1.4545Å

cell defined in Fig. 7.10 (maximum scattering angle of 54 mrad). Partial coherence was modeled using the transmission cross coefficient (see Sect. 5.4.3). The positions of the silicon atoms are initially black in Fig. 7.12a but reverse contrast as the defocus is changed, becoming white in Fig. 7.12b,c. Figure 7.13 shows the Scherzer focus image for two different thickness of the specimen. The apparent contrast of the silicon atom positions has reversed (black has become white). The sign of the contrast will changed periodically with defocus for a given thickness and also periodically with thickness for a given defocus, making image interpretation very difficult. Image simulation is one means of sorting out what is going on in the image.

An aberration corrected instrument will be corrected to some maximum angle or to some maximum order of aberration. The corrector will add a large negative C_{S3} to balance a large positive C_{S3} from the round objective lens. If the corrector is good to third order there will still be fifth order aberrations (C_{S50} plus all of the nonrotational aberrations), which may be minimized to some extent. Just as Scherzer used a lower order aberration (defocus) to partially compensate for the higher order C_{S3} aberration, the fifth order aberrations can be partially offset by the third order aberration. For simplicity ignore all of the rotational aberration like C_{32} etc. (which may be large and very important in practice). A third order corrector should be able to drive the total third order spherical aberration ($C_{S3} = C_{30}$) negative to partially offset the fifth order spherical aberration ($C_{S5} = C_{50}$). Similar pairs can be found in most of the rotational aberrations (left out from this discussion for simplicity). For example,

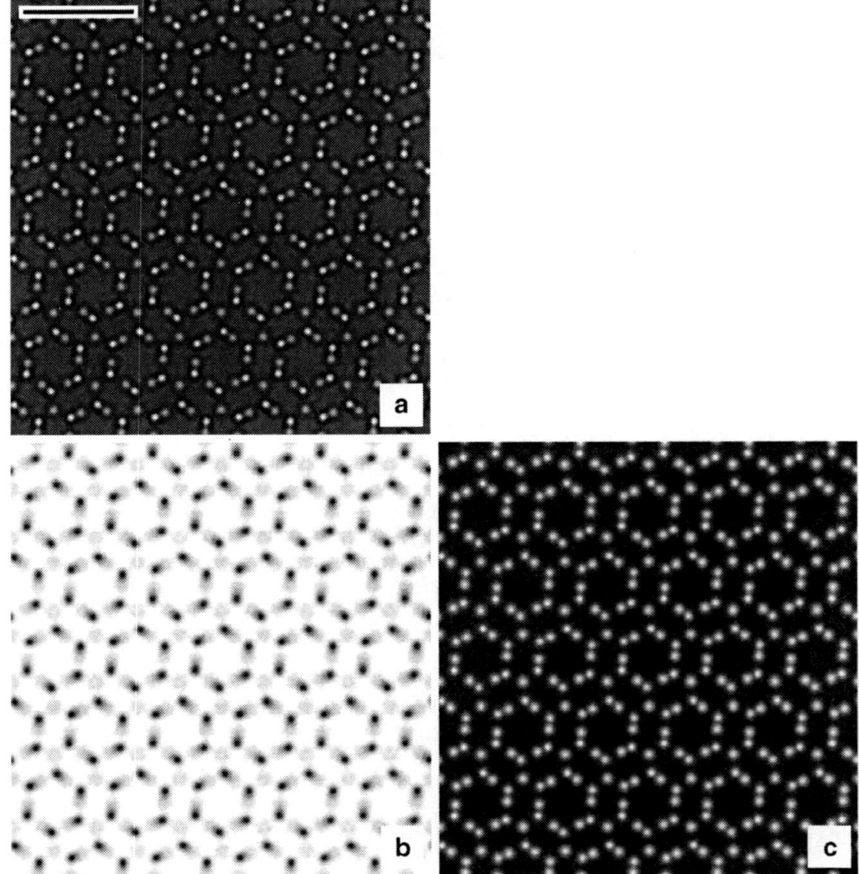

Fig. 7.11 The 200 keV electron wave function after passing through 49.5Å of 001 silicon nitride (β-Si_3N_4). (a) Magnitude $|\psi(x,y)|$ (b) Real part of $\psi(x,y)$ and (c) imaginary part of $\psi(x,y)$. The area of each image corresponds to 5×3 unit cells of the type shown in Fig. 7.10. The numerical range of each image is (a) 0.43–2.84, (b) −1.93–1.04 and (c) 0.01–2.27 (*white is a larger positive number*) The scale bar in (a) is 10Å

if $C_{S5} = 50$ mm is present at 200 kV then making a total $C_{S3} = -0.02$ mm allows the transfer function to go out to about 1Å (30 mrad) and look similar to that of Scherzer focus (Fig. 3.6).

Figure 7.14 shows Si_3N_4 calculated for an aberration corrected BF-CTEM with three different thickness using 512 by 512 pixels (maximum angle 108 mrad) and a slice thickness of 1.4545Å. The corrector is assumed to be good to third order out to an angle of 30 mrad. Figure 7.14a) is very thin (may not be possible in practice) and yields a good representation of the actual specimen structure. As the specimen gets thicker the atoms switch from black (expected) to white, which may be caused by an increase in the total phase as the electron wave passes through the specimen. At this high resolution the depth of focus is only about 30–40 Å (approximately

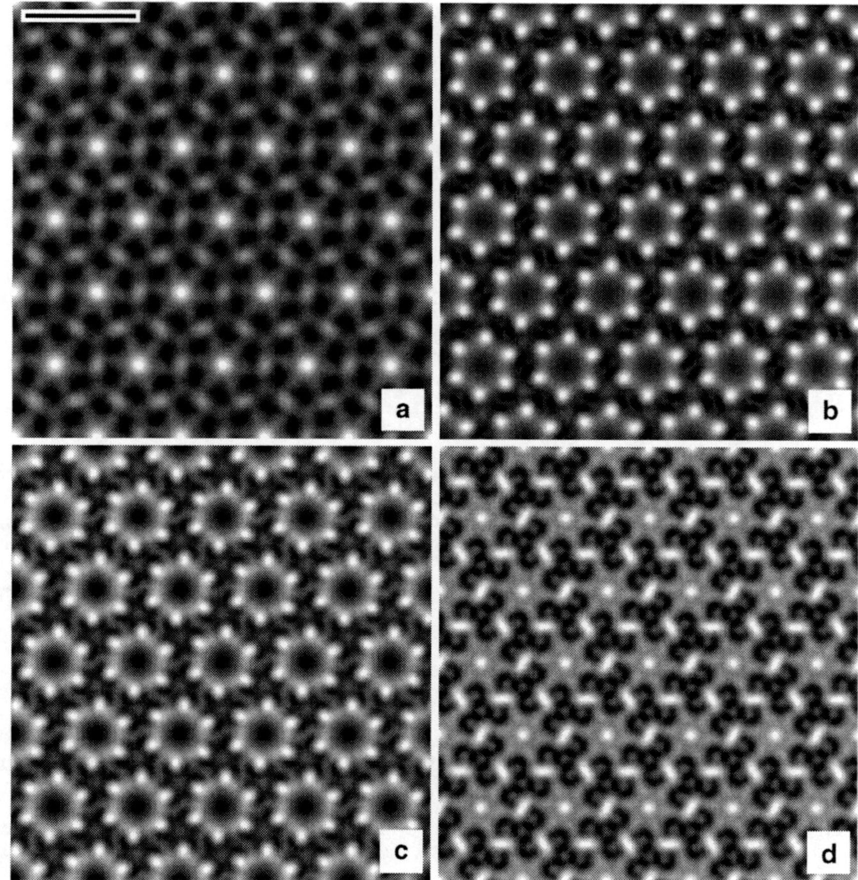

Fig. 7.12 Simulated CTEM defocus series of 001 silicon nitride (β-Si_3N_4), 49.5Å thick at a beam energy of 200 keV (C_s = 1.3 mm., obj. apert.=12 mrad, condenc. apert = 0.75 mrad, defocus spread = 100Å). The defocus values are (a) 700Å (Scherzer focus), (b) 900Å, (c) 1100Å and (d) 1300Å. The scale bar in (a) is 10Å . The area of each image corresponds to 5×3 unit cells of the type shown in Fig. 7.10. Silicon atom positions appear *black* in (a)

the resolution divided by the angle), which is less than the specimen thickness. In this calculation defocus is referred to the exit surface. Its not clear which plane is the optimum in this situation. Depth of focus becomes a significant problem as resolution increases. Jia et al. [188], Tillmann et al. [345] and Urban [348] have discussed other possible uses for negative C_{S3}.

Figure 7.15 shows a simulated ADF-STEM defocus series of silicon nitride under similar conditions to that in Fig. 7.12 except that the objective aperture was fixed at 10.37 mrad consistent with Scherzer conditions for a focused probe. The super cell size was $4a_0 \times 2b_0$ (or 30.4 × 26.3Å) with 512 × 512 pixels to allow a maximum scattering angle of 140 mrad (increased from the CTEM case in Fig. 7.12). The electron wave function was also sampled with 512 × 512 pixels. To save computer time

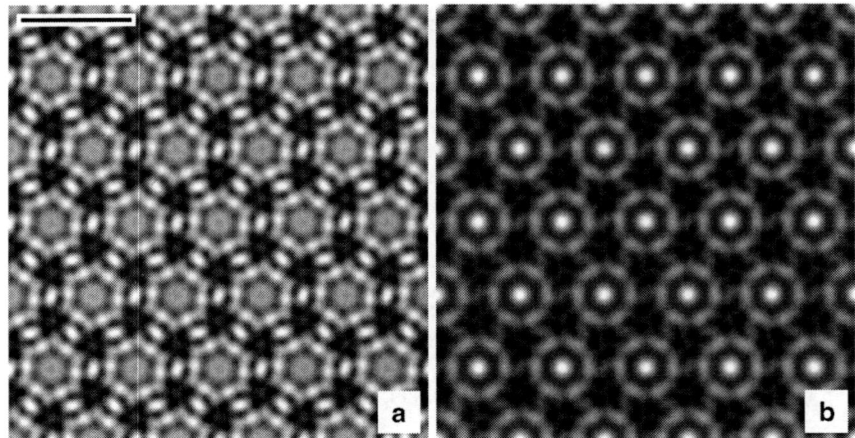

Fig. 7.13 Simulated CTEM images for two different thickness of 001 silicon nitride (β-Si_3N_4), at a beam energy of 200 keV (defocus of 700Å, $C_s = 1.3$ mm., obj. apert. = 12 mrad, condenc. apert = 0.75 mrad, defocus spread = 100Å). The specimen thickness was (a) 102Å (35(ab) layers), and (b) 151Å (52(ab) layers) The scale bar in (a) is 10Å . The area of each image corresponds to 5×3 unit cells of the type shown in Fig. 7.10

only the image in one rectangular unit cell was calculated with 32×64 pixels and replicated to fill the same area Fig. 7.12 for comparison. The silicon atom columns appear as white dots in Fig. 7.15a and do not reverse contrast unlike the BF phase contrast image, however there are still significant artifacts in the image that appear as the defocus is increased from Scherzer defocus ($\Delta f = 700$Å). The artifact are consistent with the tails of the probe. ADF-STEM is less sensitive to changes in defocus (than BF phase contrast) but there are still significant artifacts produced when defocus is changed. Figures 7.12 and 7.15 clearly indicate that the interpretation of a high-resolution image is not always straightforward in either BF-CTEM or ADF-STEM. Image simulation is one approach to verify that the image is interpreted correctly.

7.3 CBED Simulations

The CBED or Convergent Beam Electron Diffraction pattern is formed when a small focused probe is incident on the specimen. The diffraction pattern is from a microscopic area of the specimen and can also be referred to as the microdiffraction pattern. When the illuminating radiation is convergent (as required to focus on a small area of the specimen) then each diffraction spot is enlarged to the same size as the illumination cone and becomes a disk instead of a spot. The angular diameter of each disk is the same as the angular spread of the incident beam. In the case of the STEM each diffraction disk is the same size as the objective aperture.

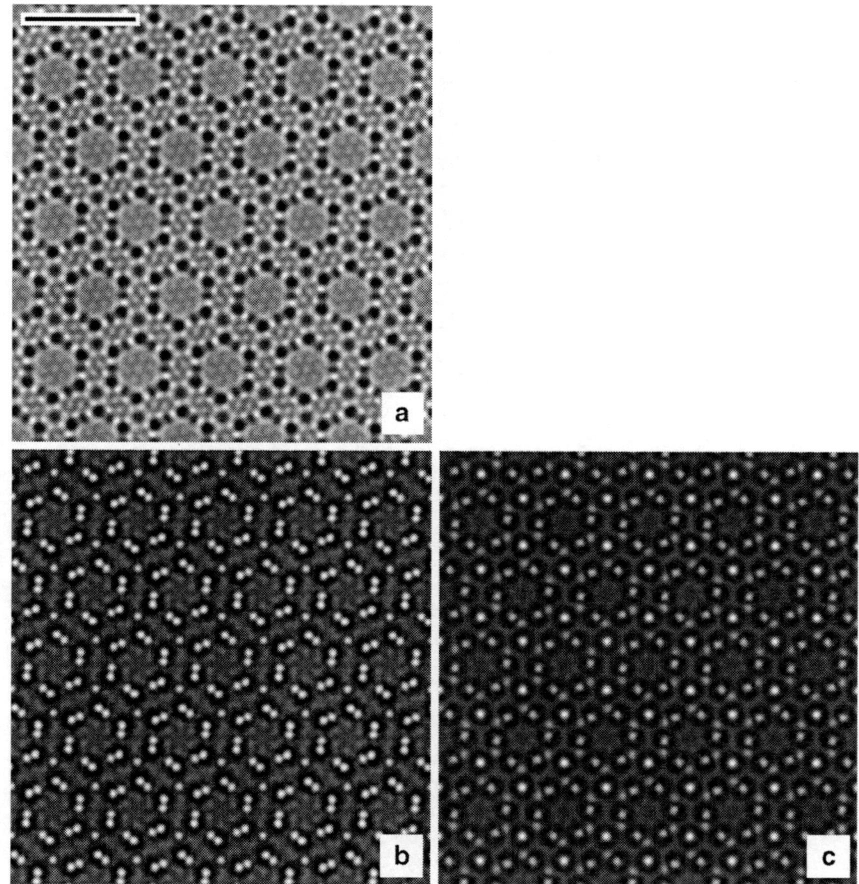

Fig. 7.14 Calculated aberration corrected BF-CTEM images for three different thickness of 001 silicon nitride (β-Si_3N_4), at a beam energy of 200 keV (defocus of 20Å, $C_{S5} = 50$ mm, $C_{S3} = -0.02$ mm, obj. apert. = 30 mrad, defocus spread = 10Å). The specimen thickness was (a) 22Å, (b) 100Å, and (c) 151Å. The scale bar in (a) is 10Å . The area of each image corresponds to 5×3 unit cells of the type shown in Fig. 7.10

The sampling error test (as in Table 7.2 and Fig. 7.5) are only an internal consistency test and do not test for systematic errors in the simulation theory or program implementation. The only real test of the simulation is a comparison to actual experimental data. Once the simulation software and theory have been verified by a detailed comparison to one or more experiments, then it can be used to predict other unknown situations.

Figure 7.16 shows experimentally recorded CBED patterns from 111 silicon at an electron energy of 100 keV (from Kirkland et al. [208]). In this case the probe (about 2.7Å diameter) is larger than the lattice spacing (1.92Å) so the CBED disks do not overlap and the lattice is not observable (Spence [332]). The outer

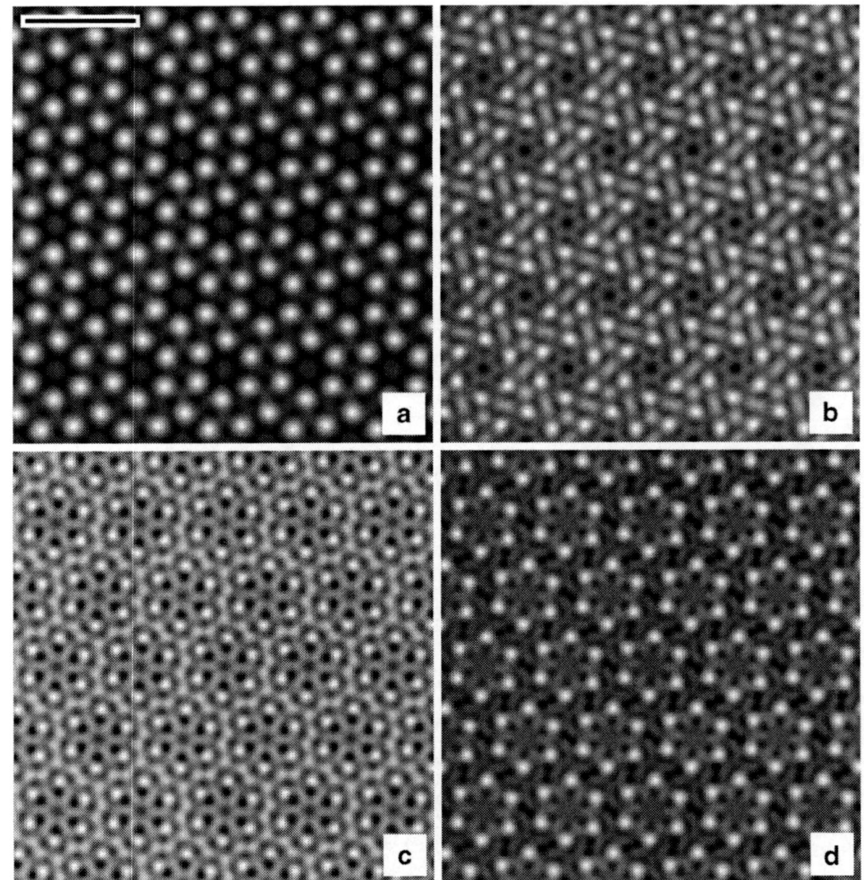

Fig. 7.15 Simulated ADF-STEM defocus series of 001 silicon nitride (β-Si_3N_4), 49.5Å thick at a beam energy of 200 keV (C_s = 1.3 mm, obj. apert. = 10.37 mrad). The defocus values are (a) 700Å (Scherzer focus), (b) 900Å, (c) 1100Å, and (d) 1300Å. The scale bar in (a) is 10Å. The area of each image corresponds to 5×3 unit cells of the type shown in Fig. 7.10. Silicon atom positions appear white in (a)

(white) ring is the HOLZ line. It is thin because the Ewald sphere intersects each diffraction disk at a steep angle. The patterns were recorded by photographing the diffraction screen (phosphor screen) with a 35 mm camera (using Kodak Plus-X film, chosen for its long linear region) with an exposure of 2 min. The dynamic range of the CBED pattern is too large to display easily so some old fashioned image processing was applied to the central seven disks (they were photographically *burned in* by a factor of six when the final prints were made in the dark room).

A CBED simulation requires only one multislice calculation like in a BF-CTEM calculation. The initial wave function is a focused probe (instead of a plane wave)

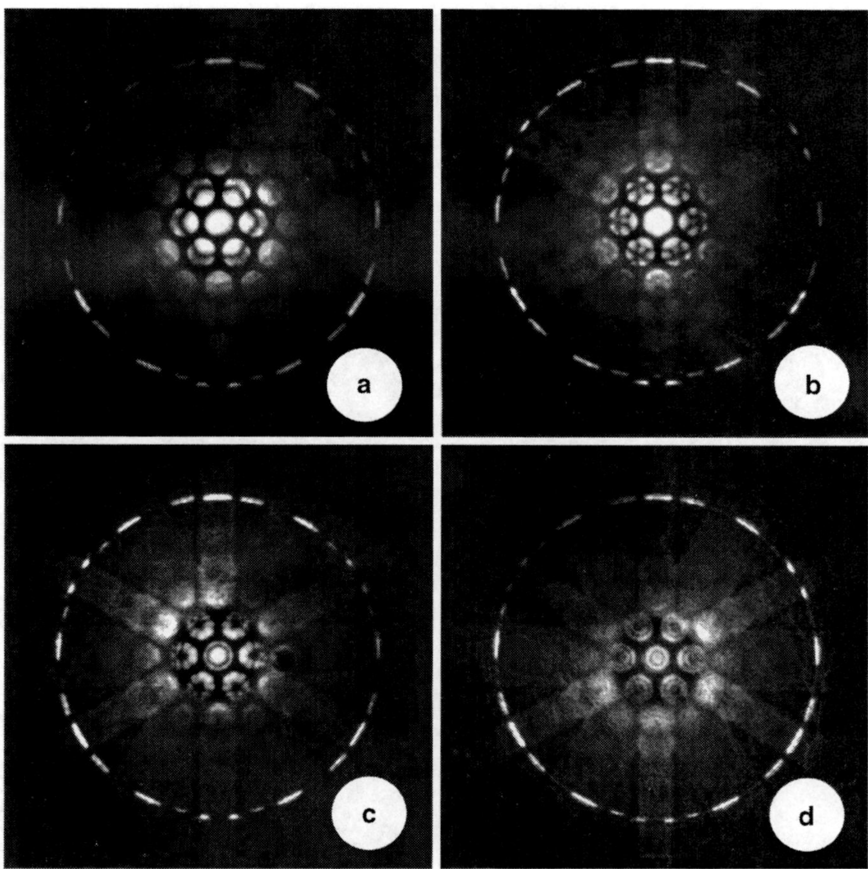

Fig. 7.16 Experimental CBED patterns of 111 silicon recorded in an HB501A STEM at 100 keV, $C_s = 3.3$ mm, with an objective aperture of 8 mrad. (Scherzer conditions). The specimen thickness was determined to be (a) 198 Å, (b) 273 Å, (c) 489 Å, (d) 1270 Å(±30 Å). Reprinted from Kirkland et al. [208] with the permission of *The Minerals, Metals and Materials Society*

and the final image is Fourier transformed into the far field to get a diffraction pattern. Silicon 111 was modeled as a layered structure with a stacking sequence of *abcabca*.... The slice thickness (one layer per slice) is 3.135 Å, with a total repeat length of 9.405 Å. The wave function and specimen potential were sampled with 512 × 512 pixels and a super cell size of 53.5Å × 53.2Å. A library of CBED patterns was calculated for various thickness in the range 100–1500 Å at intervals of 10–50Å. Each experimental CBED pattern was visually matched to a simulated CBED pattern to determine the specimen thickness. It was possible to distinguish the thickness to an accuracy of about ±30Å by observing the structure in the low order diffraction disks. The best match is shown in Fig. 7.17. Only the center two thirds of the diffraction pattern (342 × 342 pixels) is shown to avoid displaying the portion that must be set to zero to eliminate aliasing. Note that the sampling in Fig. 7.17d) is not sufficient (the total integrated intensity dropped below 0.9) but it still reproduces the general features of the experiment.

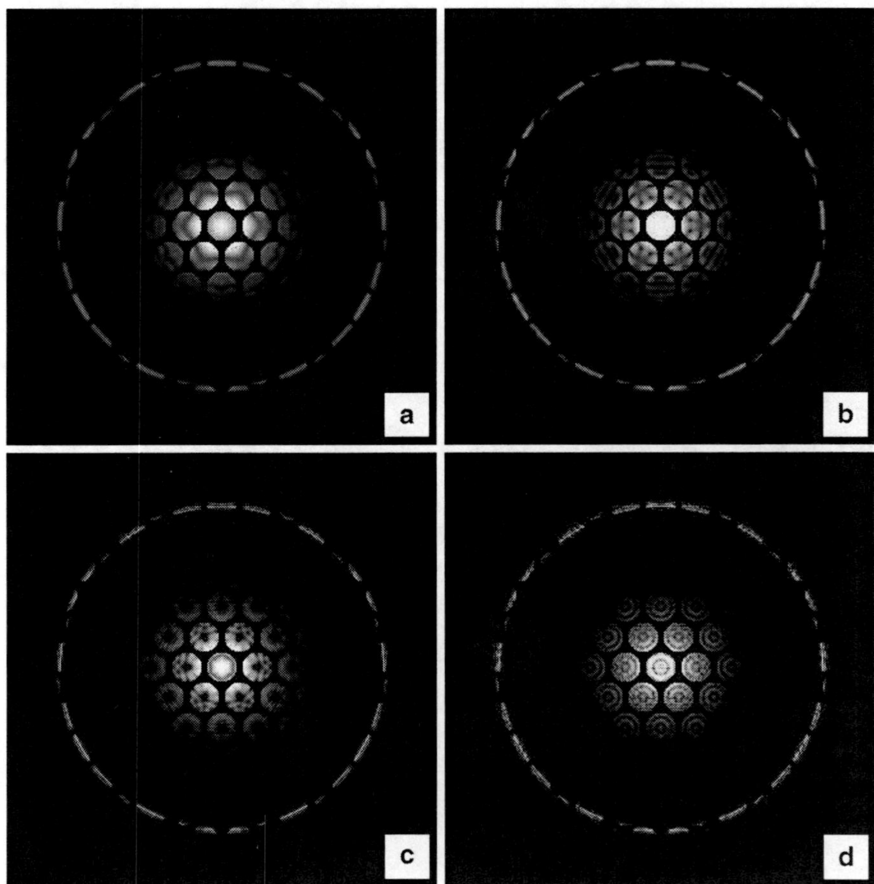

Fig. 7.17 Simulated CBED patterns of 111 silicon at 100 keV, $C_s = 3.3$ mm, with an objective aperture of 8 mrad. (Scherzer conditions). The specimen thickness was determined to be (a) 198 Å, (b) 273 Å, (c) 489 Å, (d) 1270 Å(±30 Å). Each pattern is displayed on a logarithmic scale. The maximum scattering angle is 118 mrad. Reprinted from Kirkland et al. [208] with the permission of *The Minerals, Metals and Materials Society*

An EELS (Electron Energy Loss Spectroscopy) spectra was recorded for the same region that each CBED pattern was recorded. The ratio of the integrated signal in the first plasmon peak to the intensity in the zero loss peak is a good measure of the thickness of the specimen. Plotting the EELS ratio vs. the thickness determined by matching the CBED simulation yielded a straight line with a slope indicating a plasmon mean free path of 1297 ± 25Å (with an EELS spectrometer collection angle of 8 mrad). The subjective agreement between Figs. 7.16 and 7.17 (overall features and the pattern of dark lines in the low order disks, taking into account the different intensity display scales used) and the measurement of the plasmon mean free path indicate a good agreement between the multislice simulation theory and an actual experiment. This particular experiment ignored the thermal motion of the atoms in the specimen. Later experiments by Loane et al. [228] yielded a plasmon mean free path of 1207Å including the effects of thermal atomic vibrations. The faint radial bands are called the Kikuchi bands and are also absent when thermal vibrations are ignored (as in Fig. 7.17).

7.4 Thermal Vibrations of the Atoms in the Specimen

All of the simulations considered so far have treated the atoms in the specimen as completely stationary. Most electron microscopy is done at room temperature of about 300°K (some microscopes can be equipped with heating and cooling stages). At room temperature the atoms in the specimen vibrate slightly. These atomic vibrations are quantized and the quantum unit of energy is called a phonon similar to the quantum unit of electromagnetic energy, the photon. Atomic vibrations are small compared to a typical interatomic distance so this effect is expected to be small but can lead to some interesting effects. In particular the thermal vibrations lead to a diffuse background intensity (in the diffraction pattern) in between the normal allowed diffraction positions. This type of scattering may be referred to as thermal diffuse scattering or simply TDS.

Typical optical phonons have a frequency no greater than about 10^{12} to 10^{13} Hz (Kittel [211]), and acoustic phonons are significantly lower frequencies. The imaging electrons in the microscope are traveling at about one half the speed of light (1.5×10^{10}cm/sec). At this speed it takes only about 0.7×10^{-16} s. to traverse the specimen which is much smaller than the period of oscillation of the atoms in the specimen. While the imaging electron is inside the specimen the atoms do not change their position significantly. The imaging electrons see the atoms as stationary but slightly offset from their normal lattice positions. The final image or diffraction pattern is made up of the average of many different imaging electrons. The typical current in the microscope is small enough that the time between two successive imaging electrons passing through the specimen is long compared to the period of oscillation of the thermal phonons in the specimen. Therefore, each successive imaging electron sees a slightly different configurations of atoms in the specimen. Each configuration of atoms is uncorrelated with all other configurations so the average over atomic configurations should be done incoherently.

The final image is the time average of many oscillations of the phonons in the specimen. However, only the imaging electron intensity and not the wave function can be averaged. This means that the time average must be performed in the image plane (or diffraction plane if simulating a diffraction pattern) and not in the specimen plane. The detection process (film, CCD, etc.) converts the electron wave function into an intensity (square magnitude of the wave function). It is not appropriate to replace the atomic potential of the specimen by a time averaged potential (or equivalently apply a Debye-Waller factor to the atomic potential) because the phase of the imaging electron wave function must be carried through to the detector plane. It is only in the detector plane that the wave function can be converted to an intensity.

A general theory of imaging and diffraction in the presence of thermal vibrations can be rather involved (for example: Hall and Hirsch [142], Cowley and Pogany [65], Cowley and Murray [64], Rez et al. [301], Cowley [61], Allen and Rossouw [9], Wang and Cowley [369, 370], Wang [365–368], Dinges and Rose [79], Amali and Rez [12], Mitsuishi et al. [245], Dwyer and Etheridge [85], Croitoru et al. [70]). The theory of TDS scattering of X-rays has been thoroughly reviewed by Warren [371] and is very similar to TDS scattering of electrons. Although a theoretical analysis may be complicated there is however a simple if somewhat brute force approach to numerically simulate the effects of thermal vibrations in the specimen. Given a list of atomic coordinates in the specimen, offset the position of each atom by a small random amount and then perform a normal multislice simulation to get an image or diffraction pattern. Next repeat this process with a different random offset for each atomic coordinate (each random offset should start from the original unperturbed atom position, so the random offsets are not cumulative). The final image or diffraction pattern is the intensity averaged over several different configurations of atoms with different random offsets (average $|\psi|^2$ and not ψ). This approach is called the frozen phonon approximation (Loane et al. [227], Hillyard and Silcox [158]). The random offsets can be generated using a random number generator with a Gaussian distribution which is then equivalent to the Einstein model of the density of states for phonons (see for example Kittel [211]). This method can also be labeled a Monte-Carlo method because it uses a sequence of computer generated random numbers to perform a simulation. Fan [96], Dinges and Rose [79], and Amali and Rez [12] have proposed slightly different methods.

When all of the atoms in the specimen have a slightly different offset the specimen is no longer periodic in any direction. The specimen is technically amorphous although there is still an approximate periodicity. This simulation requires a different type of multislice simulation. Each slice of the specimen is different so there is no advantage to precalculating the slices, storing them and reusing the transmission function of each slice (they cannot be reused). This type of multislice simulation first reads in all of the coordinates (in practice the program reads in only the coordinates for one unit cell and then replicates them for many unit cells) and then adds a random offset for each coordinate. Then the atomic coordinates are sorted by depth and cut into thin slices. The transmission function for each successive slice is calculated (one at a time), applied to the transmitted wave function and then discarded because it cannot be used again.

7.4.1 Silicon 111 CBED with TDS

Thermal vibrations have the most visible effects on the diffraction pattern, generating a diffuse background intensity in between the normal diffraction spots. The steps in a frozen phonon calculation are shown in Fig. 7.18. Each atom was allowed to deviate from its normal lattice position with a Gaussian distribution and a standard deviation of 0.075Å in each of three directions, consistent with a measured Debye-Waller factor at room temperature (Sears and Selly [314]). This simulation models a CBED pattern ($|\Psi(k)|^2$) of the 111 projection of silicon. The wave function and potentials were sampled with 512×512 pixels in an area of 34.6×33.3Å (maximum scattering angle 183 mrad.). Figure 7.18a shows the CBED pattern without thermal displacements. Figure 7.18b has one particular set of random offsets. The number of different configurations in the average increases to 16 in figure 7.18d. Surprisingly, this is enough configurations to produce a smooth pattern. This calculation is completely elastic and there is no inelastic scattering.

This simulation in Fig. 7.18d should match the experimental CBED pattern in Fig. 7.16c. However, the scale was changed slightly to improve the sampling. The faint white bands that travel radially outward from the center of the CBED pattern are called the Kikuchi bands and are noticeably absent in the CBED simulation in Fig. 7.17 without thermal vibrations, but are reproduced appropriately in Fig. 7.18. Figure 7.19 shows the average intensity vs. scattering angle including thermal atomic vibrations. Notice that the intensity in between the diffraction peaks has become nonzero. The main effect of thermal vibrations is to reduce the intensity in the HOLZ lines and redistribute it more uniformly over the whole range of scattering angles. The ADF-STEM signal integrates over this whole region so there is only a small qualitative effect in the ADF-STEM image. The TDS should be included in ADF-STEM for a good quantitative calculation but usually does not have a qualitative effect on ADF-STEM image (Hillyard and Silcox [158]). Möbus et al. [249] have also added TDS to simulations of BF-CTEM images and conclude that there is no significant effect for thin specimen as are typically used. Recently, Muller et al. [257] have performed a more detailed simulation using a set of phonons accurately generated from the measured band structure of the crystal and found that there was no significant deviation from the simple Einstein model with a Gaussian distribution of atomic coordinate offsets. Debye-Waller factors for many different materials can be found in the literature (for example, Reid [293], Krishna and Sirdeshmukh [218] and Schowalter et al. [313])

7.4.2 Silicon 110 ADF-STEM with TDS

Generally speaking, the TDS scattering in crystalline specimens reduces the peak intensity in the Bragg peaks on the ADF detector. A significant portion reappears

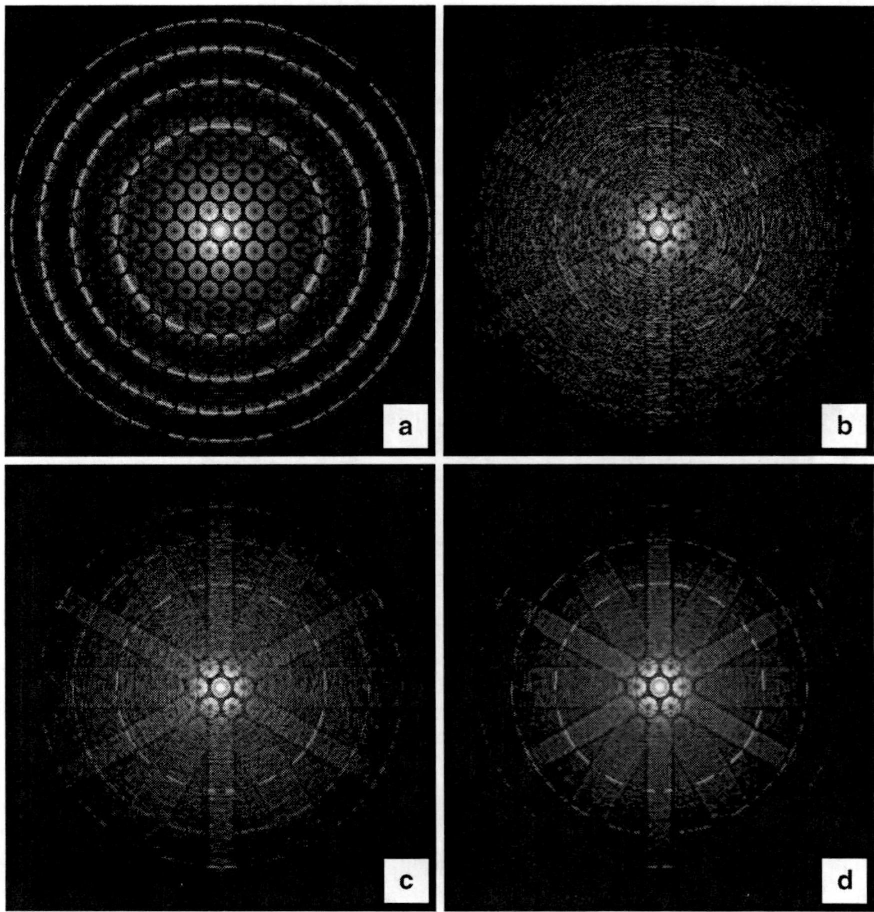

Fig. 7.18 Monte Carlo simulation of the CBED pattern of 111 silicon (489 Å thick) including thermal atomic vibrations (0.075Å in each direction) at 100 keV ($C_s = 3.3$ mm, obj. aperture 8 mrad. Scherzer conditions). (a) Has no thermal vibrations and the other images are the average of successively more sets of random displacements (b) one set, (c) four sets, and (d) 16 sets. Each pattern is displayed on a logarithmic scale. The maximum scattering angle is 183 mrad

as a diffuse background in between the Bragg peaks, which is still integrated by the ADF detector. Unless there are one or more strong Bragg peaks near the edge of the detector then TDS scattering should not have a large effect on the ADF-STEM image, although it should be included for a good quantitative comparison. Figure 7.20 shows a line scan through the so-called dumbbels (atom col. pair) in the 110 projection of Si (1.36 Å spacing) calculated with and without thermal vibrations (TDS scattering) for an aberration corrected 100 keV STEM. This calculation used 512 by 512 pixels with a super cell size of approximately 20 Å in each direction and a specimen thickness of about 100 Å. This super cell size includes scattering

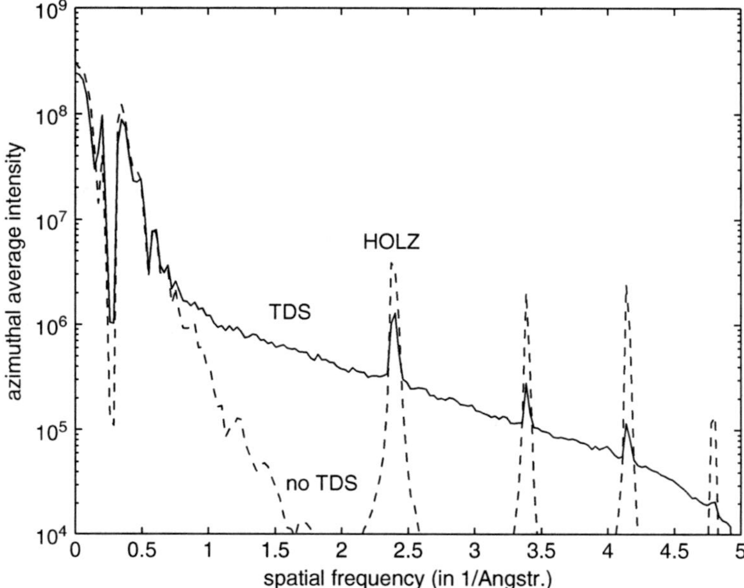

Fig. 7.19 Azimuthal average of the scattered intensity in the simulated CBED diffraction pattern including thermal atomic vibrations in 111 silicon as in Fig. 7.18d. The *solid line* includes thermal vibrations and the *dashed line* does not (Fig. 7.18a)

angles (on the ADF detector) to 290 mrad. The TDS calculation was averaged over 32 phonon configurations. Both calculations include a source size of 0.5 Å (full width half max.), calculated by convolving a 2D image (64 by 256 pixels) with a Gaussian of the appropriate width.

In the top graph (of Fig. 7.20) the atom pairs are nicely resolved and the TDS scattering does not have much effect (overall curve moved up slightly). There is a slight asymmetry due to the stacking order in the two columns, which will likely disappear in practice with surface relaxation of the crytal. The systematic error in this calculation may be about the size of the difference between these two curves so there may not be a real difference in the two curves. The bottom graph includes only high angles on the detector and the TDS scattering has the effect of reducing the overall signal which is a little unexpected. It is likely that there is a strong set of Bragg peaks near the inner angle of the detector (the FOLZ line) whose intensity gets scattered off the detector and disappears. It is not always a good idea to restrict the detector to very high angles. Usually, it's best to get the FOLZ nicely centered on the detector.

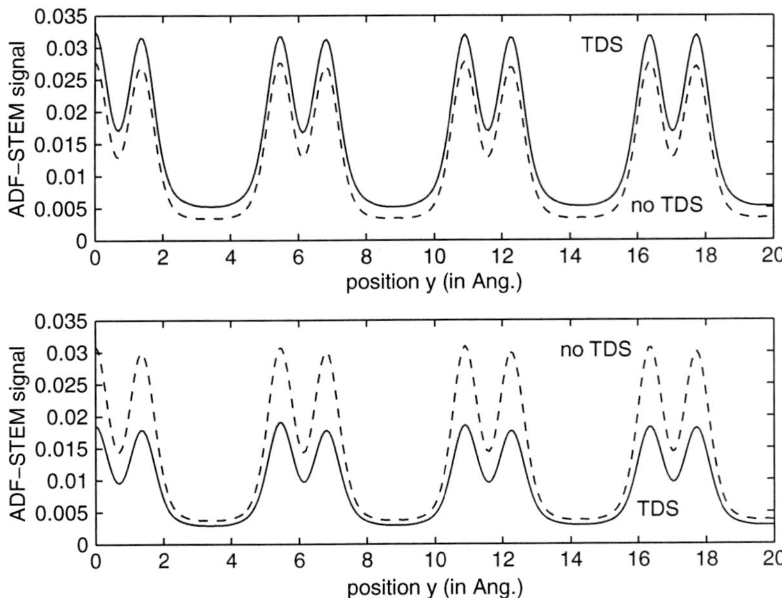

Fig. 7.20 Calculated aberration corrected ADF-STEM images for the 110 projection of silicon with and without TDS scattering, at a beam energy of 100 keV (defocus of 0Å, $C_{S5} = 0$, $C_{S3} = 0$, obj. apert. = 35 mrad). The specimen thickness was 100Å. The area of each image corresponds to 5×4 unit cells of the type shown in Fig. 5.15. The *top graph* has a detector of 80–200 mrad. and the bottom graph has a detector of 120–280 mrad

7.5 Specimen Edges or Interfaces

Calculating images of edges and interfaces presents some special problems. The specimen is no longer really periodic. The so-called wrap-around error causes an extra edge or interface to be introduced (see Fig. 6.13). Figure 7.21 shows the image of a sharp edge of copper. There is a pure crystal of copper in the left half and vacuum in the right half with an atomically sharp edge in the middle. Copper is a simple FCC (face centered cubic) structure with a unit cell size of 3.61Å.The edge could also be an interface between two different materials, with the same problem. There is one intended edge (or interface) in the center and also an unintended interface on the left and right edges due to the wrap around effect. The effects of this edge are just visible along the right edge of each image. Usually only the middle interface is of interest and typically only the middle portion of the image (with left and right edges removed) is shown. However, the whole image is shown here to help understand the situation.

The trick to calculating an edge or interface which is not strictly periodic using an FFT based multislice method in which the specimen is required to be periodic in the two-dimensional image plane is to make the specimen extra long in the nonperiodic direction. The specimen does not have to be periodic along the beam direction (into

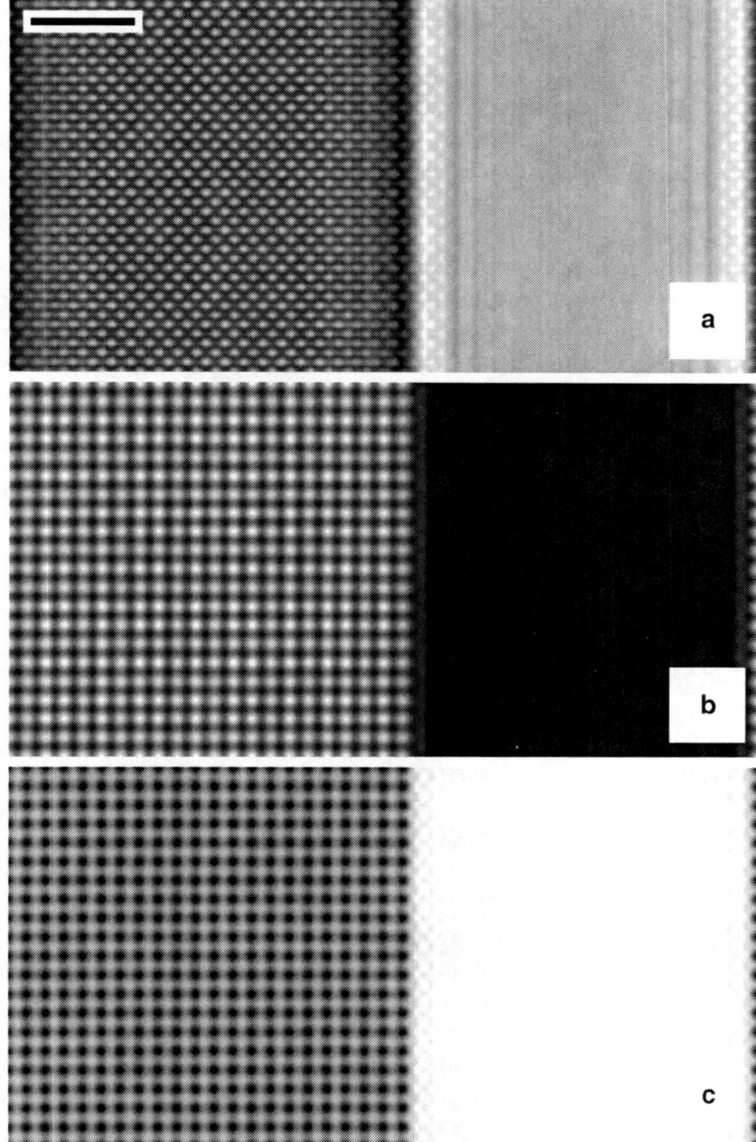

Fig. 7.21 Calcualted images of an edge of copper at 200 kV. (a) BF-CTEM image, (b) ADF-STEM image, (c) confocal STEM with a detector radius of 2Å, and the same paraemters for the collector lens as the objective lens. $C_s = 0.7$ mm. A defocus of 700Å, and an objective aperture of 12 mrad. The scale bar is 10Å

the page in this example). If it is long enough, the two interfaces are far apart and do not interfere with each other. The new unit cell is called a super cell. In this example the super cell is 72.2 by 36.1 Å and was sampled with 512 by 256 pixels for BF-CTEM and 1024 by 512 pixels for ADF and confocal STEM. The specimen was 50.54 Å thick (thin to reduce thickness effect for simplicity which also reduces the computer time). The STEM probe was sampled with 512 by 512 pixels (which effectively slides around on the larger specimen). Using a smaller sampling for the probe reduces the required computer time (which is significant for this calculation). The final image has 512 by 256 pixels. The atomic columns should appear dark in (a) and (c) and white in (b).

The ADF-STEM image (Fig. 7.21b) is incoherent and the edge appears sharp (white dots at the atom positions). The BF-CTEM image (Fig. 7.21a) is coherent and a well known Fresnel fringe (oscillations) appears at the edge (both edges) which makes it difficult to determine the exact location of the edge. This particular confocal image is well behaved although confocal is frequently not well behaved (in part because there are twice as many parameters to go wrong).

Figure 7.22 show an ADF-STEM experimental image of $SrTiO_3$ grown by pulsed laser deposition in a manner similar to that described by Ohtomo et al. [266] The growth direction is left to right (or right to left) across the page. Single atomically sharp layers of La were deposited and appear as two vertical layers (bright vertical row of dots in the image), in a demonstration of the remarkable control of the growth process using this technique. La has a higher atomic number ($Z = 57$) than Sr ($Z = 38$) so the Z-contrast feature of ADF-STEM produces a larger signal at the La positions ($Z = 22$ and 8 for Ti and O). This image was recorded on a JEOL 2010F at an energy of 200 keV.

To calculate this image, a set of coordinates were generated in three dimensions for 43 by 1 by 1 unit cells of $SrTiO_3$ with the perovskite structure (using a separate short program). One row of Sr atoms was replaced with La atoms at two positions. In reality there will probably be some relaxation of the atomic coordinates about this layer but this has been ignored. When the program was run it expanded in thickness (z) and height (y) to 22 by 26 unit cells to get a total thickness of about 100 Å. For the ADF-STEM calculation (Fig. 7.23b), the specimen was sampled with 4096 by 2048 pixels and the probe was sampled with 512 by 512 pixels. The ADF detector extended from 40–200 mrad. and the image was calculated for 512 by 256 pixels. The final image was convolved with a Gaussian low pass filter to approximate a 1.5Å source size (in the specimen plane). The BF-CTEM image (Fig. 7.23a) was calculated with 2048 by 1024 pixels. Figure 7.23 shows the calculated images to match Fig. 7.22. There is reasonable qualitative agreement. The ADF-STEM image yields a slightly more directly interpretable image at the La layers.

7.6 Biological Specimens

The multislice method is capable of handling specimens that are nearly amorphous. Each slice of the specimen is calculated independently of the other slices so the specimen can be completely amorphous in the beam direction (z as used in this

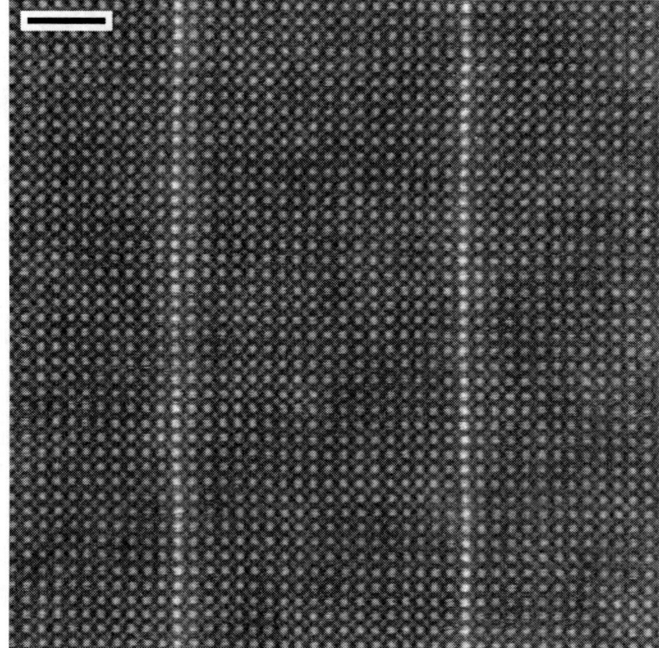

Fig. 7.22 Experimental 200 keV ADF-STEM image of single layers of La in strontium titanate ($SrTiO_3$) $C_s = 1.0$ mm and an objective aperture of 10 mrad. The scale bar is 20Å. (Courtesy of D. A. Muller, previously unpublished)

book). Using a discrete Fourier transform (the FFT) in two-dimensions perpendicular to the beam direction forces the unit cell to be repeated infinitely in those two direction (x, y as used in this book). If the specimen is somehow bounded (such as a nano particle or macromolecule) and embedded in a larger supercell such that the repeated copies of the specimen are far apart so they do not interfere with each other then amorphous objects may be calculated using the multislice method (see Fig. 6.12).

Biological macromolecules or microorganisms are good examples of amorphous particles (the method discussed next will also apply to inorganic nano particles as well). The Protein Data Base (PDB) is an on-line depository of structure data for proteins and related molecules, many of which come from X-ray diffraction studies of crystallized specimens. Each structure data file in the PDB contains a list of atomic coordinates (in three dimensions) that can be converted to a format used here for a multislice calculation (as done by Wall [363]). The image of α-hemolysin (PDB identification 7AHL.pdb, Song et al. [328]) is calculated in Fig. 7.24. This molecule has 22,778 atoms and is projected along the z axis as defined by the PDB file (which does not seem to be along the primary symmetry axis, but tilted slightly). Several major effects will be ignored for simplicity. Radiation damage is frequently the primary limiting factor in images of biological specimens. The total beam dose

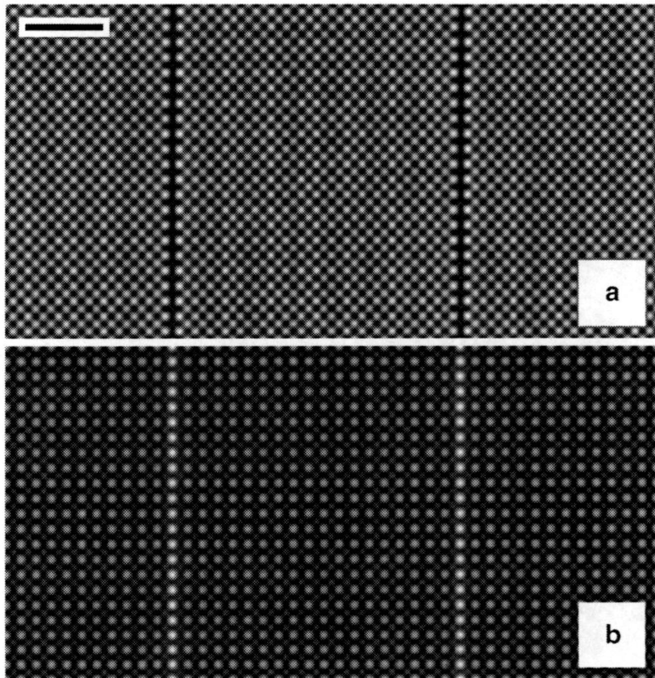

Fig. 7.23 Calculated 200 keV ADF-STEM image of single layers of La in strontium titanate ($SrTiO_3$) $C_s = 1.0$ mm and an objective aperture of 10 mrad. (a) BF-CTEM and (b) ADF-STEM with a 1.5Å source size. The scale bar is 20Å

must be limited to some small value. Part of the specimen may be moved around by the interaction with the beam if it has too large of a current. The low dose generates at lot of noise in the image. Both of these effect are ignored here for simplicity, so this result is a little bit of a fantasy.

Figure 7.24a is a BF-CTEM image (512 by 512 pixels) of the molecule with no support (magically suspended in space). Figure 7.24b is a BF-CTEM image of the molecule with a 20Å amorphous carbon support, simulated by generating uniformly distributed random numbers inside a rectangular slab on the exit surfaces used as atomic coordinates for the carbon atoms maintaining a minimum separation of 1Å and the known density of carbon. There were a little over 53,000 atoms in the carbon support. There is essentially nothing visible in the BF image, which is why biological specimens are usually stained with some heavy material (no stain here). In principle multislice can handle a surrounding stain if given a list of coordinates for the atoms in the stain (not easy to get or calculate). Figure 7.24c is an ADF-STEM (40–100 mrad detector) image and Fig. 7.24d is a confocal image. The specimen transmission functions (one per 2Å slice) were sampled with 2048 by 2048 pixels and the probe was 512 by 512 pixels. There are 256 by 256 pixels in the

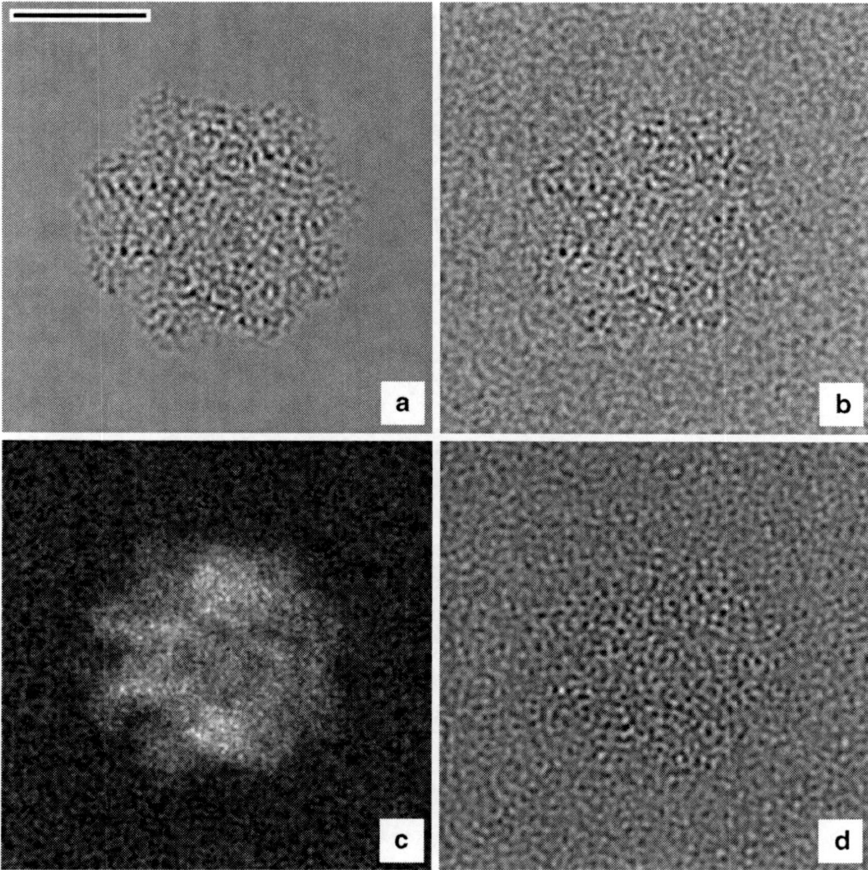

Fig. 7.24 Calculated images of α-Hemolysin(Song et al. [328] (a,b) BF-CTEM, (c) ADF-STEM and (d) SCEM (confocal). (a) Has no carbon support (not physically possible) and (b–d) have a 20Å amorphous carbon support. Electron beam energy 200 keV, $C_S = 1.3$ mm, $\Delta f = 700$Å, obj. apert. 10 mrad, confocal collector lens was similar to obj. lens. and used a 2Å diameter detector. Radiation damage and low beam dose noise are ignored for simplicity. The scale bar is 50Å

final image. It is interesting that the structure should be vivible even without staining in ADF. There might be some reason to try this with a low dose technique and a cold stage to reduce radiation damage. For this particular specimen and imaging parameters the simple incoherent image model (3.66) yields an image that is subjectively the same as the image in Fig. 7.24c, which can be calculated in seconds rather than hours. Engel and Colliex [94] and Engel [93] (amongst many) have reviewed STEM imaging of biological specimens. BF-CTEM imaging of biological specimens has been reviewed by a great many authors.

7.7 Quantitative Image Matching

Most comparisons between theoretically simulated electron micrographs and experimentally recorded electron micrographs are somewhat subjective. The two images are just displayed side by side and pronounced as being in good agreement after subjective visual inspection (Sect. 7.3 is also guilty of this practice). In principle it is possible (indeed recommended) to be more quantitative in comparing simulated and recorded images (Barry [19,20]). An easily definable figure of merit is the Chi-Squared measure of the difference between two images:

$$\chi^2 = \frac{1}{N_x N_y} \sum_{i,j} [f_{\rm exp}(x_i, y_j) - f_{\rm sim}(x_i, y_j)]^2 / \sigma_{ij}^2, \qquad (7.1)$$

where $N_x N_y$ is the number of pixels in the image, σ_{ij} the error associated with pixel (i, j), $f_{\rm exp}(x_i, y_j)$ the experimental image, and and $f_{\rm sim}(x_i, y_j)$ is the simulated image. (This symbol χ should not be confused with the same symbol used for the aberration function.) This definition of χ^2 is technically called the reduced χ^2 because it is normalized to the total number of data points. A value of $\chi^2 \sim 1$ indicates a good fit. The r-factor figure of merit (R_1, or R_2) commonly used in X-ray diffraction could also be used.

$$R_1 = \int |f_{\rm exp}(x,y) - f_{\rm sim}(x,y)| {\rm d}x{\rm d}y \Big/ \int |f_{\rm exp}(x,y)| {\rm d}x{\rm d}y$$

$$R_2 = \int |f_{\rm exp}(x,y) - f_{\rm sim}(x,y)|^2 {\rm d}x{\rm d}y \Big/ \int |f_{\rm exp}(x,y)|^2 {\rm d}x{\rm d}y. \qquad (7.2)$$

It would be very nice to be able to quote a value of χ^2 or the r-factor for the agreement between a simulated image and an experimentally recorded image. However, there are considerable obstacles to overcome to perform a quantitative image match. In practice the two images to be compared can be in different orientations (translation and rotation) and they will never be at exactly the same magnification. Just to begin a quantitative comparison requires fitting these four degrees of freedom. These properties of the image must be found in spite of the fact that the image may be noisy.

Next the overall scale and background level of the experimental image must be found. Most image detectors (film is particularly difficult to quantify) are designed to provide a linear image which is all that is required for human vision. This means that there are two additional degrees of freedom. The recorded image intensity may have an arbitrary additive and multiplicative constant:

$$f_{\rm exp}(x,y) = a_{\rm det} f_{\rm ideal}(x,y) + b_{\rm det}, \qquad (7.3)$$

where $a_{\rm det}$ and $b_{\rm det}$ are constants unique to each detector. These scaling constants can be found by recording the image intensity through a hole in the specimen and the intensity with the beam turned off (ideally at the same time that the image is recorded), but this additional measurement is rarely done. Film and plates notoriously vary with development time and temperature etc. and are difficult to quantify but more modern CCD imaging systems may make this much easier. With a lot of care an experimental image can be recorded with sufficient detail to quantitatively compare to a simulated image but this extra burden is rarely accepted in practice, so most comparisons are subjective in nature.

Thust and Urban [344] and Möbus and Rühle [250] have also proposed using the cross correlation coefficient such as:

$$C_{\text{cor}}(f_{\text{exp}}, f_{\text{sim}}) = \frac{\sum_{xy}(f_{\text{exp}}(x,y) - f_{e0})(f_{\text{sim}}(x,y) - f_{s0})}{\sqrt{\sum_{x,y}(f_{\text{exp}}(x,y) - f_{e0})^2 \sum_{xy}(f_{\text{sim}}(x,y) - f_{s0})^2}}, \qquad (7.4)$$

where f_{e0} and f_{s0} are the average values of $f_{\text{exp}}(x,y)$ and $f_{\text{sim}}(x,y)$, respectively. The cross correlation coefficient has the advantage of eliminating the dependence on the scaling parameters of the detector.

Minimizing χ^2 (or maximizing the cross correlation coefficient) with respect to some parameter of the simulation is a method for extracting that parameter from the recorded image. The specific program implementation of a minimization procedure can become rather involved and may involve a multislice simulation for each iteration. Kirkland [210] has used this approach to determine the defocus of the electron micrographs. Wilson et al. [382] have used a semiquantitative matching technique to determine the optical parameters of the microscope such as spherical aberration C_s and defocus Δf.

Ourazd et al. [278] have used a quantitative pattern matching technique to map the stoichiometry of their specimen. A precalculated set of possible specimen types was quantitatively compared to each unit cell of the specimen to determine its chemical composition. The best match determined the chemical composition of each unit cell. This requires that there be a small number of different unit cells.

King [198] and Möbus and Rühle [250] have perform nonlinear least squares fitting to extract specimen parameter such as tilt and defocus as well as the atomic coordinates. Zhang et al. [392] have uses a quantitative fitting procedure to refine the atomic coordinates at an interface. Möbus [247] and Möbus and Dehm [248] have recently proposed maximizing the cross correlation coefficient instead of minimizing χ^2 to refine the specimen parameters and coordinates. Möbus et al. [251] have presented a general structure retrieval program using image matching.

For the last decade or so many authors have found a significant discrepancy of about a factor pf 2X to 3X (principally in the lattice fringe amplitude) between calculated and experimentally measured images that has become known as the "Stobbs factor" (Hÿtch and Stobbs [167], Boothroyd [33]). As discussed earlier this is a difficult measurement, usually requiring a measurement of the incident beam intensity on an absolute scale, an accurate specimen model and a well characterized electron optical column (aberrations and partial coherence) and detector response function. A possible amorphous contamination layer (Mkoyhan et al. [246]) on the specimen (usually organic material from the air or diffusion pumps), crystal tilt (Maccagnano et al. [232]) or inelastic scattering may also confuse the issue. Thust [343] has recently obtained good quantitative agreement of BF-CTEM images using an accurate model of the CCD transfer function. Klenov et al. [213], and LeBeau et al. [223] have also obtained good agreement between measured and calculated ADF-STEM images using Bloch waves and multislice with the frozen phonon approximation. There is good reason to believe that the current theory is quantitatively correct with careful attention to experimental details.

7.8 Troubleshooting (What Can Go Wrong)

There are a large number of things that can go wrong in an image simulation. The proposed specimen structure must be specified in some detail, usually in the form of a list of atomic coordinates and atomic numbers in a unit cell. Even less well known is the thickness of the specimen. Usually, a large sequence of possible specimen thicknesses is calculated and compared to experiment.

The instrumental (optical) parameters such as the aberration constants (C_s, etc.) and aperture size of the objective lens and lens defocus must be known. Usually defocus is not know very well (particularly in bright field phase contrast). Frequently, a *defocus series* is calculated for comparison to experiment. There are also a variety of parameters such as defocus spread, illumination angle, etc. that are hard to estimate but can influence the image.

There are also many parameters that are solely related to the calculation and have very little to do with the microscope or specimen but can dramatically affect the calculation. These parameters include the sampling size (pixel size) in the image and slices and the slice thickness itself.

Multislice almost always uses an FFT to reduce the total CPU time. The FFT is a discrete Fourier transform which repeats the image infinitely in all directions. Although the image is only displayed as a single image you should remember that it is really an infinite array of identical side-by-side images. This produces a strange effect called the *wrap-around error*. The left side of the image in essence touches the right side of the image (and vice versa) and the top of the image touches the bottom of the image (see Fig. 6.12). To use the FFT each image and slice must obey periodic boundary conditions or be an integer number unit cells of the specimen (called a super cell). Interfaces and defects must be imbedded inside a large super cell.

In summary, some of the things that need to be specified correctly are:

Specimen parameters: atomic coordinates and numbers of the specimen and thickness of specimen

Instrumental parameters: defocus, C_s, objective aperture, etc.

Sampling size: number of pixels in the image and slice and the slice thickness. Ensure that the total integrated intensity is at least 0.9 or higher (1.0 to start). Calculations with slightly higher or lower sampling should yield the same result if the sampling is adequate.

Slice thickness: usually the slices should correspond to the existing atomic layers in the specimen. If the slices are too thick then the total integrated intensity will decline too much (as in the sampling size issue) and may produce false high order Laue zones corresponding to the slice thickness.

Wrap-around error: each slice must obey periodic boundary conditions.

Chapter 8
The Programs

Abstract This chapter gives some details about the programs used in this book, which can be obtained on the associated web site. This can be thought of as an illustration of some possible approaches to implementing the calculations or as a description of how to use the existing programs.

This chapter describes how to use the programs that are on the associated web site. These programs were also used to calculate many of the figures shown in the text. This group of programs will be referred to as the TEMSIM package (for conventional and scanning Transmission Electron Microscope image SIMulation). The general organization of the programs and some specific examples of the running the programs are given. It gives the information required to prepare the input data files with the specimen description and how to actually run the programs. Alternately this discussion can be viewed as a description of some possible approaches to implementing programs to perform the calculation described earlier in this book.

8.1 Program Organization

The TEMSIM programs are written in a generic no-frills manner. Their main purpose is to perform the numerical simulation in an efficient manner without being dependent on any specific type or brand of computer. They do not have a very elegant user interface by today's standards, but they are as nearly machine independent as they can be. The programs are intended for people who are comfortable using the simple command line interface available on most operating systems. There is some error checking, but very inappropriate input may cause to the programs to crash in an uncivilized manner. They have been compiled and run on a variety of different computers without requiring any changes to the program. They will probably evolve with time and may eventually get a more user friendly graphical user interface and other computational features.

E.J. Kirkland, *Advanced Computing in Electron Microscopy*,
DOI 10.1007/978-1-4419-6533-2_8, © Springer Science+Business Media, LLC 2010

The TEMSIM programs are a group of loosely coupled programs. The basic simulation steps are split into a small number of separate programs with the output of one program being used as the input for the next program. For example, there is one program called *atompot* (for atomic potential) to calculate the projected atomic potential of a two dimensional slice through the specimen. This program should be run for each slice in the specimen. The output of *atompot* can be read by the program *mulslice* to calculate the wave function that would be transmitted through the specimen using the multislice method. A third program called *image* can then use the results of *mulslice* to calculate a defocused CTEM image. The advantage of splitting the programs this way is that *atompot* and *mulslice* take more computer time than *image* but their output can be reused several times. The program *image* can use the same results of *atompot* and *mulslice* several times to generate a series of images at different defocus in an efficient manner. Furthermore, the output of *atompot* can also be used by *stemslic* to simulate a STEM image. The programs in the TEMSIM group can be combined in different ways to produce different results without having to reprogram each specific case or image simulation operation. This obviously requires using a standard disk data file format to store the intermediate images.

There are two possible strategies for describing the specimen structure. One strategy, applicable to specimens with a few repeating layers (periodic along z or the beam axis), is to manually decompose the specimen into a sequence of repeated slices. Simple crystals frequently divide into a small number of repeated layers. The atomic coordinates for each layer are generated by hand and the projected atomic potential of the layer (or slice) is calculated and stored in a disk file. The potentials are then read into memory and used over and over again in the multislice method. The programs *atompot*, *mulslice*, and *stemslic* use this approach. If there are only a few different layers that are repeated many times then this is the most efficient method because it avoids recalculating the projected atomic potential of the same layer over and over again. In principle it might be possible to automate this decomposition, but the programs still require this to be done by the human operator. A second strategy, applicable to amorphous specimen with no obvious repeating structure (along the beam direction), is to read in a list of all of the (three dimensional) coordinates of the atoms in the specimen and calculate the projected potential for each slice as it is needed and then discard it. If the specimen is completely amorphous then the potentials cannot be reused so there is no loss of efficiency, however this calculation can take substantially longer than reusing a small number of repeated layers or slices. The programs *autoslic* and *autostem* use this approach. This can also be used for periodic or semiperiodic structures if computer time is readily available.

8.2 Image Display

The numerical portion of the simulation programs can be written to be independent of the specific computer they are running on but the simple operation of displaying the image on the computer screen is very different on each type of computer and is usually also very difficult to program properly. However, there are currently many different image display and manipulation programs available on most popular computer platforms. These programs range from simple display programs (some are distributed free) to complex commercial package with elaborate image manipulations capabilities. Even simple word processing programs can include images if they are in the correct format. There is a rather strong incentive to somehow use the existing image manipulation programs to display and manipulate the results of the image simulation without having to write a display program on each different type of computer. Using an image file format that the available image display programs recognize will allow the TEMSIM results to be feed into the existing image display programs. Each image program has its own constraints on which format the image data should be stored in to use the program. There are a wide variety of image file formats that can be used, ranging from proprietary formats from specific manufacturers to well described standard formats published in publicly available manuals. Almost all of the available image display software supports the TIFF (or Tagged Image File Format) format. TIFF has been around for many years and in many ways is a defacto standard in the computer world. As of version 6.0, TIFF also has enough features to be usable for multislice image simulation. The TEMSIM packages uses the TIFF standard image file format (with the standard extensions to support both 8 bit grey scale images for display and 32 bit floating point images for calculations, see discussion later). The results of the image simulation can be easily manipulated and displayed by a wide variety of available software.

8.3 Programming Language

Some people become quit fanatical about the choice of computer operating system and programming languages. The opposite philosophy adopted here is to try to be as generic as possible so that the programs will run on as many different types of computers as possible. The programs are nearly independent of the operating system used, but there is no way around choosing a specific programming language.

There are currently two common (but not only) programming language for numerical simulation, FORTRAN and C/C++. FORTRAN has been around for a very long time and is specifically designed for numerical calculation. C (Kernigham and Richie [192]) is a relative new comer but has recently become quit popular. FORTRAN's biggest advantage for image simulation is that it has complex variables and complex arithmetic built in. Operations such as FFT's uses complex variables extensively, and are easier to program in FORTRAN. Previous versions of these image simulation programs were written in FORTRAN for this reason. However FORTRAN has several draw backs. It lacks dynamic memory allocation (FORTRAN-90 has added some memory allocation, so this comment is specifically directed at FORTRAN-77). This means that the array sizes (image sizes) must be fixed at compile time and the program must be recompiled to increase the image array size. The C/C++ language specifically includes subroutines to allocate memory and change the image array sizes dynamically at run time. This means that the image sizes can be specified when the program is run and do not require recompilation to increase the image size. Dynamic memory allocation is very useful in image simulation because the programs may be written independent of the image size. The ANSI-C library also has a variety of useful functions (for example to get the time and date) that are well defined and behave consistently from one compiler to the next. Many functions are missing or are not well defined in the standard FORTRAN libraries and tend to vary from one compiler to the next. Probably the worst problem with the FORTRAN library functions is the RECL or record length keyword in the open file subroutine. The original standard did not explicitly define the units for this important parameter. On some compilers it is in units of numeric storage units (equal to the length of a single precision floating point number which is typically four bytes long) and on others it may be in units of bytes. Past experience has found that it is almost impossible to write an image processing or simulation program in FORTRAN that does not require modifications to run correctly on different machines or even different compilers on the same machine. The C language does not have complex variables (complex variables are possible in C++ and C99 but the performance varies at the time this was written) however it is possible to write out the complex operation the long way and produce the same results. C was not really designed to be a numerical simulation language but it can be adapted to this purpose reasonably well. Also the TIFF image format is designed in a manner that assumes a byte addressable file, meaning that the program can randomly access any given byte in the file. This feature is built into C but is lacking in FORTRAN. In principle you could write a FORTRAN file caching routine that would mimic a byte addressable data file (after all, that is what the C library has done) but the TIFF file format is somewhat easier to write in C. Overall the ability to write a multislice simulation program that can be run on many different computers without change (the code is literally identical) makes the C language the most useful. The TEMSIM package is written in standard ANSI-C to be the most portable between different computers.

8.3.1 Disk File Format

As described earlier the TEMSIM package uses the TIFF standard file format to store image data. TIFF was originally designed by Aldus in the late 1980s to store black and white images for desk top publishing and has evolved into a published standard supporting both black and white images and color images as well as two tone line drawings. Murray and vanRyper [259] have discussed TIFF and cataloged many other standard image formats. TIFF is more powerful and flexible than many other formats, which is both an advantage and a curse because it is also complicated. It can be very difficult to program a good TIFF subroutine library but the format can be extended to do a lot of different things. TIFF has many more features than are needed in any one applications and most applications only support a subset of TIFF features. Simply saying that a file is in TIFF format is not enough to specify the type of data format. The image must also be specified as color or black and white or a simple line drawing and the number of bits per pixel must be identified. The TEMSIM package uses a combination of 8 bit grey scale images and 32 bit floating point images (described in more detail later). When most image processing program say they support TIFF format they typically mean 8 bit grey scale images or occasionally mapped color or true color images.

The first 8 bytes or header of a TIFF file specify the byte ordering scheme the file uses and the byte offset of the first IFD or image file directory in the file. There is a threaded directory structure inside each TIFF file. TIFF files can use either the big-endian or little-endian byte ordering scheme. Byte ordering refers to the order in which the bytes (one byte is 8 bits) are addressed within a 32 bit (or four byte) data element (or word). In big-endian byte ordering (used by some microprocessors) the byte address starts at the most significant byte and advances to the least significant byte. Little-endian byte ordering (used on Intel microprocessors) is just the opposite, the address starts at the least significant byte and increase to the most significant byte. All TIFF readers are required to read both byte ordering schemes so that TIFF image files should be transportable between different types of computers. (In practice some application software does not always follow this rule.) Also contained in the TIFF 8 byte header is a pointer (or byte offset within the TIFF file) to an IFD or image file directory. The IFD contains a list of data about the image, such as the number of bits per pixel, the number of pixels in x and y, etc. Each data item is identified with a unique tag or 8 bit code. TIFF stands for Tagged Image File Format because of this tagged data structure. There are several dozen different tags listed in the TIFF standard but only a small subset of a dozen or so are typically used in a given application. Subroutines that read TIFF files must somehow deal with many different possible tags and this tends to make TIFF difficult to program. Also contained within the IFD is a list of byte offsets (or pointers) for the raster image data itself and a pointer to the next IFD in the file. If there are no more images in the file then the next IFD pointer is NULL. Only the 8 byte header is in a fixed position in the file. The IFD's and the image data associated with them may be anywhere in the file. The TIFF file has an internal threaded data structure. There can be more than one image, each with its own IFD. The first 8 byte header points to the first IFD and its image data. The first IFD then points to the second IFD and its image data, and so forth. To read a TIFF file the computer program must follow all of the pointers as they wind their way through the file. This also presumes that the programming language can address any specific byte within the file (i.e., this essential requires using the C programming language). Refer to the actual published TIFF standard (Aldus [55]) for a complete description of the data format.

Most software applications that support TIFF images only read the first image in the file. The stated purpose of the second and later images in the TIFF file is for storing higher resolution versions of the image. This feature is very useful for image simulation. As of version 6.0 (Aldus [55]) TIFF also supports a floating point pixel data type (each pixel is one 32 bit IEEE format floating point number). TEMSIM uses this feature to combine an 8 bit image with a 32 bit floating point image. The first image in the TIFF file is an 8 bit grey scale version of the image and the second image is the 32 bit floating point version of the image. Most display programs assume square pixels, so the 8 bit image is expanded in one direction if necessary (using bilinear interpolation as in appendix D) to get square pixels. (Compressing the image using interpolation runs the risk of leaving out sharp points so is not used.) The second image, which is ignored by many image display programs, is a 32 bit floating point image used for numerical simulation. Although it seems as if each data file looks like an 8 bit image it really has 32 bits of precision for numerical calculation. Storing both versions of the image increases the size of the file by about 25% but allows using readily available image display programs. When you look at an intermediate TIFF imaged used in the TEMSIM package you should be careful not to save it from a standard TIFF applications, because this will most likely destroy the hidden 32 bit portion of the image. Stacking the images and using 32 bit floating point pixels data is allowed within the TIFF standard so this type of format may be referred to as an extended TIFF format. When one TEMSIM program reads the results generated by another it is helpful for the image file to contain its parameters like electron energy, etc. This information is transferred in a third image in the TIFF file that is 64 pixels wide and one pixel high. This image is not for display but contains a 64 element array of image parameters. Complex images are stored side by side as two real floating point images. The file *tiffsubs.c* contains a library of TIFF subroutines for reading and writing both integer and floating point TIFF images.

Some application programs require that the name of a TIFF file end in ".tif" to be identified as a TIFF file, so it is a good practice to end all image data file names with ".tif." Also, some display programs put the origin in the upper left corner and some put it in the lower left corner, which is sometimes confusing.

8.4 BF-CTEM Sample Calculations for Periodic Specimens

There are three basic computational steps in simulating a BF-CTEM image. The first step is to calculate the projected atomic potential of each slice of the specimen using the *atompot* program. The second step is to transmit the incident electron wave (usually a plane wave) through the specimen using the program *mulslice*. The third and final step is to form the image with defocus and the objective lens aberration using the program *image*.

This example will simulate an image of strontium titanate ($SrTiO_3$), which has a cubic perovskite structure with a cubic unit cell size of $a = 3.905$Å. There are Sr atoms on all of the corners of the cubic and O atoms on center of each face, with a single Ti atom in the center of the cube. (There is an alternate description of this structure with the positions of Sr swapped with that of Ti and the O atoms on the edges instead of the faces.) With the optic axis of the electron microscope along one of the three primary cubic axis this specimen naturally divides into two rectangular slices. One slice is a face of the cube and the other is through the center of the cube.

The examples later list the computer response while each program is running. Each line containing information that the user must supply has a ">" at the beginning of the line. This is to separate the computer response from the user response and should not be entered when the program is actually run. Multiple numbers should be entered separated with spaces and without commas.

8.4.1 Atomic Potentials

The projected atomic potential of the slices in the specimen are calculated using *atompot*. This program calculates the structure factor in reciprocal space using the scattering factors in the first Born approximation and then does an inverse FFT to get the projected atomic potential [see (5.21)]. The result stored in the file does not contain the leading factor of $\lambda m/m_0$ so that it is independent of electron energy. The programs *mulslice* and *stemslic* add this factor later when the potential is used.

The *atompot* program can also apply symmetry operations to the atomic position using the following symmetry operation:

$$x_{\mathrm{new}} = S_{xai} x_{\mathrm{old}} + S_{xbi} \tag{8.1}$$

$$y_{\mathrm{new}} = S_{yai} y_{\mathrm{old}} + S_{ybi} \tag{8.2}$$

where the parameters S_{xai}, S_{xbi}, S_{yai}, and S_{ybi} are supplied for each symmetry operation. $(x,y)_{\mathrm{old}}$ are the specified atomic coordinates and $(x,y)_{\mathrm{new}}$ are the coordinates generated by the symmetry operation. If the specimen has a high degree of symmetry this feature can greatly reduce the amount of data the you have to type in.

The format of the input data file for *atompot* is shown in abbreviated form in Table 8.1. The first line of the file has three numbers (ax, by, cz) that are the dimensions of the super cell of one slice of the specimen in Angstroms. The slice thickness is cz. The next line has the number of symmetry operations (Ns) for each coordinate, followed by the symmetry operations themselves. There is one line for each symmetry operation and the number of lines must match Ns. If Ns = 0 then no symmetry operations are listed. After the symmetry operations the atomic coordinates of atoms are listed in groups with the same atomic number Zatom. Each line of coordinate data has the reduced coordinates xpos = x/ax and ypos = y/by of each atom. The first number on the coordinate data line is the occupancy (occ) of each atom. The occupancy is typically one but may be set to a fractional value (for example the symmetry operations may generate several identical atoms at the same location and the occupancy may be used to correct for this duplication). At the end of each coordinate data line is a number labeled wobble signifying the rms random displacement (in Angstroms) for simulating thermal phonons (typically wobble = 0). This value is the rms deviation in each direction and not the 3D rms value. *atompot* uses a random number generator with a Gaussian distribution to simulate random thermal (phonon) displacements. The initial seed for the random number generator is obtained from the clock() function, so that each run should produce a different (random) results. This feature is mainly an historical artifact (may be removed in future) and is usually not used (*autoslic* and *autostem* do this much better). When all of the atoms for a particular atomic number have been listed there is a single blank line. Following this blank line is the next atomic number (Zatom2) and the coordinates for this atomic number are listed in the same format. The input is terminated by two successive blank lines. Two different atomic numbers are shown but there may be any number of different atomic numbers (limited by the amount of computer memory available).

Table 8.1 The format of the input data for *atompot.c*

```
ax      by     cz
Ns
Sxa1    Sxb1   Sya1   Syb1
Sxa2    Sxb2   Sya2   Syb2
:       :      :      :
SxaNs   SxbNs  SyaNs  SybNs
<blank line>
Zatom1
occ1    xpos1  ypos1  wobble1
:       :      :      :
occn    xposn  yposn  wobblen
<blank line>
Zatom2
occ1    xpos1  ypos1  wobble1
:       :      :      :
occm    xposm  yposm  wobblem
<blank line>
<blank line>
```

There are two different atomic numbers Zatom1 and Zatom2 with n and m coordinates each. There may be an arbitrary number of different atomic numbers and an arbitrary number of coordinates for each atomic number

The input coordinates for both of the layers of the $SrTiO_3$ specimen are shown in Tables 8.2 and 8.3. Layer *a* has two atoms and no symmetry operations and layer *b* has three atoms and no symmetry operations. The input data files may be prepared using standard text editors or word processing programs if the file is saved as "text only." The input data specifies the coordinates for one unit cell and *atompot* will

Table 8.2 The *atompot* input data file srta.dat for the a layer of strontium titanate

```
    3.9051   3.9051   1.9525
  0
 38
    1.0000   0.0000   0.0000

 8
    1.0000   0.5000   0.5000
```

This is one face of the cube and has one strontium atom (Z = 38) in the corner and one oxygen atom (Z = 8) in the center

Table 8.3 The *atompot* input data file srtb.dat for the a layer of strontium titanate This is a plane through the center of the cube and has one titanium atom (Z = 22) in the center and two oxygen atoms (Z = 8) on the edges

```
   3.9051   3.9051   1.9525
 0
22
   1.0000   0.5000   0.5000

 8
   1.0000   0.5000   0.0000
   1.0000   0.0000   0.5000
```

expand the final potential to an arbitrary (positive) integer number of unit cells. The rms random deviation is left blank because this feature will not be used in this example. The results of running *atompot* on the data is shown below.

```
<---- run program atompot ---->

atompot version dated 3-jul-2008 EJK
Copyright (C) 1998, 2008 Earl J. Kirkland
This program is provided AS-IS with ABSOLUTELY NO WARRANTY
  under the GNU general public license

calculate projected atomic potentials

Name of file with input crystal data :
>srta.dat
Name of file to get binary output of atomic potential :
>srtapot.tif
Real space dimensions in pixels Nx, Ny :
>512 512
Replicate unit cell by NCELLX,NCELLY,NCELLZ  :
>8 8 1
Do you want to add thermal displace. to atomic coord.? (y/n)  :
>n
2D lattice constants= 3.905100 x 3.905100 Angstroms
 and propagation constant= 1.952500 Angstroms
Unit cell replicated to a= 31.2408, b= 31.2408, c=1.9525  Ang.
Maximum symmetrical resolution set to 0.122034 Angstroms

   64.00 atoms with Z=  38 (Sr)
   64.00 atoms with Z=   8 ( O)

   for a grand total of     128.00 atoms
 pix range 0.253791 to 47.6074
103184 fourier coeff. calculated in right half plane
The average real space value was 0.986566
CPU time (excluding set-up) = 0.495000 sec.

<---- run program atompot ---->
```

```
atompot version dated 3-jul-2008 EJK
Copyright (C) 1998, 2008 Earl J. Kirkland
This program is provided AS-IS with ABSOLUTELY NO WARRANTY
  under the GNU general public license

calculate projected atomic potentials

Name of file with input crystal data :
>srtb.dat
Name of file to get binary output of atomic potential :
>srtbpot.tif
Real space dimensions in pixels Nx, Ny :
>512 512
Replicate unit cell by NCELLX,NCELLY,NCELLZ :
>8 8 1
Do you want to add thermal displace. to atomic coord.? (y/n) :
>n
2D lattice constants= 3.905100 x 3.905100 Angstroms
 and propagation constant= 1.952500 Angstroms
Unit cell replicated to a= 31.2408, b= 31.2408, c=1.9525  Ang.
Maximum symmetrical resolution set to 0.122034 Angstroms

   64.00 atoms with Z=  22 (Ti)
  128.00 atoms with Z=   8 ( O)

   for a grand total of     192.00 atoms
 pix range 0.0406337 to 30.2206
103184 fourier coeff. calculated in right half plane
The average real space value was 0.834076
CPU time (excluding set-up) = 0.469000 sec.
```

8.4.2 Multislice

After the projected atomic potential has been calculated for each layer in the specimen (and stored in a file), the *mulslice* program is run. This program performs a multislice calculation to transmit the electron wave function through the specimen. A sample output using the strontium titanate potentials from Sect. 8.4.1 is shown below. This program allows the incident beam and the crystal to be tilted. The incident beam tilt should obey periodic boundary conditions (i.e., not all angles are allowed) and be small. The crystal tilt is calculated by adding a phase factor to the propagator (6.99) and is only valid for small angles of no more than about one degree.

mulslice can also print out a table of values (real and imaginary part) for selected beams vs. thickness. The beam (or Fourier coefficient) can be specified by its crytallographic index (h,k). However, this it the index in the super cell which is not necessarily the same as the primitive unit cell. This feature is not used in this example.

```
<---- run program mulslice ---->

mulslice version dated 3-jul-2008 ejk
```

```
Copyright (C) 1998, 2008 Earl J. Kirkland
This program is provided AS-IS with ABSOLUTELY NO WARRANTY
  under the GNU general public license

perform traditional multislice calculation

Type in the stacking sequence :
>12(ab)

Type in the name of 2 atomic potential layers :

Name of file with input atomic potential a :
>srtapot.tif
Name of file with input atomic potential b :
>srtbpot.tif
Name of file to get binary output of multislice result:
>srtmul.tif
Do you want to include partial coherence (y/n) :
>n
NOTE, the program image must also be run.
Do you want to start from previous result (y/n) :
>n
Incident beam energy in kev:
>400
Crystal tilt x,y in mrad.:
>0 0
Incident beam tilt x,y in mrad.:
>0 0
Do you want to record the (real,imag) value
 of selected beams vs. thickness (y/n) :
>n
Wavelength = 0.016439 Angstroms
layer a, cz = 1.952500
layer b, cz = 1.952500
Size in pixels Nx x Ny= 512 x 512 = 262144 beams
Lattice constant a =      31.2408, b =      31.2408
Total specimen thickness = 46.86 Angstroms
Bandwidth limited to a real space resolution of 0.183052 Ang.
   (= 89.81 mrad)  for symmetrical anti-aliasing.
Number of symmetrical non-aliasing beams = 91529
slice    1, layer = a, integrated intensity = 0.999959
slice    2, layer = b, integrated intensity = 0.999935
slice    3, layer = a, integrated intensity = 0.999851
slice    4, layer = b, integrated intensity = 0.999811
slice    5, layer = a, integrated intensity = 0.999658
slice    6, layer = b, integrated intensity = 0.999593
slice    7, layer = a, integrated intensity = 0.999351
slice    8, layer = b, integrated intensity = 0.999252
slice    9, layer = a, integrated intensity = 0.998905
slice   10, layer = b, integrated intensity = 0.998767
slice   11, layer = a, integrated intensity = 0.998303
slice   12, layer = b, integrated intensity = 0.998121
slice   13, layer = a, integrated intensity = 0.997534
slice   14, layer = b, integrated intensity = 0.997306
```

```
slice   15, layer = a, integrated intensity = 0.996597
slice   16, layer = b, integrated intensity = 0.996322
slice   17, layer = a, integrated intensity = 0.995501
slice   18, layer = b, integrated intensity = 0.995186
slice   19, layer = a, integrated intensity = 0.994268
slice   20, layer = b, integrated intensity = 0.993912
slice   21, layer = a, integrated intensity = 0.992916
slice   22, layer = b, integrated intensity = 0.992537
slice   23, layer = a, integrated intensity = 0.991511
slice   24, layer = b, integrated intensity = 0.991112
pix range -3.95399 to 0.996049 real,
          -1.85484 to 1.76015 imag
Total CPU time = 1.520000 sec.
```

The stacking sequence of the slices or layers of the specimen is entered in symbolic form as 12(ab) in this example. Each layer is given a unique single character name in the order abcd...xyzABCD...XYZ. There are 52 possible layer names (it is case sensitive). Pairs of parenthesis denote a group of slices and the leading number is the number of times that the group in parenthesis is repeated. 12(ab) means that there are twelve repeats of the sequence ab. Parenthesis may be nested up to 100 levels and structures such as 5(2(ab)3(ca)) are possible.

8.4.3 Image Formation

The multislice calculation produces the wave function at the exit surface of the specimen. The objective lens images this wave function. The program *image* adds the effects of the objective lens aberrations and defocus and produces an image as it would be observed in the electron microscope. *image* can perform this calculation in a completely coherent mode or add partial coherence. The example shown below is a coherent image. *image* does not change the input data file (from *mulslice*) so it can be run several times with the same input file (without running *mulslice* again) to produce a whole defocus series.

```
<---- run program image ---->

image version dated 3-jul-2008 (ejk)
Copyright (C) 1998, 2008 Earl J. Kirkland
This program is provided AS-IS with ABSOLUTELY NO WARRANTY
  under the GNU general public license

calculate TEM images with defocus

Name of file with input multislice result:
srtmul.tif
Type 0 for coherent real space image,
  or 1 for partially coherent real space image,
  or 2 for diffraction pattern output:
>0
```

```
Name of file to get defocused output:
>srtimg.tif
Spherical aberration in mm.:
>1.3
Defocus in Angstroms:
>566
Objective aperture size in mrad:
>9.33
Mag. and angle of two-fold astig. (in Angst. and degrees):
>0 0
Mag. and angle of three-fold astig. (in Angst. and degrees):
>0 0
Objective lens and aperture center x,y in mrad
 (i.e. non-zero for dark field):
>0 0
Starting pix energy =   400.00 keV
Starting pix range -3.95399 0.996049 real
                   -1.85484 1.76015 imag
Pix range 0.051091 to 1.437282
Elapsed time = 0.106000 sec
```

8.4.4 Partial Coherence

Partial coherence may be added in two different ways. If the specimen is thin then the image may be calculated using the transmission cross coefficient as in Sect. 5.4.3 as shown below. The actual simulated image is shown in Fig. 8.1. The black dots correspond to the oxygen atom positions and there is a contrast reversal in the image.

```
<---- run program image ---->

image version dated 3-jul-2008 (ejk)
Copyright (C) 1998, 2008 Earl J. Kirkland
This program is provided AS-IS with ABSOLUTELY NO WARRANTY
  under the GNU general public license

calculate TEM images with defocus

Name of file with input multislice result:
>srtmul.tif
Type 0 for coherent real space image,
  or 1 for partially coherent real space image,
  or 2 for diffraction pattern output:
>1
Name of file to get defocused output:
srtimg2.tif
Spherical aberration in mm.:
>1.3
Defocus in Angstroms:
```

```
>566
Objective aperture size in mrad:
>12
Illumination semi-angle in mrad:
>0.6
Defocus spread in Angstroms:
>50
Starting pix energy =    400.00 keV
Starting pix range -3.95399 0.996049 real
                   -1.85484 1.76015 imag
Pix range 0.150198 to 1.332640
Elapsed time = 0.404000 sec
```

Fig. 8.1 The results of the program *image* for a partially coherent BF-CTEM image of strontium titanate. The scale bar in the upper left corner is 10 Å

A more accurate type of calculation (for thick specimen) is shown next. This method performs a multislice calculation for each incident angle (from the condenser illumination system) and sums the resulting images incoherently as described in (6.104). The program *mulslice* does this calculation as shown below. The allowed incident angles must satisfy periodic boundary conditions so only a few discrete illumination angles can be used.

```
<---- run program mulslice ---->

mulslice version dated 3-jul-2008 ejk
Copyright (C) 1998, 2008 Earl J. Kirkland
This program is provided AS-IS with ABSOLUTELY NO WARRANTY
  under the GNU general public license

perform traditional multislice calculation

Type in the stacking sequence :
>12(ab)

Type in the name of 2 atomic potential layers :

Name of file with input atomic potential a :
>srtapot.tif
Name of file with input atomic potential b :
>srtbpot.tif
Name of file to get binary output of multislice result:
>srtmul2.tif
Do you want to include partial coherence (y/n) :
>y
Illumination angle min, max in mrad:
>0 0.6
Spherical aberration (in mm.):
>1.3
Defocus, mean, standard deviation, and sampling size (in Ang.)=
>566 50 10
Objective aperture (in mrad) =
>12
Magnitude and angle of 2-fold astig. (in Ang. and degrees):
>0 0
Magnitude and angle of 3-fold astig. (in Ang. and degrees):
>0 0
Incident beam energy in kev:
>400
Crystal tilt x,y in mrad.:
>0 0
Wavelength = 0.016439 Angstroms
layer a, cz = 1.952500
layer b, cz = 1.952500
Size in pixels Nx x Ny= 512 x 512 = 262144 beams
Lattice constant a =      31.2408, b =      31.2408
Total specimen thickness = 46.86 Angstroms
Bandwidth limited to a real space resolution of 0.183052 Ang.
   (= 89.81 mrad)  for symmetrical anti-aliasing.
Number of symmetrical non-aliasing beams = 91529
Illumination angle sampling (in mrad) = 0.526217, 0.526217

Illum. angle =   0.000,  -0.526 mrad, total intensity= 0.991150
```

```
Illum. angle =  -0.526,    0.000 mrad, total intensity= 0.991150
Illum. angle =   0.000,    0.000 mrad, total intensity= 0.991112
Illum. angle =   0.526,    0.000 mrad, total intensity= 0.991149
Illum. angle =   0.000,    0.526 mrad, total intensity= 0.991149
Total number of illumination angle = 5
Total number of defocus values = 25
pix range 0.196332 to 1.45425 real,
          0 to 0 imag
Total CPU time = 10.089000 sec.
```

8.5 ADF-STEM Sample Calculations for Periodic Specimens

The program *stemslic.c* performs a multislice simulation for a STEM image. It generates an incident focused probe wave function at each position of the final image, transmits the wave function through the specimen (using multislice) and then integrates the electron intensity on the detector. It can generate the signal for several detectors at the same time. This is obviously a significant increase in computer time relative to a CTEM image because a STEM simulation requires a multislice simulation at each point in the image whereas a whole CTEM image is calculated in parallel. In response to this problem *stemslic* can generate a one-dimensional line scan or a two-dimensional image. A line scan is relatively quicker to calculate and can yield a significant amount of information about the specimen. *stemslic* uses the same projected atomic potential (produced by *atompot*) for each slice as is required for *mulslice*. Modeling interfaces may require a very large number of pixels in the transmission function to avoid periodic interference between interfaces. *stemslic* allows for the number of pixels in the probe wave function to be less than the number of pixels in the transmission function to decrease the required computer time. The pixel size in each must be the same however. A smaller probe wave function slides around inside a larger transmission function (with the periodicity of the transmission function not the probe wave function).

There are two examples shown below. Each uses the projected atomic potential as calculated using *atompot* for the CTEM simulation (Sect. 8.4.1). The first simulation later is a 1D line scan image and the second simulation is a 2D image simulation. As in *mulslice* crystal tilt is calculated by adding a phase factor to the propagator (6.99) and is only valid for small angles of no more than about one degree. Both an inner an outer angle may be specified for the objective aperture to model hollow cone and apodization effects (Loane and Silcox [226]) to minimize the probe size. The initial point of the line scan is (x_i, y_i) and the final point is (x_f, y_f). The simulated ADF-STEM image is shown in Fig. 8.2. The white spots correspond to the Sr atom positions. The grey dot in the middle of four white dots are the Ti atom positions and the black dots are the oxygen positions.

```
<---- run program stemslic ---->

stemslic(e) version dated 3-jul-2008 (ejk)
Copyright (C) 1998, 2008 Earl J. Kirkland
This program is provided AS-IS with ABSOLUTELY NO WARRANTY
  under the GNU general public license

perform STEM multislice

Type in the stacking sequence :
>12(ab)

Type in the name of 2 atomic potential layers :

Name of file with input atomic potential a :
>srtapot.tif
Name of file with input atomic potential b :
>srtbpot.tif
STEM probe param., V0(kv), Cs(mm) df(Ang.), apert1,2(mrad) :
>100 1.3 850 0 11.43
Magnitude and angle of 2-fold astig. (in Ang. and degrees):
>0 0
Magnitude and angle of 3-fold astig. (in Ang. and degrees):
>0 0
wavelength = 0.037014 Angstroms
Size of probe wavefunction Nx,Ny in pixels :
>512 512
Crystal tilt x,y in mrad. :
>0 0
Do you want to calculate a 1D line scan (y/n) :
>y
Number of detector geometries :
>2
Name of file to get output of 1D STEM multislice result :
>srt100.1d
Detector   1: Type, min,max angles(mrad) of collector :
>50 200
Detector   2: Type, min,max angles(mrad) of collector :
>100 200
xi, xf, yi, yf, nout :
>0 12 0 0 25
layer a, cz = 1.952500
layer b, cz = 1.952500
Size in pixels Nx x Ny = 512 x 512 = 262144 total pixels,
 lattice constants a,b = 31.240801 x 31.240801
Total specimen thickness = 46.86 Angstroms
Number of symm. anti-aliasing beams in trans. function = 91529
  with a resolution of 0.183052 Angstroms.
Number of symmetrical anti-aliasing beams in probe = 91529
    1           0              0    0.08650974    0.06920083
    2         0.5              0    0.07222058    0.05741175
    3           1              0    0.04217643    0.03276586
    4         1.5              0    0.01854871    0.01359941
    5           2              0    0.01147204   0.007921558
```

```
      6             2.5               0    0.02186334     0.01627105
      7               3               0    0.04796352     0.03749446
      8             3.5               0    0.07684164     0.06122042
      9               4               0    0.08594899     0.06873755
     10             4.5               0    0.06700292      0.0531163
     11               5               0     0.0366695     0.02827674
     12             5.5               0    0.01585812     0.01143615
     13               6               0     0.0120988    0.008422802
     14             6.5               0    0.02578121     0.01943735
     15               7               0     0.0538918     0.04234877
     16             7.5               0    0.08070254     0.06440548
     17               8               0    0.08428839      0.0673659
     18             8.5               0     0.0613654     0.04848158
     19               9               0    0.03155807     0.02412053
     20             9.5               0    0.01379981    0.009784867
     21              10               0    0.01335468    0.009428199
     22            10.5               0    0.03026422     0.02307026
     23              11               0    0.05980234     0.04719782
     24            11.5               0    0.08366199      0.0668486
     25              12               0    0.08159152     0.06513921
The total integrated intensity range was:
   0.968401 to 0.996714

CPU time = 30.65 sec.
```

A two-dimensional image can be calculated as:

```
<---- run program stemslic ---->

stemslic(e) version dated 3-jul-2008 (ejk)
Copyright (C) 1998, 2008 Earl J. Kirkland
This program is provided AS-IS with ABSOLUTELY NO WARRANTY
  under the GNU general public license

perform STEM multislice

Type in the stacking sequence :
>12(ab)

Type in the name of 2 atomic potential layers :

Name of file with input atomic potential a :
>srtapot.tif
Name of file with input atomic potential b :
>srtbpot.tif
STEM probe param., V0(kv), Cs(mm) df(Ang.), apert1,2(mrad) :
>100 1.3 850 0 11.43
Magnitude and angle of 2-fold astig. (in Ang. and degrees):
>0 0
Magnitude and angle of 3-fold astig. (in Ang. and degrees):
>0 0
wavelength = 0.037014 Angstroms
Size of probe wavefunction Nx,Ny in pixels :
>512 512
```

```
Crystal tilt x,y in mrad. :
>0 0
Do you want to calculate a 1D line scan (y/n) :
>n
Number of detector geometries :
>1
Detector   1: Type, min,max angles(mrad) of collector :
>50 200
Name of file to get output of result for this detector:
>srtadf.tif
xi,xf,yi,yf, nxout,nyout :
>0 15.376 0 15.376 64 64
layer a, cz = 1.952500
layer b, cz = 1.952500
Size in pixels Nx x Ny = 512 x 512 = 262144 total pixels,
 lattice constants a,b = 31.240801 x 31.240801
Total specimen thickness = 46.86 Angstroms
Number of symm. anti-aliasing beams in trans. function = 91529
  with a resolution of 0.183052 Angstroms.
Number of symmetrical anti-aliasing beams in probe = 91529
output file size in pixels is 64 x 64
iy= 0, min[0]= 0.011394, max[0]= 0.086510
iy= 1, min[0]= 0.011394, max[0]= 0.086510
iy= 2, min[0]= 0.011394, max[0]= 0.086510
iy= 3, min[0]= 0.011394, max[0]= 0.086510
 :                          :
iy= 56, min[0]= 0.011394, max[0]= 0.086510
iy= 57, min[0]= 0.011394, max[0]= 0.086510
iy= 58, min[0]= 0.011394, max[0]= 0.086510
iy= 59, min[0]= 0.011394, max[0]= 0.086510
iy= 60, min[0]= 0.011394, max[0]= 0.086510
iy= 61, min[0]= 0.011394, max[0]= 0.086510
iy= 62, min[0]= 0.011394, max[0]= 0.086510
iy= 63, min[0]= 0.011394, max[0]= 0.086510
output pix range : 0.0113938 to 0.0865097
The total integrated intensity range was:
   0.968401 to 0.996743

CPU time = 4885.65 sec.
```

8.6 NonPeriodic Specimens

Specimens that have no obvious periodic structure along the beam or z direction require another approach to image simulation. The programs *autoslic.c* and *autostem.c* implement this strategy. The specimen is described as a sequence of (x, y, z) coordinates with an associated atomic number $Zatom$. This sequence is first sorted by depth (along z) and then is automatically sliced into layers of a specified thickness Δz. The projected atomic potential of each slice is calculated, the electron wave function is transmitted through the slice and propagates to the next slice. The atomic

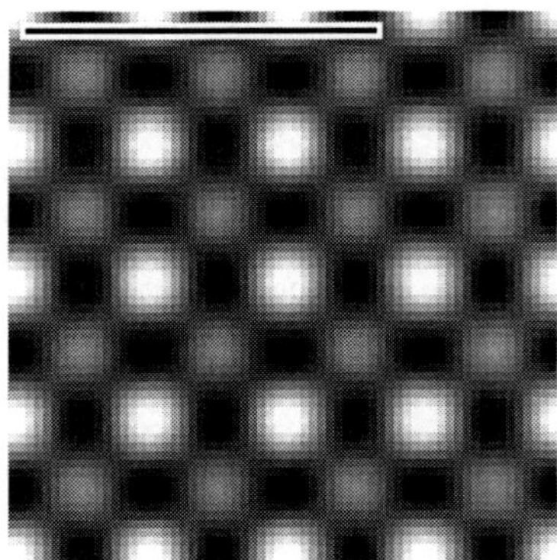

Fig. 8.2 The results of *stemslic* for an ADF-STEM image of strontium titanate. The scale bar in the *upper left corner* is 10 Å. This image corresponds to the *lower left corner* of the BF-CTEM image in figure 8.1. The Sr atom is in the lower left corner and should be *white*

potential is discarded after each slice and the process is repeated for each successive slice until the electron wave function has reached the exit surface of the specimen. This approach does not require a possibly tedious effort on the part of the user to decompose the specimen into repeating layers but will take much more computer time than calculations using repeating slices (as in *atompot* and *mulslice*). However, if there is no repetitive structure along the beam direction (z) then there is no significant difference in computer time. *autoslic* and *autostem* can be dramatically easier to use than *mulslice* for specimens with defects or interface or completely amorphous specimens.

This approach also uses a different method to calculate the projected atomic potential of each slice (as compared to *atompot*). The computer time for this non-periodic approach to image simulation is dominated by the calculation of the projected atomic potentials of the slices, and this step should be optimized if possible. The atomic potential is very localized in real space (see Fig. 5.5 for example) but it is very extended in reciprocal space. It is more efficient to calculate the projected atomic potential in real space using (5.19) because there are much fewer pixels to fill in (Pan et al. [279]). Each atomic potential is assumed to have a range of no more that 3Å in real space and the individual projected potentials are calculated from expression 5.11. The transcendental functions involved are time consuming to calculate so a look up table of cubic spline interpolation coefficients (using the quasi-Hermite spline method of Akima [3,4]) is generated as needed (i.e., only for the specific atomic numbers needed). The potential is sampled on a logarithmic grid to get more points near the origin where the potential in rapidly changing. Using the spline look up table reduces the CPU time by about a factor of three to four on typical specimens. The total projected potential (integrated from minus infinity to plus infinity) so each slice must be thicker than the range of the potential for a single atom (about one Angstroms).

The input data format to describe a specimen structure is shown in Table 8.4. The first line is a comment line with a brief description of the specimen (ignored by the program). The second line has the unit cell dimensions of the specimen in units of Angstroms. Each following line is the coordinates for one atom. The first number Zatom is the atomic number for the atom. The next three numbers are its three dimensional coordinates (xpos,ypos,zpos). All of the coordinates are assumed to be between (0,0,0) and (ax,by,cz) (i.e., the coordinates are positive). The fifth number is the occupancy for the atom (as in *atompot*). The last number on the line, wobble, is the standard deviation (in Angstroms) of the rms displacement if random thermal displacements are used. (As in *atompot* this is the rms value in each direction and not the 3D rms value.) These displacements are generated using a random number generator with a Gaussian distribution. The initial seed for the random number generator is obtained from the clock() function, and should produce a different (pseudo-random) result each time the program is run.

Table 8.4 The format of the input data for *autoslic.c* and *autostem.c*

```
<comment line>
ax       by      cz
Zatom1  xpos1  ypos1  zpos1  occ1  wobble1
Zatom2  xpos2  ypos2  zpos2  occ2  wobble2
:       :      :      :      :     :
ZatomN  xposN  yposN  zposN  occN  wobbleN
-1
```

The first line is a comment line (ignored by the program). The second line is the unit cell dimensions of the specimen. Each line following the second line is the atomic number and position of one atom in the specimen. There may be an arbitrary number of different atoms, each with it own atomic number, position and thermal vibration amplitude.

The specimen can be expanded to an arbitrary number of unit cells using the unit cell dimensions. The specimen would then be periodic and it would be better to use *mulslice*. However this feature is useful for simulating the effects of random thermal vibration, because random displacements can be added to the replicated unit cell coordinates to generate an essentially nonperiodic structure. The random thermal displacements can be scaled with a semiclassical temperature law as:

$$\text{wobble} = \text{wobble}_0 \sqrt{\frac{T}{300}}, \tag{8.3}$$

where wobble_0 is the value appearing in the input file. Although this is not very rigorous, it provides a simple methods of changing all of the thermal displacements with a single control variable. The apparent temperature can be set to some nonphysical value to get the actual scaling for a more appropriate scaling law if necessary.

autoslic can print out a table of values for selected beams vs. thickness. The beam (or Fourier coefficient) is specified by its crytallographic index (h,k), however this is the index in the super cell which is not necessarily the same as the primitive unit cell.

8.6.1 Fixed Beam Calculation

The previous examples were calculations of real space images. This example will use *autoslic* to calculate a convergent beam diffraction (CBED) pattern of a specimen with thermal vibrations. This program is not limited to diffraction patterns and can calculate images as well, just as *atompot* and *mulslice* can calculate diffraction patterns as well as images.

Table 8.5 The *autoslic.c* input data for 100 silicon

```
one unit cell of 100 silicon
      5.43     5.43     5.43
  14  0.0000   0.0000   0.0000   1.0   0.078
  14  2.7150   2.7150   0.0000   1.0   0.078
  14  1.3575   4.0725   1.3575   1.0   0.078
  14  4.0725   1.3575   1.3575   1.0   0.078
  14  2.7150   0.0000   2.7150   1.0   0.078
  14  0.0000   2.7150   2.7150   1.0   0.078
  14  1.3575   1.3575   4.0725   1.0   0.078
  14  4.0725   4.0725   4.0725   1.0   0.078
  -1
```

The program *probe* calculates a focused probe wave function. This wave function can be used as the initial starting point for *autoslic* or *mulslice*. The focused probe is transmitted through the specimen and the square modulus of the Fourier transform of the exit wave function is the CBED pattern. A conventional electron diffraction pattern can also be calculated if the incident wave function were a plane wave (this is possible with these programs but is not shown here). There are three programs that must be run to simulate a CBED pattern. First run *probe* to calculate the incident probe wave function, then run *autoslic* to transmit the wave function through the specimen, and finally run *image* to calculate the diffraction pattern from the exit wave function. This sequence is shown later and the resulting CBED pattern is shown in Fig. 8.3. The input data for this run is in the file si100.xyz shown in Table 8.5. Each silicon atom is given an rms random displacement of 0.078 Å, to simulate thermal diffuse scattering. This calculation should be repeated several times and averaged over many different sets of random displacements to get an appropriate CBED pattern (using the program *sumpix*, not discussed here). Only one run is shown for simplicity.

```
<---- run program probe ---->

probe version dated 23-nov-2008 ejk
Copyright (C) 1998, 2008 Earl J. Kirkland
This program is provided AS-IS with ABSOLUTELY NO WARRANTY
  under the GNU general public license
```

```
calculate focused probe wave function

Name of file to get focused probe wave function:
>siprobe.tif
Desired size of output image in pixels Nx,Ny:
>512 512
Size of output image in Angstroms ax,by:
>32.50 32.50
Probe param., V0(kv), Cs3(mm), Cs5(mm), df(Ang.), apert(mrad):
>100  1.3  0.0  850  11.43
Magnitude and angle of 2-fold astig. (in Ang. and degrees):
>0 0
Magnitude and angle of 3-fold astig. (in Ang. and degrees):
>0 0
Type 1 for smooth aperture:
>0
Probe position in Angstroms:
>20 20
electron wavelength = 0.0370144 Angstroms
there were 317 pixels inside the aperture
Pix range       -1.503085 to       3.224012 real,
      and       -2.465231 to        13.6526 imaginary

CPU time = 0.193000 sec

<---- run program autoslic ---->

autoslic(e) version dated 3-jul-2008 ejk
Copyright (C) 1998, 2008 Earl J. Kirkland
This program is provided AS-IS with ABSOLUTELY NO WARRANTY
  under the GNU general public license

perform multislice with automatic slicing

Name of file with input atomic potential in x,y,z format:
>si100.xyz
Replicate unit cell by NCELLX,NCELLY,NCELLZ :
>6 6 40
Name of file to get binary output of multislice result:
>simul.tif
Do you want to include partial coherence (y/n) :
>n
NOTE, the program image must also be run.
Do you want to start from previous result (y/n) :
>y
Name of file to start from:
>siprobe.tif
Crystal tilt x,y in mrad.:
>0 0
Slice thickness (in Angstroms):
>1.3575
Do you want to record the (real,imag) value
 of selected beams vs. thickness (y/n) :
>n
```

```
Do you want to include thermal vibrations (y/n) :
>y
Type the temperature in degrees K:
>100
Random number seed initialized to 1258514039
Do you want to output intensity vs. depth cross section (y/n) :
>n
Starting pix range -1.50308 to 3.22401 real
                   -2.46523 to 13.6526 imag
Beam voltage = 100 kV
Old crystal tilt x,y = 0, 0 mrad
electron wavelength = 0.0370144 Angstroms
11520 atomic coordinates read in
one unit cell of 100 silicon
Size in pixels Nx, Ny= 512 x 512 = 262144 beams
Lattice constant a,b =      32.5800,      32.5800
Total specimen range is
 0 to 31.2225 in x
 0 to 31.2225 in y
 0 to 215.842 in z
Range of thermal rms displacements (300K) = 0.076 to 0.076
Bandwidth limited to a real space res. of 0.190898 Angstroms
   (= 193.90 mrad)  for symmetrical anti-aliasing.
Sorting atoms by depth...
Thickness range with thermal displacement
                          is -0.125289 to 215.948 (in z)
fit from r= 0.01 to r= 5
z= 1.018125 A, 91529 beams, 72 coord.,
     aver. phase= 0.017376, total intensity = 0.999993
z= 2.375625 A, 91529 beams, 72 coord.,
     aver. phase= 0.017374, total intensity = 0.999986
z= 3.733125 A, 91529 beams, 72 coord.,
     aver. phase= 0.017377, total intensity = 0.999981
z= 5.090625 A, 91529 beams, 72 coord.,
     aver. phase= 0.017374, total intensity = 0.999945
z= 6.448125 A, 91529 beams, 72 coord.,
     aver. phase= 0.017374, total intensity = 0.999933
z= 7.805625 A, 91529 beams, 72 coord.,
     aver. phase= 0.017377, total intensity = 0.999911
z= 9.163125 A, 91529 beams, 72 coord.,
     aver. phase= 0.017371, total intensity = 0.999905
z= 10.520625 A, 91529 beams, 72 coord.,
     aver. phase= 0.017372, total intensity = 0.999858
z= 11.878125 A, 91529 beams, 72 coord.,
     aver. phase= 0.017376, total intensity = 0.999845
z= 13.235625 A, 91529 beams, 72 coord.,
     aver. phase= 0.017370, total intensity = 0.999828
z= 14.593125 A, 91529 beams, 72 coord.,
     aver. phase= 0.017375, total intensity = 0.999820
z= 15.950625 A, 91529 beams, 72 coord.,
     aver. phase= 0.017378, total intensity = 0.999742
    :                             :
z= 208.715625 A, 91529 beams, 72 coord.,
     aver. phase= 0.017370, total intensity = 0.988574
```

```
z= 210.073125 A, 91529 beams, 72 coord.,
     aver. phase= 0.017375, total intensity = 0.988555
z= 211.430625 A, 91529 beams, 72 coord.,
     aver. phase= 0.017375, total intensity = 0.988384
z= 212.788125 A, 91529 beams, 72 coord.,
     aver. phase= 0.017375, total intensity = 0.988357
z= 214.145625 A, 91529 beams, 72 coord.,
     aver. phase= 0.017373, total intensity = 0.988294
z= 215.503125 A, 91529 beams, 72 coord.,
     aver. phase= 0.017374, total intensity = 0.988274
z= 216.860625 A, 91529 beams, 72 coord.,
     aver. phase= 0.017374, total intensity = 0.988163
pix range -6.31375 to 4.76657 real,
          -29.1689 to 8.22712 imag
Total CPU time = 25.277000 sec.


<---- run program image ---->

image version dated 3-jul-2008 (ejk)
Copyright (C) 1998, 2008 Earl J. Kirkland
This program is provided AS-IS with ABSOLUTELY NO WARRANTY
  under the GNU general public license

calculate TEM images with defocus

Name of file with input multislice result:
>simul.tif
Type 0 for coherent real space image,
  or 1 for partially coherent real space image,
  or 2 for diffraction pattern output:
>2
Name of file to get diffraction pattern:
>sicbed.tif
Do you want to include central beam (y/n) :
>y
Do you want to impose the aperture (y/n) :
>n
Type 0 for linear scale,
  or 1 to do logarithmic intensity scale:
  or 2 to do log(1+c*pixel) scale:
>1
Starting pix energy =   100.00 keV
Starting pix range -6.31375 4.76657 real
                   -29.1689 8.22712 imag
Pix range 9.531965 to 19.695755
Elapsed time = 0.120000 sec
```

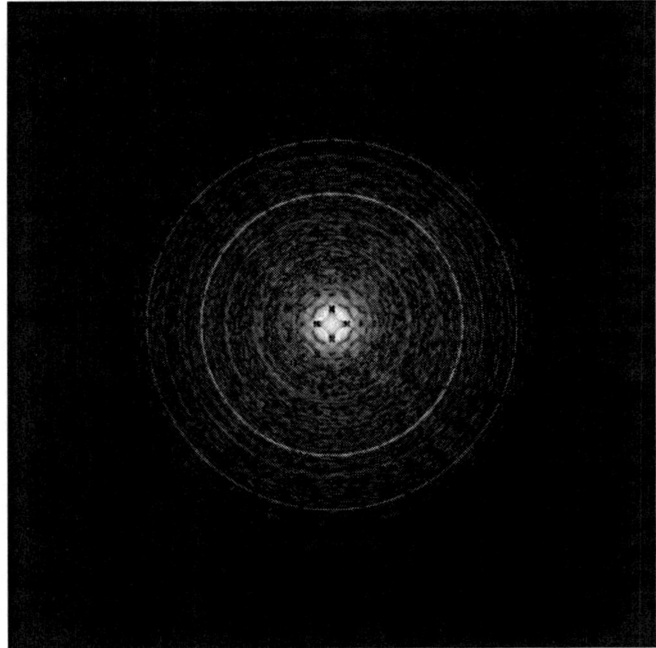

Fig. 8.3 The results of the *autoslic* for a convergent beam diffraction pattern of 100 silicon. The image intensities are shown on a logarithmic scale

8.6.2 Scanned Beam Calculation

The program *autostem* calculates a high-resolution ADF-STEM or scanning confocal image of a specimen that is described by a list of atoms and their positions. It is a combination of *autoslic* and *stemslic* with a few extra tricks to dramatically reduce the CPU time and make it easier to produce ADF STEM images with the frozen phonon approximation. A large fraction of the CPU time is used in the calculation of the transmission functions for each slice, which are all different with phonons. *stemslic* reuses many identical slices to reduce CPU time, but this does not work with the frozen phonon method or amorphous specimens. *autostem* reduces the CPU time by reusing the transmission function for several probes (using lots of CPU memory). For an image it will do a whole line of the image at one time using the same transmission function. The statistics seem to work out if there are enough phonon configurations (i.e., adjacent probe positions are not really independent for phonons). This produces a large improvement in speed, making this calculation practical (close to the same time for one phonon configuration as *stemslic*). *autostem* also does all of the averaging (over phonon configurations) in memory with a single simple output file rather than requiring the user to calculate the average with a separate program (as with *autoslic*). This process is generally transparent to the user (except for being much faster). For example, if you calculate a 128×128 image, *autostem* will propagate 128 probes (each 512×512 typically) at once (using a lot more memory) and repeat for 128 lines and also repeat both steps for each phonon configuration.

This program can save multiple thickness in one run to avoid multiple runs for intermediate thickness and includes higher order spherical aberration (Cs3 and Cs5) for an aberration corrected microscope. The thickness levels are measured from the beginning of the specimen (unit cell). It propagates many probes at once which is much faster but uses much more memory (its not hard to run out of memory on a 32 bit computer). In addition this program is multi-threaded. If running on a computer with more than one processor in a shared memory configuration (SMP) the program will use all available CPUs (up to one per probe) to reduce the total computation time (using openMP). Although it may still take a few days of CPU time for some specimens it is much faster than other approaches (for ADF-STEM with TDS) and can run by itself most of the time.

Confocal mode has a lot in common with ADF-STEM so this program can calculate confocal results by just adding an extra lens on the exit.

8.6.2.1 Atomic Coordinates

This program reads in a 3D atomic coord. in *xyz* format (Table 8.4), automatically slices the specimen (as *autoslic*), calculates ADF-STEM images and line scans (as *stemslice*), and integrates over phonon configurations. The atomic coordinates are randomly displaced to simulate thermal motion (the frozen phonon approx.) and then sorted by z as in *autoslic*. The calculation starts slightly before the beginning of the specimen (lowest z) with the specified defocus value, and continues until slightly past the specimen (covering a distance slightly larger that the super cell of the specimen).

8.6.2.2 Source Size

autostem can also include the effects of a nonzero source size as part of the Monte-Carlo averaging process. The source size is the physical size of the image of the tip on the specimen (actually the virtual source size appropriately demagnified) not the aberration limited probe size (which is already taken care of with Cs3, Cs5, etc). *autostem* just adds a random offset to the probe positions during the phonon averaging process. This approach is not really that efficient and may require a large number of samples to get a smooth function. It is probably better to calculate a whole 2D image and convolve with a Gaussian source distribution (low pass filter) but this also requires a lot of computer time (2D vs. 1D).

8.6.2.3 Sample Run

Below is an example of running *autostem* in one mode. Lines beginning with > are typed in by the user (without the >). The other lines are printed by the program. There are many (almost too many) different modes of this program. It can produce a single line scan or a whole image. There is also an xz mode in which the line scan is saved at all thickness levels and output in image form, which is a convenient visual representation of the fringe contrast vs. depth. This example uses only two phonon configurations for simplicity which is not sufficient. In practice something like 10–20 configuration should be used (averaged over). Figure 8.4 shows the results for xz mode (other figures in this book use other modes). Two detector geometries are for ADF-STEM and one is for a confocal detector, which adds some extra questions on the collector lens and a little CPU time for this extra step. The initial point of the line scan is (x_i, y_i) and the final point is (x_f, y_f). Several output files are generated (different thickness levels, different detectors etc.). The output ".txt" file lists which files correspond to which parameters. The ".dat" file has the line scan data and the ".tif" files have image data.

```
autostem (ADF,confocal) version dated 4-aug-2009 (ejk)
Copyright (C) 1998,2008,2009 Earl J. Kirkland
This program is provided AS-IS with ABSOLUTELY NO WARRANTY
  under the GNU general public license

multithreaded using openMP
calculate STEM images

Name of file with input atomic potential in x,y,z format:
>si100.xyz
Replicate unit cell by NCELLX,NCELLY,NCELLZ :
>5 5 38
STEM probe parameters, V0(kv), Cs3(mm), Cs5(mm),
                          df(Angstroms), apert1,2(mrad) :
>200 1 0 450 0 10
Magnitude and angle of 2-fold astig. (in Ang. and degrees):
>0 0
Magnitude and angle of 3-fold astig. (in Ang. and degrees):
>0 0
wavelength = 0.025079 Angstroms
Size of specimen transmission function Nx,Ny in pixels :
>512 512
Size of probe wave function Nx,Ny in pixels :
>512 512
Crystal tilt x,y in mrad. :
>0 0
Do you want to calculate a 1D line scan (y/n) :
>y
Do you want to save all depth inform. as xz image (y/n) :
>y
File name prefix to get output of STEM multislice result
                                          (no extension):
>test1dbig
```

```
Number of detector geometries (>=1):
>3
Detector   1, type: min max angles(mrad) or radius(Ang.)
 followed by m or A
>50 200 m
normal ADF detector
Detector   2, type: min max angles(mrad) or radius(Ang.)
 followed by m or A
>80 200 m
normal ADF detector
Detector   3, type: min max angles(mrad) or radius(Ang.)
 followed by m or A
>0  3 A
confocal detector
Collector lens parameters, Cs3(mm), Cs5(mm),
                           df(Angstroms), apert1,2(mrad)  :
>1.3 0 700 0 10.37
Magnitude and angle of 2-fold astig. (in Ang. and degrees):
>0 0
Magnitude and angle of 3-fold astig. (in Ang. and degrees):
>0 0
xi, xf, yi, yf, nout :
>1 11 5.43 5.43 64
Slice thickness (in Angstroms):
>1.36
Do you want to include thermal vibrations (y/n) :
>y
Type the temperature in degrees K:
>300
Type number of configurations to average over:
>2
Random number seed initialized to 1259269207
Type source size (FWHM in Ang.):
>0.1
electron wavelength = 0.0250793 Angstroms
7600 atomic coordinates read in
one unit cell of 100 silicon
Lattice constant a,b,c =
             27.1500,      27.1500,     206.3400
save up to 152 thickness levels
Total specimen range is
 0 to 25.7925 in x
 0 to 25.7925 in y
 0 to 204.982 in z
Range of thermal rms displacements (300K) = 0.076 to 0.076
Bandwidth limited to a real space res. of 0.159082 Angstroms
   (= 157.65 mrad)  for symmetrical anti-aliasing.
Number of symmetrical anti-aliasing beams in probe = 91529
configuration # 1
The new range of z is -0.163809 to 205.137
specimen range is 0 to 206.34 Ang.
slice ending at z= 1.02 Ang. with 50 atoms
fit from r= 0.01 to r= 5
save ADF/confocal signals, thickness level 0
```

```
slice ending at z= 2.38 Ang. with 50 atoms
save ADF/confocal signals, thickness level 1
slice ending at z= 3.74 Ang. with 50 atoms
save ADF/confocal signals, thickness level 2
slice ending at z= 5.1 Ang. with 50 atoms
save ADF/confocal signals, thickness level 3
slice ending at z= 6.46 Ang. with 50 atoms
save ADF/confocal signals, thickness level 4
slice ending at z= 7.82 Ang. with 50 atoms
save ADF/confocal signals, thickness level 5
slice ending at z= 9.18 Ang. with 50 atoms
save ADF/confocal signals, thickness level 6
slice ending at z= 10.54 Ang. with 50 atoms
save ADF/confocal signals, thickness level 7
slice ending at z= 11.9 Ang. with 50 atoms
save ADF/confocal signals, thickness level 8
   :                 :
save ADF/confocal signals, thickness level 148
slice ending at z= 203.66 Ang. with 52 atoms
save ADF/confocal signals, thickness level 149
slice ending at z= 205.02 Ang. with 47 atoms
save ADF/confocal signals, thickness level 150
slice ending at z= 206.38 Ang. with 16 atoms
save ADF/confocal signals, thickness level 151
configuration # 2
The new range of z is -0.0966492 to 205.187
specimen range is 0 to 206.34 Ang.
slice ending at z= 1.02 Ang. with 50 atoms
   :                      :
slice ending at z= 200.94 Ang. with 52 atoms
save ADF/confocal signals, thickness level 147
slice ending at z= 202.3 Ang. with 50 atoms
save ADF/confocal signals, thickness level 148
slice ending at z= 203.66 Ang. with 52 atoms
save ADF/confocal signals, thickness level 149
slice ending at z= 205.02 Ang. with 45 atoms
save ADF/confocal signals, thickness level 150
slice ending at z= 206.38 Ang. with 18 atoms
save ADF/confocal signals, thickness level 151
output file= test1dbig.dat
output file= test1dbig000.tif
test1dbig000.tif: output pix range : 2.44786e-005 to 0.0437663
output file= test1dbig001.tif
test1dbig001.tif: output pix range : 8.29768e-006 to 0.0188693
output file= test1dbig002.tif
test1dbig002.tif: output pix range : 0.182288 to 0.621157
Number of symm. anti-aliasing beams in trans. function = 91529
The total integrated intensity range was:
   0.992999 to 0.997616

CPU time = 1101.8 sec.
wall time = 1101.86 sec.
```

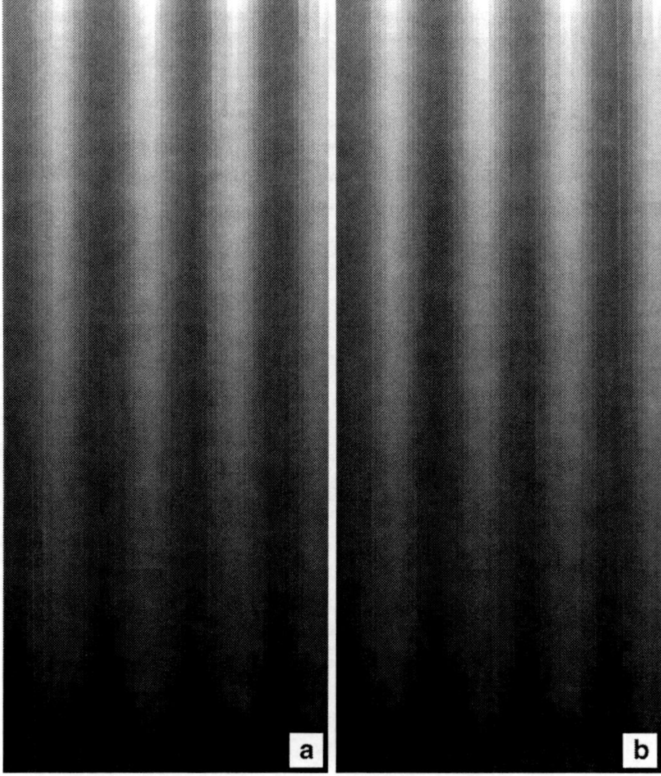

Fig. 8.4 Example of *autostem* in *xz* mode (line scan vs. thickness), with the start of the specimen (*thin*) at the bottom and the end (*thick part*) at the top. This illustrates how the lattice intensity increases with thickness. The horizontal and vertical scales may be different. In this case, the scale is 10Å in the horizontal directions and 206Å in the vertical direction. (a) First ADF detector and (b) second ADF detector

8.7 Program Display

The program *display* can read the data files produced by the TEMSIM programs and convert the image into readable ASCII data or postscript format for printing. It can extract a single line scan through the image and print it out in a form than many data plotting programs can read or it can display a 2D image. The 2D ASCII form generates a large file but the data an be read by other data plotting programs. An example is shown below.

```
<---- run program display ---->

display version dated 3-jul-2008 (ejk)
Copyright (C) 1998, 2008 Earl J. Kirkland
This program is provided AS-IS with ABSOLUTELY NO WARRANTY
  under the GNU general public license
```

```
generate display of TEMSIM images
The available image output modes are:
    code  mode
     1    print single (1D) line of image data as text
     2    ASCII text file of 2D image
     3    postscript (EPS) greyscale in file
     4    postscript (EPS) contour plot in file
Enter code number:
>3
Name of file that has binary data to display:
>sicbed.tif
Name of output file:
>sicbed.eps
Size of old image in pixels, Nx, Ny = 512, 512
  created 2009:11:17 22:29:59
Pix goes from 9.428522 to 19.675783
Lattice constants are: ax, by = 15.715163, 15.715163
Do you want to rescale greyscale range (y/n) :
>n
Do you want to complement image (y/n) :
>n
Please print file sicbed.eps.
```

8.8 Program Slicview

The program *slicview* will draw a 3D hard sphere model of the specimen to help debug the data files describing the specimen. The method used is described in appendix E and a sample run is shown later. The program can create a TIFF file or an EPS file. The viewing distance is in the same units as the atomic coordinates and controls the visual depth. The sphere size is relative to the width of the final image (0.10 means 10% of the width of the final image).

```
sliceview C version dated 3-jul-2008 ejk
Copyright (C) 1998, 2008 Earl J. Kirkland
This program is provided AS-IS with ABSOLUTELY NO WARRANTY
  under the GNU general public license

Create a 3d perspective view of a multislice specimen

Type 1 for atompot input format,
  or 2 for XYZ autoslic input format:
>1

Name of output file to get 3D perspective view of
                                    multislice specimen:
>srt3d.eps
Type 1 for TIFF output or 2 for EPS output.
>2
EPS size in inch. (real) xsize ysize :
```

```
>4 4
Type viewing distance and sphere size:
>70 0.10
Type rotation, tilt angle in degrees:
>0 0
maximum number of slices = 1000
Type in the stacking sequence :
>3(ab)
Type in the names of 2 atompot description files.
Name of file with input crystal data for layer a :
>srta.dat
replicate layer a unit cell by ncellx,ncelly:
>3 3
Name of file with input crystal data for layer b :
>srtb.dat
replicate layer b unit cell by ncellx,ncelly:
>3 3
Lattice constants = 3.905100  3.905100  1.952500 Angstroms
Lattice constants = 3.905100  3.905100  1.952500 Angstroms
Total specimen thickness = 11.715000
Total number of atoms = 135
with a total occupancy of 135.000000
Sorting atoms by depth...
Drawing atoms...
CPU time = 0.017000 sec
```

Appendix A
Plotting Transfer Functions

Abstract This appendix gives several simple MATLAB scripts to interactively plot the transfer function for CTEM and STEM. This is a simple and relatively easy to use approach to investigate the transfer function.

The following scripts interactively calculate and plot transfer functions for the CTEM and STEM on the screen or in publication quality hardcopy using Matlab (distributed and trademarked by: The MathWorks, Inc., www.mathworks.com). Matlab is relatively easy to use and provides a graphical output on many popular computers in a nearly machine independent manner. It is a complete programming language and has a variety of sophisticated mathematical functions and procedures. Matlab's ease of use comes at a price however. It is mostly an interpreted language (newer versions have a just-in-time compiler that improves performance) with the inherent speed penalty. However Matlab's fundamental operations are on matrices and vectors. If the problem is vectorized (i.e., the operation are on a whole array or vector of numbers at one time) then the performance penalty typically associated with an interpreted language is partially overcome. Matlab is well suited for small to medium calculations (such as calculating and plotting transfer functions) but probably should not be used for large numerical simulation.

Each of the three MATLAB programs ctemtf.m, stempsf.m and stemmtf.m should be called directly from the MATLAB command line. All files should be in the default directory. The MATLAB functions do not need to be called directly but are called from the other three programs. Each program first asks for the electron optical parameters. Then it calculates and plots the appropriate function on the screen. The STEM programs may take a significantly longer time because they must calculate a Fourier-Bessel transform. Once the graph appears on the screen it can be printed from the command line using the standard Matlab print command.

E.J. Kirkland, *Advanced Computing in Electron Microscopy*,
DOI 10.1007/978-1-4419-6533-2_9, © Springer Science+Business Media, LLC 2010

A.1 CTEM

There is one Matlab programs and one Matlab functions (each is a separate file). A sample output of each program is shown later (Fig. A.1) along with the source listing.

- **ctemtf.m** Plot CTEM transfer function (calls ctemh.m below).
- **ctemh.m** Calculate the CTEM transfer function (3.40).

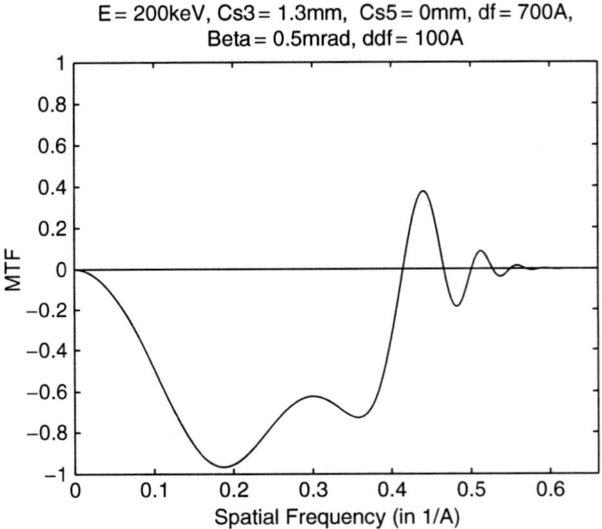

Fig. A.1 Example of CTEM transfer function from the MATLAB program **ctemtf.m**

ctemtf.m

```
%
%  MATLAB script ctemtf.m to plot CTEM transfer functions
%  this script calls ctemh.m
%
%  Cs3,5 = Spherical Aberration
%  df = defocus
%  kev = electron energy in keV
%  ddf = chromatic aberation defocus spread
%  beta =  spread in illumination angles
%
disp( 'Plot CTEM transfer function' );
p.kev  = input('Type electron energy in keV : ');
p.Cs3  = input('Type spherical aberration Cs3 in mm : ');
p.Cs5  = input('Type spherical aberration Cs5 in mm : ');
p.df   = input('Type defocus df in Angstroms : ');
p.ddf  = input('Type defocus spread ddf in Angs. : ');
```

```
p.beta = input('Type illumination semiangle in mrad : ');
%
% electron wavelength
wav = 12.3986/sqrt((2*511.0+p.kev)*p.kev);
Cs = abs(p.Cs3);
if( Cs < 0.1 )
    Cs = 0.1;
end
ds = sqrt( sqrt( Cs*1.0e7*wav*wav*wav ));
kmax = 2.5/ds;
k = 0.:(kmax/500):kmax;  % 500 points
sinw = ctemh( k, p, 0 );
plot( k, sinw  );
axis([0, kmax, -1, +1]);
xlabel( 'Spatial Frequency (in 1/A)');
ylabel( 'MTF' );
s1 = sprintf('E= %gkeV, Cs3= %gmm, ', p.kev, p.Cs3);
s2 = sprintf(' Cs5= %gmm, df= %gA,  ', p.Cs5, p.df);
s3 = sprintf('Beta= %gmrad, ddf= %gA', p.beta, p.ddf);
title([s1 s2 s3]);
hold on;         % plot line through zero
x = [0, kmax];
y = [0, 0];
plot( x, y );
hold off;
```

ctemh.m

```
function y = ctemh(k,params,type)
%
%  MATLAB function ctemh.m to calculate CTEM bright
%  field phase contrast transfer function with partial
%  coherence for weak phase objects
%    input array k has the spatial freq. values (in 1/A)
%    input array params has the optical parameters
%          params = [Cs, df, kev, ddf, beta]
%    input type = 0 for phase contrast
%           and 1 for amplitude contrast
%    output array contains the transfer function vs k
%
%  params.Cs3,5  = spherical aberration (in mm)
%  params.df    = defocus (in Angstroms)
%  params.kev   = electron energy (in keV)
%  params.ddf   = chrom. aberr. def. spread (in Angst.)
%  params.beta  = spread in illum. angles (in mrad)
%
%  reference
%  R. H. Wade and J. Frank, Optik 49 (1977) p.81
%
Cs3 = params.Cs3*1.0e7;
Cs5 = params.Cs5*1.0e7;
df = params.df;
kev = params.kev;
```

```
ddf = params.ddf;
beta = params.beta*0.001;
mo = 511.0;       % electron rest mass in keV
hc = 12.3986;     % in keV-Angstroms
wav = (2*mo)+kev;
wav = hc/sqrt(wav*kev);
wavsq = wav*wav;
w1 = pi*Cs3*wavsq*wav;
w2 = pi*wav*df;
w3 = pi*Cs5*wavsq*wavsq*wav;
e0 = (pi*beta*ddf)^2;
k2 = k .* k;
wr = ((w3.*k2+w1).*k2-w2).*k*beta/wav;
wi = pi*wav*ddf.*k2;
wi = wr.*wr + 0.25.*wi.*wi;
wi = exp(-wi./(1+e0.*k2));
wr = w3*(1-2.0*e0.*k2)/3.0;
wr = wr.*k2 + 0.5*w1.*(1-e0.*k2);
wr = (wr.*k2 - w2).*k2./(1+e0.*k2);
if type == 0
    y = sin(wr).* wi;
else
    y = cos(wr).* wi;
end;
```

A.2 STEM

There are two Matlab programs and three Matlab functions (each is a separate file). The first integral over the lens aberration function is not well behaved so a Matlab adaptive quarature routine is used to gain efficiency and accuracy (in stmhr.m). A sample output of each program is shown later (Figs. A.2, A.3) along with the source listing.

- **stempsf.m** Plot the STEM probe profile (calls stemhr.m below).
- **stemtf.m** Plot the STEM transfer function (calls stemhr.m and stemhk.m below).
- **stemhr.m** Calculate the STEM probe profile (3.69, calls lens.m).
- **stemhk.m** Calculate the STEM transfer function(3.70, calls stemhr.m).
- **lens.m** Calculate the lens aberration function to use for the adaptive quadrature function.

stempsf.m

```
%  stempsf.m
%  Matlab file to plot the STEM probe profile
%  this script calls stemhr.m
%
clear;
clf;
disp( 'Plot STEM probe intensity' );
```

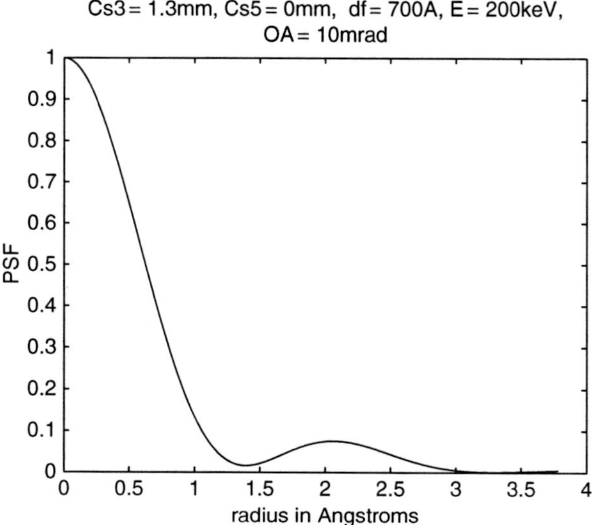

Fig. A.2 Example of STEM probe intensity profile the output from the MATLAB program **stempsf.m**

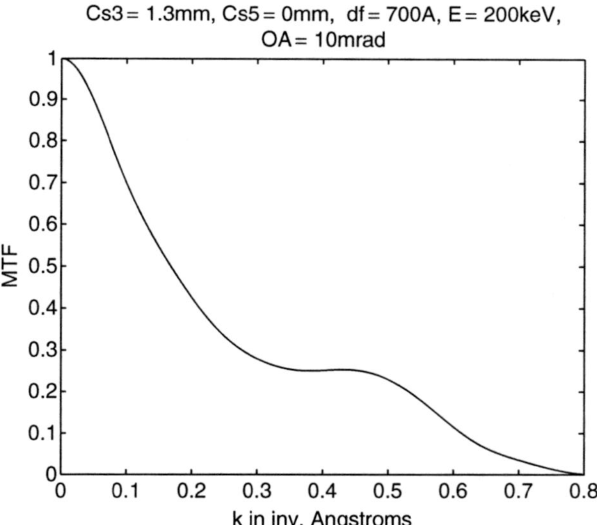

Fig. A.3 Example of the STEM transfer function output from the MATLAB program **stemtf.m**

```
p.kev  = input( 'Type electron energy in keV : ');
p.Cs3  = input( 'Type spherical aberration Cs3 in mm : ');
p.Cs5  = input( 'Type spherical aberration Cs5 in mm : ');
p.df   = input( 'Type defocus df in Angstroms : ');
p.amax = input( 'Type obj. apert. semiangle in mrad : ');
%
% electron wavelength
```

```
wav = 12.3986/sqrt((2*511.0+p.kev)*p.kev);
Cs = abs(p.Cs3);
if( Cs < 0.1 )
    Cs = 0.1;
end
rmax = sqrt( sqrt( Cs*1.0e7*wav*wav*wav ));
npts = 300;   %  number of points in curve
r = 0:(rmax/npts):rmax;
psf = stemhr( r, p );
plot( r, psf );
xlabel( 'radius in Angstroms');
ylabel( 'PSF' );
s1 = sprintf('Cs3= %gmm, Cs5= %gmm, ', p.Cs3, p.Cs5 );
s2 = sprintf(' df= %gA, ', p.df);
s3 = sprintf('E= %gkeV, OA= %gmrad', p.kev, p.amax);
title([s1 s2 s3]);
```

stemtf.m

```
%  stemtf.m
%  Matlab file to plot the STEM probe mtf
%  this script calls stemhk.m
%
clear;
clf;
disp( 'Plot STEM transfer function' );
p.kev  = input( 'Type electron energy in keV : ');
p.Cs3  = input( 'Type spherical aberration Cs3 in mm : ');
p.Cs5  = input( 'Type spherical aberration Cs5 in mm : ');
p.df   = input( 'Type defocus df in Angstroms : ');
p.amax = input( 'Type obj. apert. semiangle in mrad : ');
%
% electron wavelength
wav = 12.3986/sqrt((2*511.0+p.kev)*p.kev);
kmax = 2*0.001*p.amax/wav;
npts = 500;  % number of points
k = 0:(kmax/npts):kmax;
mtf = stemhk( k, p );
plot( k, mtf );
xlabel( 'k in inv. Angstroms');
ylabel( 'MTF' );
s1 = sprintf('Cs3= %gmm, Cs5= %gmm, ', p.Cs3, p.Cs5 );
s2 = sprintf(' df= %gA, ', p.df);
s3 = sprintf('E= %gkeV, OA= %gmrad', p.kev, p.amax);
title([s1 s2 s3]);
```

stemhr.m

```
function psf = stemhr(r,params)
%
%  MATLAB function stemhf.m to calculate
```

```
%            STEM probe profile vs. r
%    input array r has the radial positions (in Angs.)
%    input variable params has the optical parameters
%          <Cs, df, kev, amax> as elements
%    output array contains the transfer function
%
%  param.Cs3  = third order spherical aberration (in mm)
%  param.Cs5  = fifth order spherical aberration (in mm)
%  param.df   = defocus (in Angstroms)
%  param.kev  = electron energy (in keV)
%  param.amax = objective aperture  (in mrad)
%
global w2 w4 w6 intr;  % constants for lens.m
df = params.df;
kev = params.kev;
amax = params.amax*0.001;
% electron wavelength
wav = 12.3986/sqrt((2*511.0+kev)*kev);
kmax = amax/wav;
w2 = wav*pi*df;
w4 = 0.5*pi*params.Cs3*1.0e7*wav*wav*wav;
w6 = pi*params.Cs5*1.0e7*wav*wav*wav*wav*wav /3.0;
nr = length( r );
for ir=1:nr,
    intr = 2*pi*r(ir);
    tol = 1.0e-7/(0.01*ir); % sliding accuracy to speed up
    % use adaptive quadrature because integrand
    %     not well behaved
    hr(ir) = quad( @lens, 0, kmax, tol );
end;
% a little faster than abs()
psf = real(hr).^2 + imag(hr).^2;
a = max(psf);
psf = psf/a;   % norm. probe intensity to a max. of 1
```

stemhk.m

```
function mtf = stemhk( k, params )
%
%  MATLAB function stemhk.m to calculate STEM mtf vs. k
%    input array k has the spatial freq. (in inv. Angs.)
%    input variable params has the optical parameters
%          [Cs, df, kev, amax] as elements
%    output array contains the transfer function
%
%  param.Cs3  = third order spherical aberration (in mm)
%  param.Cs5  = fifth order spherical aberration (in mm)
%  param.df   = defocus (in Angstroms)
%  param.kev  = electron energy (in keV)
%  param.amax = objective aperture  (in mrad)
%
Cs3 = params.Cs3;
```

```
df = params.df;
kev = params.kev;
amax = params.amax*0.001;
%  first calculate the psf using stemhr()
nr = 500;  % number of points in integral over r
wav = 12.3986/sqrt((2*511.0+kev)*kev); % elect. wavelength
Cs = abs(Cs3);
if( Cs < 0.1 )  % guess if Cs3=0
    Cs = 0.1;
end
rmax = 2.0*sqrt( sqrt( Cs*1.0e7*wav*wav*wav) );
r = 0:(rmax/nr):rmax;
psf = stemhr( r, params );
% next inverse psf to get mtf
nk = length( k );
for ik=1:nk,
h = psf .* besselj( 0, 2*pi*r*k(ik) ) .*r;
mtf(ik) = sum(h);
end;
a = mtf(1);
mtf = mtf/a;  % normalize mtf(0)=1
```

lens.m

```
function expchi = lens(k)
%
%  dummy function to integrate (used by stempsf.m)
%  MATLAB function lens.m to calculate complex
%                               aberr. function
%    input k (in 1/Angs.), wav = electron wavelength
%
%  chi = pi*wav*k^2*[ 0.5*Cs3*wav^2*k^2
%                    + (1/3)*Cs5*wav^4*k^4 - df ]
%  return exp( -i*chi )
%
% globals:
%   w2 = pi*defocus*wav
%   w4 = 0.5*pi*Cs3*wav^3
%   w6 = (1/3)*pi*Cs5*wav^5
%   intr = 2*pi*r
%
global w2 w4 w6 intr;  % constants from lenhr.m
k2 = k.*k;
w = ( (w6.*k2 + w4) .*k2 - w2 ).*k2;
expw = exp( -i*w );
expchi = expw .* besselj( 0, intr.*k ).*k;
```

Appendix B
The Fourier Projection Theorem

Abstract This appendix derives a property of a two-dimensional Fourier transforms that can be used to generate the projected atomic potential in two dimensions of a three-dimensional atomic potential.

It seems like a two-dimensional inverse Fourier transform should never be applied to a function of three dimensions. However, the inverse two-dimensional transform of a three-dimensional object will end up producing the desired inverse transform with respect to two of the dimensions and integrating (or projecting) along the third dimension.

To see that this is true, define a function $f(x,y,z)$ that is a function of three spatial coordinates x,y,z. The three-dimensional Fourier transform of this function will be $F(k_x,k_y,k_z)$ which is a function of three reciprocal space coordinates k_x,k_y,k_z.

$$\begin{aligned} F(k_x,k_y,k_z) &= \mathrm{FT}_{3D}[f(x,y,z)] \\ &= \int_{x,y,z} f(x,y,z)\exp[2\pi \mathrm{i}(xk_x+yk_y+zk_x)]\mathrm{d}x\mathrm{d}y\mathrm{d}z \end{aligned} \quad \text{(B.1)}$$

Next calculate the inverse Fourier transform in two dimensions of the three-dimensional object $F(k_x,k_y,k_z)$:

$$\mathrm{FT}_{2D}^{-1}[F(k_x,k_y,k_z)] = \int_{k_x,k_y} F(k_x,k_y,k_z)\exp[-2\pi \mathrm{i}(xk_x+yk_y)]\mathrm{d}k_x\mathrm{d}k_y \quad \text{(B.2)}$$

and then substitute the original expression for $F(k_x,k_y,k_z)$ from (B.1) using dummy variable r,s,t in place of x,y,z for the second occurrence of the spatial coordinates inside the integrand:

$$\begin{aligned} \mathrm{FT}_{2D}^{-1}[F(k_x,k_y,k_z)] = \int_{k_x,k_y} &\left\{ \int_{r,s,t} f(r,s,t)\exp[2\pi \mathrm{i}(rk_x+sk_y+tk_z)]\mathrm{d}r\mathrm{d}s\mathrm{d}t \right\} \\ &\times \exp[-2\pi \mathrm{i}(xk_x+yk_y)]\mathrm{d}k_x\mathrm{d}k_y \end{aligned}$$

E.J. Kirkland, *Advanced Computing in Electron Microscopy*,
DOI 10.1007/978-1-4419-6533-2_10, © Springer Science+Business Media, LLC 2010

$$= \int_{r,s,t} f(r,s,t) \left\{ \int_{k_x} \exp[2\pi \mathrm{i}(r-x)k_x]\mathrm{d}k_x \right\}$$
$$\times \left\{ \int_{k_y} \exp[2\pi \mathrm{i}(s-y)k_y]\mathrm{d}k_y \right\} \exp[2\pi \mathrm{i}tk_z]\mathrm{d}r\mathrm{d}s\mathrm{d}t$$
$$= \int_{r,s,t} f(r,s,t)\delta(r-x)\delta(s-y)\exp[2\pi \mathrm{i}tk_z]\mathrm{d}r\mathrm{d}s\mathrm{d}t$$
$$= \int_t f(x,y,t)\exp[2\pi \mathrm{i}tk_z]\mathrm{d}t \tag{B.3}$$

where $\delta(x)$ is the Dirac delta function. Finally, if $k_z = 0$ and z is substituted for the dummy variable t this leaves:

$$\mathrm{FT}_{2D}^{-1}[F(k_x,k_y,k_z=0)] = \int_z f(x,y,z)\mathrm{d}z. \tag{B.4}$$

Therefore, if a two-dimensional inverse Fourier transform is applied to a function of three dimensions the result is an inverse transform over the appropriate two dimensions and a projection (or integral) over the third dimension if the missing third reciprocal space coordinate (k_z above) is set to zero.

Appendix C
Atomic Potentials and Scattering Factors

Abstract This appendix gives a detailed listing of the atomic scattering factors of all atoms in the periodic chart. This data is used extensively to calculate electron microscope images.

The projected atomic potential of the atoms in the specimen is a necessary starting point for the calculation of an electron microscope image. The basic principles of quantum mechanics enable a well defined calculation of the atomic potentials for single isolated atoms. The potentials for all of the atoms in the periodic chart have been calculated and the results are tabulated later. Treating the specimen as a collection of single isolated atoms neglects the change in electronic structure due to bonding etc. in the solid. This should be a small effect because electron scattering is mainly from the nucleus with the core and valence electrons screening the nucleus. However the low angle scattering may be in error due to this approximation. This might give rise to problems with phase contrast bright field images however high angle annular dark field (STEM) images should be more accurate. A rigorous calculation of the electronic structure of solids including bonding is currently an active area of research. Including the effects of bonding in a rigorous and general manner in the calculation of electron microscope images is beyond the capability of generally available computers at present and is not considered here.

Herman and Skillman [155] tabulated the electron wave functions resulting from Hartree-Fock calculations. The atomic scattering factors for X-rays and electrons have been tabulated in many places and a representative sample is given in Table C.1. The data given by Doyle and Turner [82] and Doyle and Cowley [81] (using the program of Coulthard [56]) is generally considered to be the most accurate. However, Doyle and Turner do not calculate the whole periodic chart and Doyle and Cowley do not calculate the scattering factors for high angles. Rez et al. [298–300] have recently published a new set of scattering factors that covers the whole periodic chart and includes high angle scattering. The scattering factor table presented later is most similar to that of Doyle and Turner [82], Doyle and Cowley [81] and Rez et al. [298, 299] (i.e., all of these use a relativistic Hartree-Fock calculation) and

E.J. Kirkland, *Advanced Computing in Electron Microscopy*,
DOI 10.1007/978-1-4419-6533-2_11, © Springer Science+Business Media, LLC 2010

covers the whole periodic chart including high scattering angles. Bonham and Fink [32] have also reviewed electronic structure calculations for electron scattering in an energy range appropriate for electron microscopy.

Table C.1 Some tabulations of atomic potentials and scattering factors

Author	Year	Quantity	Z	Comments	
Bragg and West [38]	1929	fx	8,9,11–14, 17,19,20,26	TF,H	
James and Brindley [185]	1931	fx	1–22,29,37	TF,H	
McWeeny et al. [238]	1951	fx	1–10	analytical	
Vand et al. [359]	1957	fx	1–100	param.	
Ibers [168]	1958	fe	1–12,18, 20,25–80	TFD, H, HF	
Freeman [112, 113]	1959	fx	3,4,6–11,13,14 17,19,20,22 23,25,26,31,81	H, HF	
Forsyth and Wells [108]	1959	fx	1–92	param.	
Dawson [74]	1961	fe	9–20,36	HF	
Ibers and Vainshtein [169]	1962	fx	1–104	HF,TF	
Smith and Burge [326]	1962	fe	1–18,20-104	param.	
Cromer and Waber [72, 73]	1965	fx	2–103	RHF, TF	TF = Thomas Fermi
Mott and Massey [256]	1965	pce	3–36, 47,74,80	HF	
Cox and Bonham [67]	1967	fe	1–54	param.	
Cromer and Mann [71]	1968	fx	2–103	HF	
Doyle and Turner [82]	1968	fx,fe	2–38,42, 47–56,63,79, 80–83, 86,92	RHF	
Hasse [136–138]	1968	fe	1–92	HF, TF	
Doyle and Cowley [81]	1974	fe	2–98	HF, RHF	
Fox et al. [109]	1989	fx	2–98	param.	
Weickenmeier and Kohl [376]	1991	fx	2-98	param.	
Rez et al. [298–300]	1994	fx	2–92	RHF	
Peng et al. [281]	1996	fe	1–98	param.	
Wang et al. [364]	1996	fx	2–18	RHF	
Su and Coppens [338]	1997	fx	1–54	RHF, param.	

TF = Thomas Fermi, TFD = Thomas-Fermi-Dirac, H = Hartree, HF = Hartree-Fock, RHF = Relativistic Hartree, fx = X-ray, fe = electron scattering factor, pce = partial cross section(electrons), param. = parameterized

C.1 Atomic Charge Distribution

The distribution of charge in an atom must be found from a quantum mechanical description of the electrons and nucleus of the atom. Unfortunately, the hydrogen atom is the only element in the periodic chart that can be solved analytically. With more than one electron the atom becomes a many-body problem and requires a numerical solution with suitably approximations. The Hartree-Fock procedure (see for example Hartree [147], Froese-Fischer [117], Froese-Fischer et al. [106] and Cowan [57]) is a method of calculating the electron wave functions of all of the electrons in an atom assuming a central potential model. It is a variational calculation to minimize the total energy of the many-electron atomic system, and includes the interaction of the electrons with each other and with the nucleus. The total wave function is fashioned so that it is antisymmetric on the exchange of identical electrons, which leads to the so-called exchange terms. Hartree-Fock starts with an initial guess of the electron wave functions (such as the hydrogenic analytical form). From this guess a net potential for each electron orbital is calculated due to the other electrons and the nucleus (each orbital sees a slightly different potential because it does not interact with itself). From this potential a new set of wave functions is calculated and the process repeats until a self consistent answer is obtained. Each orbital produces one coupled integral-differential equation (one equation per orbital). With the assumption of a central potential the angular integrations may be done analytically, so it is only the radial portion of the wave function that need to be calculated. The wave function is sampled on a discrete grid vs. radius.

The kinetic energy of the inner shell electrons of the heavier elements such as gold is of the order of 100 keV, which produces some relativistic effects. A relativistic form of the Hartree-Fock procedure uses the Dirac relativistic wave equation instead of the nonrelativistic Schrödinger equation. Grant [130, 131] has given a thorough discussion of relativistic Hartree-Fock theory. With the Dirac equation the wave function for each electron has two components $P(r)$ and $Q(r)$ instead of one and there is a separate orbital for each spin state of the electron. This produces roughly twice as many sets of coupled integro-differential equations to solve and generally increases the amount of computation significantly. Desclaux [77], Grant et al. [132], and Dyall et al. [86] have published extensive Dirac-Fock programs.

A program based on the average configuration theory (Grant [131]) was used to calculate the relativistic wave functions for all atoms in the periodic chart (atomic number Z = 3 to Z = 103) except hydrogen (which is known analytically). Helium (Z = 2) was calculated nonrelativistically. The atomic radial charge distribution, X-ray and electron scattering factors and the projected atomic potential were then calculated from the electron wave functions. The configurations of the electrons in each atom were obtained from standard tables (for example, Wiese and Martin [378], appendix 5 of Morrison et al. [253], Table 19.3a of Haken and Wolf [140]). The wave functions vary rapidly near the nucleus, so it is more efficient to change the independent variable from r to $t = \log r$ with equal spacings in t. This produces a fine sampling (small grid size) at small r where the wave function is changing rapidly but keeps the total number of grid points manageably small by increasing the grid size at large r where the wave function varies slowly. Each wave function was sampled with 500 points in each component (1,000 points all together). The wave functions are initially set to the relativistic hydrogenic wave functions that are known analytically (Burke and Grant [41]). At each iteration the effective electron-electron interaction is calculated from the current electron wave functions and a new set of electron wave functions is then calculated. The calculation proceeded until the energy eigenvalues changed by less than one part in 1×10^6. The minimum radius was set to $r = 1 \times 10^{-6} a_0$ and the maximum radius varied between $8a_0$ and $15a_0$

(where $a_0 = 0.5292\text{Å}$ is the Bohr radii). The energy and size ($< r^2 >$) of each orbital are in good agreement with those tabulated by Desclaux [76].

For historical reasons it was more convenient to use the radial charge density (calculated from the electron wave functions) than to produce the atomic potentials directly. First the X-ray scattering factors were calculated from the wave functions. Then the electron scattering factors and hence the projected atomic potential were obtained from the X-ray scattering factors via the Mott-Bethe formula (i.e., the electron scattering factor in the first Born approximation is the Fourier transform of the atomic potential).

The relativistic electron wave function of each atomic orbital i consists of two components $Q_i(r)$ and $P_i(r)$ of the radial portion of the Dirac wave equation and an occupancy c_i for each atomic orbital. The radial distribution of the electron charge $\rho(r)$ is calculated from the wave function as:

$$4\pi r^2 \rho(r) = \sum_i c_i \left[|Q_i(r)|^2 + |P_i(r)|^2\right]. \tag{C.1}$$

Both sides of this equation have units of electrons per Angstrom. r is the three-dimensional radial coordinate. The wave functions are sampled on an exponential grid to get more points near the nucleus where the wave function is changing rapidly. The r coordinates are defined on a grid as:

$$r_n = R_{\text{min}} e^{n\Delta t} \quad ; \quad n = 0, 1, 2, ..., (N_r - 1) \tag{C.2}$$

R_{min} is the minimum radius and and Δt is a logarithmic spacing ($t = \log(r)$) in the radial coordinate r. This is equivalent to changing the independent variable from r to $\log(r)$. Remember that $Q_i(0) = P_i(0) = 0$.

An example of the radial charge distribution of mercury (Z = 80) is shown in Fig. C.1 for both a relativistic and nonrelativistic calculation (using the average configuration theory as defined by Cowan [57]). The main effect of relativity is that the inner electron shell moves closer to the nucleus. This changes the effective screening of the nucleus and can have a large effect on the valence shell energies for heavy elements.

C.2 X-ray Scattering Factors

The X-ray scattering factor for a spherically symmetric charge distributions is defined as:

$$fx(q) = 4\pi \int r^2 \rho(r) \frac{\sin(2\pi qr)}{2\pi qr} \mathrm{d}r, \tag{C.3}$$

where $q = \sin(\alpha)/\lambda$ is the magnitude of the three dimensional wavevector that is the difference between the incident and scattered X-ray. λ is wavelength and α is the scattering semiangle. $f_x(q)$ is a dimensionless quantity corresponding to the number of electrons. Most numerical integration formulas for tabulated data

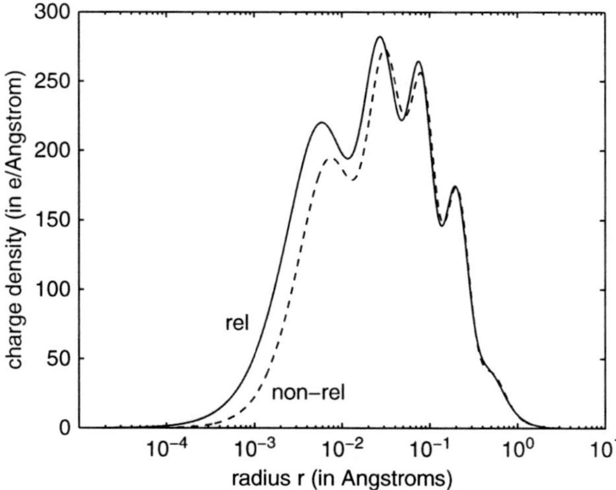

Fig. C.1 The radial electron charge density $4\pi r^2\rho(r)$ vs. radius for Hg (Z = 80) calculated relativistically (*solid line*) and nonrelativistically (*dashed line*)

assume that the data is sampled on a regularly spaced grid and not an exponential grid as used here. The integral can however be converted to a regular spacing in the independent variable by changing the integration variable from dr to $d(\ln r) = \mathrm{d}r/r$.

$$f_X(q) = \int \left[4\pi r^2\rho(r)\right] \frac{\sin(2\pi qr)}{2\pi q} d(\ln r). \tag{C.4}$$

The integrand is now sampled at regularly spaced intervals of $\Delta t = \Delta ln(r)$ and can be easily integrated numerically. Simpson's three point formula was used here.

For neutral atoms $f_X(0) = Z$ where Z is the atomic number so integrating the charge distribution to find $f_X(0)$ is only a test of the program.

C.3 Electron Scattering Factors

It is traditional to tabulate the electron scattering factor $f_e(q)$ in the first Born approximation. The scattering factor is the amplitude for scattering of a single electron by a single atom. The first Born approximation is totally inadequate for calculating electron scattering and electron microscope image (Zeitler [388, 389], Glauber and Shoemaker [121]). However, the first Born approximation is useful because it is the three dimensional Fourier transform of the atomic potential (see for example Sect. 38 of Schiff [312]):

$$\begin{aligned} f_e(\mathbf{q}) &= \frac{2\pi m_0 e}{h^2} \int V(\mathbf{r}) \exp(2\pi i \mathbf{q}\cdot\mathbf{r}) \mathrm{d}^3 r \\ &= \frac{1}{2\pi e a_0} \int V(\mathbf{r}) \exp(2\pi i \mathbf{q}\cdot\mathbf{r}) \mathrm{d}^3 r, \end{aligned} \tag{C.5}$$

where $V(\mathbf{r})$ is the 3D atomic potential of the atom, m_0 the rest mass of the electron, e the charge of the electron, h Planck's constant, and $a_0 = \hbar^2/m_0e^2 = 0.5292\text{Å}$ is the Bohr radius. $f_e(q)$ is in units of Å. For the case where the atom is spherically symmetric this reduces to:

$$f_e(q) = \frac{1}{\pi e a_0 q}\int_0^\infty V(r)\sin(2\pi qr)r\mathrm{d}r. \tag{C.6}$$

The electron scattering factor in the first Born approximation is also simply related to the X-ray scattering factor for an atom with atomic number Z using the Mott-Bethe [25, 27, 255, 256] formula:

$$f_e(q) = \frac{2m_0e^2}{h^2}\left(\frac{Z - f_X(q)}{q^2}\right) = \frac{1}{2\pi^2 a_0}\left(\frac{Z - f_X(q)}{q^2}\right) \tag{C.7}$$

(Bethe [26] has recently given an English language translation of his original German publication.) There is a singularity at $q = 0$ so the Mott-Bethe formula must be replaced by the following expression due to Ibers [168]:

$$f_e(0) = \frac{4\pi^2 m_0 e^2}{3h^2} Z < r^2 >= \frac{Z}{3a_0} < r^2 > \tag{C.8}$$

$$< r^2 > = \frac{\int_0^\infty r^2[4\pi r^2\rho(r)]\mathrm{d}r}{\int_0^\infty [4\pi r^2\rho(r)]\mathrm{d}r}, \tag{C.9}$$

where $< r^2 >$ is the mean square radius of the electrons in the atom. Note that neutral atoms also satisfy:

$$Z = \int_0^\infty [4\pi r^2\rho(r)]\mathrm{d}r \tag{C.10}$$

The Mott-Bethe formula is equivalent to solving Poison's equation in reciprocal space to obtain the potential distribution from the charge distribution (including the point charge of the nucleus). For historical reasons it is more convenient to calculate the atomic potentials in a round-about manner from the Mott-Bethe formula and the X-ray scattering factors.

If the valence shell electrons are not in the s ($\mathrm{l} = 0$) angular momentum state then the charge distribution in the atom is not necessarily spherically symmetric as assumed in the expressions for the scattering factors. McWeeny [238, 239] and Freeman [112, 113] have shown that the X-ray scattering from aspherical atoms (p-state valence shells) may vary by approximately 5–10% with azimuthal angle in low Z atoms. This difference shows up mainly at low scattering angles. The Mott-Bethe formula also implies that high angle electron scattering is mainly due to the nucleus and becomes insensitive to the X-ray scattering factor at high scattering angles. Bonding in the solid should produce a similar order of magnitude error in the electron scattering factor at small scattering angles. This 5–10% error should be regarded as an estimate of the error in image simulation produced by treating a solid as a collection of isolated (nonbonded) spherically symmetric atoms (i.e., by using the values tabulated here).

C.4 Parameterization

The electron and X-ray scattering factors for all neutral atoms with atomic numbers Z = 2 through Z = 103 were calculated using the relativistic Hartree-Fock program as outlined earlier. The results were tabulated for scattering angles $0 < q < 12\text{Å}^{-1}$ (equivalent to $0 < s < 6$ in the notation of Doyle and Turner [82]) at intervals of 0.05Å^{-1} and parameterized to make it easy to hard code into the subroutine library used for the main programs. $f_e(q)$ is in units of Å and $f_x(q)$ is in dimensionless units of electron number. A few low atomic number atoms ($2 \leq Z \leq 6$) do not scatter appreciably at this high angle so were stopped when $f_x(q) < 0.010$.

Hydrogen is the only atom that can be solved analytically. The (nonrelativistic) electron distribution for hydrogen is:

$$\rho(r) = |\psi(r)|^2 = \frac{1}{\pi a_0^3} \exp(-2r/a_0). \tag{C.11}$$

Using this expression yields:

$$fx(q) = (1+\pi^2 a_0^2 q^2)^{-2} \tag{C.12}$$

$$< r^2 > = 3a_0^2 \tag{C.13}$$

$$f_e(q=0) = a_0. \tag{C.14}$$

These analytical expressions for the scattering from hydrogen were also fit to the same parameters as the rest of the atoms in the periodic chart for completeness, although not strictly necessary.

There is a relatively large amount of data to represent the scattering factors. The simulation programs could read in a large tabulation and interpolate it to obtain the potentials or scattering factors at every required point. Parameterizing the scattering factors can considerably reduce the amount of required data and also allow for easier analytical calculations although it is not necessarily required for a simulation program. Any parameterization should have the correct asymptotic form at high and low angles. Because the electron charge distribution in the atom has a nonzero size the X-ray scattering factor $f_x(q)$ must approach zero at high angles. The Mott-Bethe formula then implies that the electron scattering factor must approach $f_e(q) \propto s^{-2}$ at high angles. The parameterization must also approach a constant value at $q = 0$. Doyle and Turner [82] had some success in fitting linear combinations of Gaussians to the tabulated data. Gaussians fit well at low scattering angles but fall off too rapidly at high angles. Weickenmeier and Kohl [376] have proposed an alternate form that has the appropriate form at high angles, but is more complicated to work with analytically. The following form has the appropriate form at large and small angles and can be transformed analytically [see (C.20)]:

$$f_e(q) = \sum_{i=1}^{N_L} \frac{a_i}{q^2 + b_i} + \sum_{i=1}^{N_G} c_i \exp(-d_i q^2), \tag{C.15}$$

where $N_L = 3$ is the number of Lorenzians (first summation) and $N_G = 3$ is the number of Gaussians (second summation). The Lorenzians have the correct behavior at high angles and the Gaussians empirically fit the behavior at low angles. It is best to parameterize $f_e(q)$ and not $f_x(q)$ because of the singularity in the Mott-Bethe formula. The inverse Mott-Bethe formula:

$$f_x(q) = Z - 2\pi^2 a_0 q^2 f_e(q) \tag{C.16}$$

is however well behaved everywhere, so if $f_e(q)$ is known then it is easy to calculate $f_x(q)$. Peng and Cowley [280] have shown that serious errors may result from applying the Mott-Bethe formula to parameterized X-ray scattering factors. It is much better to parameterize the electron scattering factors than the X-ray scattering factors.

The actual parameters (a_i, b_i, c_i, d_i) are found by performing a nonlinear least squares fit of the numerical tabulation of the X-ray scattering factor f_{xj} and electron scattering factors f_{ej} at the angle q_j to the form of $f_e(q_j)$ given in (C.15). The combined reduced χ^2 fit:

$$\begin{aligned} \chi^2 &= \frac{1}{2N_q - 2N_L - 2N_G} \sum_{j=1}^{N_q} \left\{ \left[\frac{f_{xj} - f_x(q_j)}{\sigma_{xj}} \right]^2 + \left[\frac{f_{ej} - f_e(q_j)}{\sigma_{ej}} \right]^2 \right\} \\ &= \text{minimum} \end{aligned} \tag{C.17}$$

is minimized. N_q is the number of points in the q direction and $f_x(q_j)$ is calculated from the inverse Mott-Bethe formula and $f_e(q_j)$. The effective error of each data point was set to:

$$\begin{aligned} \sigma_{xj} &= 10^{-3} f_x(0) \\ \sigma_{ej} &= 10^{-3} f_e(0) \end{aligned} \tag{C.18}$$

$f_e(q)$ and $f_x(q)$ sometimes differ by a factor or two or more (at the same q) so each must be scaled before summing in the figure of merit χ^2 to give equal weight to the X-ray and electron scattering factors. This choice of errors gives a slight preference for matching the scattering factors at low scattering angle but in practice does not seem to hinder the fit at high scattering angles either. A value of $\chi^2 \sim 1$ means than the parameterization is good to about three significant figures on average at low scattering angle (where f_e and f_x are a maximum).

The actual fit was performed with the Levenberg-Marquardt algorithm (for example see Press et al. [288]). The parameters were constrained to be positive by fitting the square root of the parameters (and then squaring them in (C.15). The

Levenberg-Marquardt algorithm is reasonably robust if it is started relatively close to the correct values of the parameters. With a large number of degrees of freedom (i.e., many free parameters) the algorithm frequently converges to an incorrect answer (i.e., it is not globally convergent). The fit was greatly improved by trying several different starting points and keeping only the best one. In practice several hundred different starting points were generated for each atomic number using a random number generator (with appropriate scaling). Also in many cases the correct parameters are very close to the parameters for other atomic numbers so each fit was also tried by starting from the best-fit parameters for all other atoms. There is one additional failure mode in which $b_i \gg 1$ and $a_i \gg 1$ but $a_i/b_i \sim 1$. This results in a constant value of f_e at large angles which is not correct. Therefore any fit with $|b_i| > 200$ was rejected (in practice only a small number of cases produce this response). The resulting parameters fit the tabulated values very well in most cases but it should be noted that the parameters are probably not unique. This fitting procedure is a relatively brute force solution but it only has to be done once so it is not worth optimizing this procedure.

The resulting parameterization with 12 parameters (for each neutral atom) is tabulated later. The parameters are listed for each atomic number Z at the end of this chapter in the following order:

```
Z=  6,   chisq=      0.143335
     a1       b1       a2       b2
     a3       b3       c1       d1
     c2       d2       c3       d3
```

a_i, b_i, c_i, and d_i have units of Å^{-1}, Å^{-2},Å,Å^2, respectively. Although this may seem like a large number of parameters they can be used for both X-ray and electron scattering factors as well as the atomic potential. Some atoms such as He (Z = 2) produce $b_1 \sim b_2 \sim b_3$ which means that there is really only one Lorenzian. However other atoms require all six independent functions.

The figure of merit χ^2 (C.17) is shown vs. atomic number Z in Fig. C.2 as a solid line. Most atomic numbers have $\chi^2 < 0.1$ indicating a very good fit. The error introduced by using the parameterized form instead of the tabulated form is significantly less than the error introduced by ignoring the effects of bonding in the solid.

If the previously tabulated values of $f_x(s)$ and $f_e(s)$ given by Doyle and Turner [82], Doyle and Cowley [81], Cromer and Waber [73], and Fox et al. [109] are considered as one data set then the effective χ^2 formed by comparing this data set to the new parameters is shown in Fig. C.2 as a dashed line. Each of the old tabulations were typed in from the published literature (6076 values of $f_x(s)$ and 5432 values for $f_e(s)$). The data was entered twice and compared to find any typing errors. Because the new parameters fit the new tabulation relatively well this χ^2 is a measure of the agreement between the two Hartree-Fock programs. Many atomic numbers agree to $\chi^2 < 1.0$ which is good. However some disagree significantly. The maximum disagreement occurs for Z = 84 with $\chi^2 = 367$. This corresponds to an average relative error of $0.1\sqrt{\chi^2} = 1.9\%$ (with the definitions of the error terms σ_x

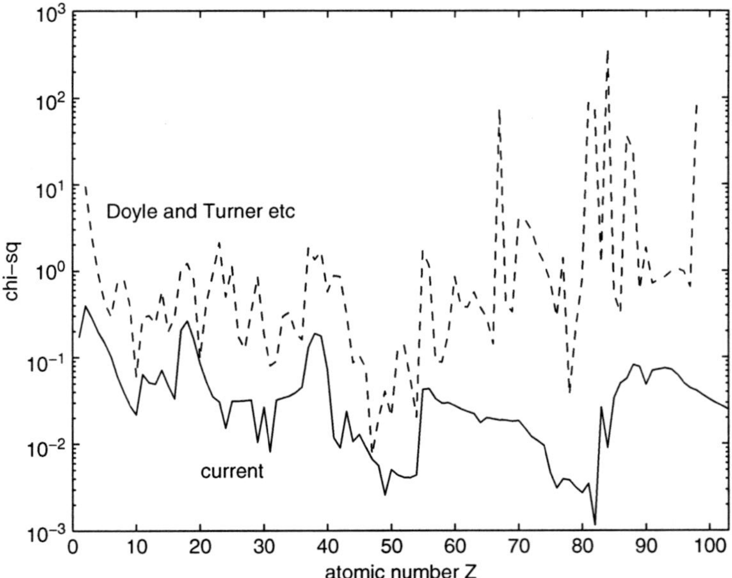

Fig. C.2 The figure of merit χ^2 of the parameterized form of the X-ray and electron scattering factors vs. atomic number Z. The parameters are compared to the data of Doyle and Turner, Doyle and Cowley and Fox et al. in the *dashed line* curve and to the current calculation in the *solid line* curve

and σ_e in (C.18). The largest percentage error (of about 10%) is in $f_e(0)$. The inverse Mott-Bethe formula constrains $f_x(0)$ to be exact. The scattering factors agree fairly well at high angles.

The particular parameterization was chosen to be a combination of relatively simple functions so that it can be inverse Fourier transformed analytically. The three-dimensional atomic potential $V(\mathbf{r}) = V(x,y,z)$ is:

$$V(x,y,z) = 2\pi a_0 e \int f_e(q) \exp(-2\pi i \mathbf{q}\cdot\mathbf{r}) d^3r =$$

$$2\pi^2 a_0 e \sum_i \frac{a_i}{r} \exp(-2\pi r\sqrt{b_i}) + 2\pi^{5/2} a_0 e \sum_i c_i d_i^{-3/2} \exp(-\pi^2 r^2/d_i) \quad \text{(C.19)}$$

$$\text{with} \quad r^2 = x^2 + y^2 + z^2$$

and the projected atomic potential is:

$$V_z(x,y) = \int_{-\infty}^{+\infty} V(x,y,z) dz =$$

$$4\pi^2 a_0 e \sum_i a_i K_0(2\pi r\sqrt{b_i}) + 2\pi^2 a_0 e \sum_i \frac{c_i}{d_i} \exp(-\pi^2 r^2/d_i) \quad \text{(C.20)}$$

$$\text{with} \quad r^2 = x^2 + y^2,$$

where $K_0(x)$ is the modified Bessel function. The integral for the first summation was helped by expression 3.387.6 of Gradshteyn and Ryzhik [128]. Abramowitz and Stegun [1] (Sect. 9.8) give a convenient numerical expression for evaluating $K_0(x)$. The right hand side of the expression for $V_z(x,y)$ (C.20) has units of the electron charge e. By combining the Rydberg constant $Ry = 0.5e^2/a_0$ ($Ry/e = 13.6$ volts) and the Bohr radius a_0=0.529Å the electron charge can be written in an unconventional set of units as $e = 14.4$ Volt-Angstroms which is more convenient for evaluating $V_z(x,y)$.

```
Z=  1,  chisq=    0.170190
  4.20298324e-003  2.25350888e-001  6.27762505e-002  2.25366950e-001
  3.00907347e-002  2.25331756e-001  6.77756695e-002  4.38854001e+000
  3.56609237e-003  4.03884823e-001  2.76135815e-002  1.44490166e+000

Z=  2,  chisq=    0.396634
  1.87543704e-005  2.12427997e-001  4.10595800e-004  3.32212279e-001
  1.96300059e-001  5.17325152e-001  8.36015738e-003  3.66668239e-001
  2.95102022e-002  1.37171827e+000  4.65928982e-007  3.75768025e+004

Z=  3,  chisq=    0.286232
  7.45843816e-002  8.81151424e-001  7.15382250e-002  4.59142904e-002
  1.45315229e-001  8.81301714e-001  1.12125769e+000  1.88483665e+001
  2.51736525e-003  1.59189995e-001  3.58434971e-001  6.12371000e+000

Z=  4,  chisq=    0.195442
  6.11642897e-002  9.90182132e-002  1.25755034e-001  9.90272412e-002
  2.00831548e-001  1.87392509e+000  7.87242876e-001  9.32794929e+000
  1.58847850e-003  8.91900236e-002  2.73962031e-001  3.20687658e+000

Z=  5,  chisq=    0.146989
  1.25716066e-001  1.48258830e-001  1.73314452e-001  1.48257216e-001
  1.84774811e-001  3.34227311e+000  1.95250221e-001  1.97339463e+000
  5.29642075e-001  5.70035553e+000  1.08230500e-003  5.64857237e-002

Z=  6,  chisq=    0.102440
  2.12080767e-001  2.08605417e-001  1.99811865e-001  2.08610186e-001
  1.68254385e-001  5.57870773e+000  1.42048360e-001  1.33311887e+000
  3.63830672e-001  3.80800263e+000  8.35012044e-004  4.03982620e-002

Z=  7,  chisq=    0.060249
  5.33015554e-001  2.90952515e-001  5.29008883e-002  1.03547896e+001
  9.24159648e-002  1.03540028e+001  2.61799101e-001  2.76252723e+000
  8.80262108e-004  3.47681236e-002  1.10166555e-001  9.93421736e-001

Z=  8,  chisq=    0.039944
  3.39969204e-001  3.81570280e-001  3.07570172e-001  3.81571436e-001
  1.30369072e-001  1.91919745e+001  8.83326058e-002  7.60635525e-001
  1.96586700e-001  2.07401094e+000  9.96220028e-004  3.03266869e-002

Z=  9,  chisq=    0.027866
  2.30560593e-001  4.80754213e-001  5.26889648e-001  4.80763895e-001
  1.24346755e-001  3.95306720e+001  1.24616894e-003  2.62181803e-002
  7.20452555e-002  5.92495593e-001  1.53075777e-001  1.59127671e+000

Z= 10,  chisq=    0.021836
  4.08371771e-001  5.88228627e-001  4.54418858e-001  5.88288655e-001
  1.44564923e-001  1.21246013e+002  5.91531395e-002  4.63963540e-001
  1.24003718e-001  1.23413025e+000  1.64986037e-003  2.05869217e-002

Z= 11,  chisq=    0.064136
  1.36471662e-001  4.99965301e-002  7.70677865e-001  8.81899664e-001
  1.56862014e-001  1.61768579e+001  9.96821513e-001  2.00132610e+001
  3.80304670e-002  2.60516254e-001  1.27685089e-001  6.99559329e-001
```

```
Z= 12,  chisq=    0.051303
  3.04384121e-001  8.42014377e-002  7.56270563e-001  1.64065598e+000
  1.01164809e-001  2.97142975e+001  3.45203403e-002  2.16596094e-001
  9.71751327e-001  1.21236852e+001  1.20593012e-001  5.60865838e-001

Z= 13,  chisq=    0.049529
  7.77419424e-001  2.71058227e+000  5.78312036e-002  7.17532098e+001
  4.26386499e-001  9.13331555e-002  1.13407220e-001  4.48867451e-001
  7.90114035e-001  8.66366718e+000  3.23293496e-002  1.78503463e-001

Z= 14,  chisq=    0.071667
  1.06543892e+000  1.04118455e+000  1.20143691e-001  6.87113368e+001
  1.80915263e-001  8.87533926e-002  1.12065620e+000  3.70062619e+000
  3.05452816e-002  2.14097897e-001  1.59963502e+000  9.99096638e+000

Z= 15,  chisq=    0.047673
  1.05284447e+000  1.31962590e+000  2.99440284e-001  1.28460520e-001
  1.17460748e-001  1.02190163e+002  9.60643452e-001  2.87477555e+000
  2.63555748e-002  1.82076844e-001  1.38059330e+000  7.49165526e+000

Z= 16,  chisq=    0.033482
  1.01646916e+000  1.69181965e+000  4.41766748e-001  1.74180288e-001
  1.21503863e-001  1.67011091e+002  8.27966670e-001  2.30342810e+000
  2.33022533e-002  1.56954150e-001  1.18302846e+000  5.85782891e+000

Z= 17,  chisq=    0.206186
  9.44221116e-001  2.40052374e-001  4.37322049e-001  9.30510439e+000
  2.54547926e-001  9.30486346e+000  5.47763323e-002  1.68655688e-001
  8.00087488e-001  2.97849774e+000  1.07488641e-002  6.84240646e-002

Z= 18,  chisq=    0.263904
  1.06983288e+000  2.87791022e-001  4.24631786e-001  1.24156957e+001
  2.43897949e-001  1.24158868e+001  4.79446296e-002  1.36979796e-001
  7.64958952e-001  2.43940729e+000  8.23128431e-003  5.27258749e-002

Z= 19,  chisq=    0.161900
  6.92717865e-001  7.10849990e+000  9.65161085e-001  3.57532901e-001
  1.48466588e-001  3.93763275e-002  2.64645027e-002  1.03591321e-001
  1.80883768e+000  3.22845199e+001  5.43900018e-001  1.67791374e+000

Z= 20,  chisq=    0.085209
  3.66902871e-001  6.14274129e-002  8.66378999e-001  5.70881727e-001
  6.67203300e-001  7.82965639e+000  4.87743636e-001  1.32531318e+000
  1.82406314e+000  2.10056032e+001  2.20248453e-002  9.11853450e-002

Z= 21,  chisq=    0.052352
  3.78871777e-001  6.98910162e-002  9.00022505e-001  5.21061541e-001
  7.15288914e-001  7.87707920e+000  1.88640973e-002  8.17512708e-002
  4.07945949e-001  1.11141388e+000  1.61786540e+000  1.80840759e+001

Z= 22,  chisq=    0.035298
  3.62383267e-001  7.54707114e-002  9.84232966e-001  4.97757309e-001
  7.41715642e-001  8.17659391e+000  3.62555269e-001  9.55524906e-001
  1.49159390e+000  1.62221677e+001  1.61659509e-002  7.33140839e-002

Z= 23,  chisq=    0.030745
  3.52961378e-001  8.19204103e-002  7.46791014e-001  8.81189511e+000
  1.08364068e+000  5.10646075e-001  1.39013610e+000  1.48901841e+001
  3.31273356e-001  8.38543079e-001  1.40422612e-002  6.57432678e-002

Z= 24,  chisq=    0.015287
  1.34348379e+000  1.25814353e+000  5.07040328e-001  1.15042811e+001
  4.26358955e-001  8.53660389e-002  1.17241826e-002  6.00177061e-002
  5.11966516e-001  1.53772451e+000  3.38285828e-001  6.62418319e-001

Z= 25,  chisq=    0.031274
  3.26697613e-001  8.88813083e-002  7.17297000e-001  1.11300198e+001
```

```
  1.33212464e+000  5.82141104e-001  2.80801702e-001  6.71583145e-001
  1.15499241e+000  1.26825395e+001  1.11984488e-002  5.32334467e-002

Z= 26,  chisq=    0.031315
  3.13454847e-001  8.99325756e-002  6.89290016e-001  1.30366038e+001
  1.47141531e+000  6.33345291e-001  1.03298688e+000  1.16783425e+001
  2.58280285e-001  6.09116446e-001  1.03460690e-002  4.81610627e-002

Z= 27,  chisq=    0.031643
  3.15878278e-001  9.46683246e-002  1.60139005e+000  6.99436449e-001
  6.56394338e-001  1.56954403e+001  9.36746624e-001  1.09392410e+001
  9.77562646e-003  4.37446816e-002  2.38378578e-001  5.56286483e-001

Z= 28,  chisq=    0.032245
  1.72254630e+000  7.76606908e-001  3.29543044e-001  1.02262360e-001
  6.23007200e-001  1.94156207e+001  9.43496513e-003  3.98684596e-002
  8.54063515e-001  1.04078166e+001  2.21073515e-001  5.10869330e-001

Z= 29,  chisq=    0.010467
  3.58774531e-001  1.06153463e-001  1.76181348e+000  1.01640995e+000
  6.36905053e-001  1.53659093e+001  7.44930667e-003  3.85345989e-002
  1.89002347e-001  3.98427790e-001  2.29619589e-001  9.01419843e-001

Z= 30,  chisq=    0.026698
  5.70893973e-001  1.26534614e-001  1.98908856e+000  2.17781965e+000
  3.06060585e-001  3.78619003e+001  2.35600223e-001  3.67019041e-001
  3.97061102e-001  8.66419596e-001  6.85657228e-003  3.35778823e-002

Z= 31,  chisq=    0.008110
  6.25528464e-001  1.10005650e-001  2.05302901e+000  2.41095786e+000
  2.89608120e-001  4.78685736e+001  2.07910594e-001  3.27807224e-001
  3.45079617e-001  7.43139061e-001  6.55634298e-003  3.09411369e-002

Z= 32,  chisq=    0.032198
  5.90952690e-001  1.18375976e-001  5.39980660e-001  7.18937433e+001
  2.00626188e+000  1.39304889e+000  7.49705041e-001  6.89943350e+000
  1.83581347e-001  3.64667232e-001  9.52190743e-003  2.69888650e-002

Z= 33,  chisq=    0.034014
  7.77875218e-001  1.50733157e-001  5.93848150e-001  1.42882209e+002
  1.95918751e+000  1.74750339e+000  1.79880226e-001  3.31800852e-001
  8.63267222e-001  5.85490274e+000  9.59053427e-003  2.33777569e-002

Z= 34,  chisq=    0.035703
  9.58390681e-001  1.83775557e-001  6.03851342e-001  1.96819224e+002
  1.90828931e+000  2.15082053e+000  1.73885956e-001  3.00006024e-001
  9.35265145e-001  4.92471215e+000  8.62254658e-003  2.12308108e-002

Z= 35,  chisq=    0.039250
  1.14136170e+000  2.18708710e-001  5.18118737e-001  1.93916682e+002
  1.85731975e+000  2.65755396e+000  1.68217399e-001  2.71719918e-001
  9.75705606e-001  4.19482500e+000  7.24187871e-003  1.99325718e-002

Z= 36,  chisq=    0.045421
  3.24386970e-001  6.31317973e+001  1.31732163e+000  2.54706036e-001
  1.79912614e+000  3.23668394e+000  4.29961425e-003  1.98965610e-002
  1.00429433e+000  3.61094513e+000  1.62188197e-001  2.45583672e-001

Z= 37,  chisq=    0.130044
  2.90445351e-001  3.68420227e-002  2.44201329e+000  1.16013332e+000
  7.69435449e-001  1.69591472e+001  1.58687000e+000  2.53082574e+000
  2.81617593e-003  1.88577417e-002  1.28663830e-001  2.10753969e-001

Z= 38,  chisq=    0.188055
  1.37373086e-002  1.87469061e-002  1.97548672e+000  6.36079230e+000
  1.59261029e+000  2.21992482e-001  1.73263882e-001  2.01624958e-001
  4.66280378e+000  2.53027803e+001  1.61265063e-003  1.53610568e-002
```

```
Z= 39,  chisq=     0.174927
  6.75302747e-001  6.54331847e-002  4.70286720e-001  1.06108709e+002
  2.63497677e+000  2.06643540e+000  1.09621746e-001  1.93131925e-001
  9.60348773e-001  1.63310938e+000  5.28921555e-003  1.66083821e-002

Z= 40,  chisq=     0.072078
  2.64365505e+000  2.20202699e+000  5.54225147e-001  1.78260107e+002
  7.61376625e-001  7.67218745e-002  6.02946891e-003  1.55143296e-002
  9.91630530e-002  1.76175995e-001  9.56782020e-001  1.54330682e+000

Z= 41,  chisq=     0.011800
  6.59532875e-001  8.66145490e-002  1.84545854e+000  5.94774398e+000
  1.25584405e+000  6.40851475e-001  1.22253422e-001  1.66646050e-001
  7.06638328e-001  1.62853268e+000  2.62381591e-003  8.26257859e-003

Z= 42,  chisq=     0.008976
  6.10160120e-001  9.11628054e-002  1.26544000e+000  5.06776025e-001
  1.97428762e+000  5.89590381e+000  6.48028962e-001  1.46634108e+000
  2.60380817e-003  7.84336311e-003  1.13887493e-001  1.55114340e-001

Z= 43,  chisq=     0.023771
  8.55189183e-001  1.02962151e-001  1.66219641e+000  7.64907000e+000
  1.45575475e+000  1.01639987e+000  1.05445664e-001  1.42303338e-001
  7.71657112e-001  1.34659349e+000  2.20992635e-003  7.90358976e-003

Z= 44,  chisq=     0.010613
  4.70847093e-001  9.33029874e-002  1.58180781e+000  4.52831347e-001
  2.02419818e+000  7.11489023e+000  1.97036257e-003  7.56181595e-003
  6.26912639e-001  1.25399858e+000  1.02641320e-001  1.33786087e-001

Z= 45,  chisq=     0.012895
  4.20051553e-001  9.38882628e-002  1.76266507e+000  4.64441687e-001
  2.02735641e+000  8.19346046e+000  1.45487176e-003  7.82704517e-003
  6.22809600e-001  1.17194153e+000  9.91529915e-002  1.24532839e-001

Z= 46,  chisq=     0.009172
  2.10475155e+000  8.68606470e+000  2.03884487e+000  3.78924449e-001
  1.82067264e-001  1.42921634e-001  9.52040948e-002  1.17125900e-001
  5.91445248e-001  1.07843808e+000  1.13328676e-003  7.80252092e-003

Z= 47,  chisq=     0.006648
  2.07981390e+000  9.92540297e+000  4.43170726e-001  1.04920104e-001
  1.96515215e+000  6.40103839e-001  5.96130591e-001  8.89594790e-001
  4.78016333e-001  1.98509407e+000  9.46458470e-002  1.12744464e-001

Z= 48,  chisq=     0.005588
  1.63657549e+000  1.24540381e+001  2.17927989e+000  1.45134660e+000
  7.71300690e-001  1.26695757e-001  6.64193880e-001  7.77659202e-001
  7.64563285e-001  1.66075210e+000  8.61126689e-002  1.05728357e-001


Z= 49,  chisq=     0.002569
  2.24820632e+000  1.51913507e+000  1.64706864e+000  1.30113424e+001
  7.88679265e-001  1.06128184e-001  8.12579069e-002  9.94045620e-002
  6.68280346e-001  1.49742063e+000  6.38467475e-001  7.18422635e-001

Z= 50,  chisq=     0.005051
  2.16644620e+000  1.13174909e+001  6.88691021e-001  1.10131285e-001
  1.92431751e+000  6.74464853e-001  5.65359888e-001  7.33564610e-001
  9.18683861e-001  1.02310312e+001  7.80542213e-002  9.31104308e-002

Z= 51,  chisq=     0.004383
  1.73662114e+000  8.84334719e-001  9.99871380e-001  1.38462121e-001
  2.13972409e+000  1.19666432e+001  5.60566526e-001  6.72672880e-001
  9.93772747e-001  8.72330411e+000  7.37374982e-002  8.78577715e-002
```

```
Z= 52,  chisq=     0.004105
  2.09383882e+000  1.26856869e+001  1.56940519e+000  1.21236537e+000
  1.30941993e+000  1.66633292e-001  6.98067804e-002  8.30817576e-002
  1.04969537e+000  7.43147857e+000  5.55594354e-001  6.17487676e-001

Z= 53,  chisq=     0.004068
  1.60186925e+000  1.95031538e-001  1.98510264e+000  1.36976183e+001
  1.48226200e+000  1.80304795e+000  5.53807199e-001  5.67912340e-001
  1.11728722e+000  6.40879878e+000  6.60720847e-002  7.86615429e-002

Z= 54,  chisq=     0.004381
  1.60015487e+000  2.92913354e+000  1.71644581e+000  1.55882990e+001
  1.84968351e+000  2.22525983e-001  6.23813648e-002  7.45581223e-002
  1.21387555e+000  5.56013271e+000  5.54051946e-001  5.21994521e-001

Z= 55,  chisq=     0.042676
  2.95236854e+000  6.01461952e+000  4.28105721e-001  4.64151246e+001
  1.89599233e+000  1.80109756e-001  5.48012938e-002  7.12799633e-002
  4.70838600e+000  4.56702799e+001  5.90356719e-001  4.70236310e-001

Z= 56,  chisq=     0.043267
  3.19434243e+000  9.27352241e+000  1.98289586e+000  2.28741632e-001
  1.55121052e-001  3.82000231e-002  6.73222354e-002  7.30961745e-002
  4.48474211e+000  2.95703565e+001  5.42674414e-001  4.08647015e-001

Z= 57,  chisq=     0.033249
  2.05036425e+000  2.20348417e-001  1.42114311e-001  3.96438056e-002
  3.23538151e+000  9.56979169e+000  6.34683429e-002  6.92443091e-002
  3.97960586e+000  2.53178406e+001  5.20116711e-001  3.83614098e-001

Z= 58,  chisq=     0.029355
  3.22990759e+000  9.94660135e+000  1.57618307e-001  4.15378676e-002
  2.13477838e+000  2.40480572e-001  5.01907609e-001  3.66252019e-001
  3.80889010e+000  2.43275968e+001  5.96625028e-002  6.59653503e-002

Z= 59,  chisq=     0.029725
  1.58189324e-001  3.91309056e-002  3.18141995e+000  1.04139545e+001
  2.27622140e+000  2.81671757e-001  3.97705472e+000  2.61872978e+001
  5.58448277e-002  6.30921695e-002  4.85207954e-001  3.54234369e-001

Z= 60,  chisq=     0.027597
  1.81379417e-001  4.37324793e-002  3.17616396e+000  1.07842572e+001
  2.35221519e+000  3.05571833e-001  3.83125763e+000  2.54745408e+001
  5.25889976e-002  6.02676073e-002  4.70090742e-001  3.39017003e-001

Z= 61,  chisq=     0.025208
  1.92986811e-001  4.37785970e-002  2.43756023e+000  3.29336996e-001
  3.17248504e+000  1.11259996e+001  3.58105414e+000  2.46709586e+001
  4.56529394e-001  3.24990282e-001  4.94812177e-002  5.76553100e-002


Z= 62,  chisq=     0.023540
  2.12002595e-001  4.57703608e-002  3.16891754e+000  1.14536599e+001
  2.51503494e+000  3.55561054e-001  4.44080845e-001  3.11953363e-001
  3.36742101e+000  2.40291435e+001  4.65652543e-002  5.52266819e-002

Z= 63,  chisq=     0.022204
  2.59355002e+000  3.82452612e-001  3.16557522e+000  1.17675155e+001
  2.29402652e-001  4.76642249e-002  4.32257780e-001  2.99719833e-001
  3.17261920e+000  2.34462738e+001  4.37958317e-002  5.29440680e-002

Z= 64,  chisq=     0.017492
  3.19144939e+000  1.20224655e+001  2.55766431e+000  4.08338876e-001
  3.32681934e-001  5.85819814e-002  4.14243130e-002  5.06771477e-002
  2.61036728e+000  1.99344244e+001  4.20526863e-001  2.85686240e-001
```

```
Z= 65,  chisq=     0.020036
  2.59407462e-001  5.04689354e-002  3.16177855e+000  1.23140183e+001
  2.75095751e+000  4.38337626e-001  2.79247686e+000  2.23797309e+001
  3.85931001e-002  4.87920992e-002  4.10881708e-001  2.77622892e-001

Z= 66,  chisq=     0.019351
  3.16055396e+000  1.25470414e+001  2.82751709e+000  4.67899094e-001
  2.75140255e-001  5.23226982e-002  4.00967160e-001  2.67614884e-001
  2.63110834e+000  2.19498166e+001  3.61333817e-002  4.68871497e-002

Z= 67,  chisq=     0.018720
  2.88642467e-001  5.40507687e-002  2.90567296e+000  4.97581077e-001
  3.15960159e+000  1.27599505e+001  3.91280259e-001  2.58151831e-001
  2.48596038e+000  2.15400972e+001  3.37664478e-002  4.50664323e-002

Z= 68,  chisq=     0.018677
  3.15573213e+000  1.29729009e+001  3.11519560e-001  5.81399387e-002
  2.97722406e+000  5.31213394e-001  3.81563854e-001  2.49195776e-001
  2.40247532e+000  2.13627616e+001  3.15224214e-002  4.33253257e-002

Z= 69,  chisq=     0.018176
  3.15591970e+000  1.31232407e+001  3.22544710e-001  5.97223323e-002
  3.05569053e+000  5.61876773e-001  2.92845100e-002  4.16534255e-002
  3.72487205e-001  2.40821967e-001  2.27833695e+000  2.10034185e+001

Z= 70,  chisq=     0.018460
  3.10794704e+000  6.06347847e-001  3.14091221e+000  1.33705269e+001
  3.75660454e-001  7.29814740e-002  3.61901097e-001  2.32652051e-001
  2.45409082e+000  2.12695209e+001  2.72383990e-002  3.99969597e-002

Z= 71,  chisq=     0.015021
  3.11446863e+000  1.38968881e+001  5.39634353e-001  8.91708508e-002
  3.06460915e+000  6.79919563e-001  2.58563745e-002  3.82808522e-002
  2.13983556e+000  1.80078788e+001  3.47788231e-001  2.22706591e-001

Z= 72,  chisq=     0.012070
  3.01166899e+000  7.10401889e-001  3.16284788e+000  1.38262192e+001
  6.33421771e-001  9.48486572e-002  3.41417198e-001  2.14129678e-001
  1.53566013e+000  1.55298698e+001  2.40723773e-002  3.67833690e-002

Z= 73,  chisq=     0.010775
  3.20236821e+000  1.38446369e+001  8.30098413e-001  1.18381581e-001
  2.86552297e+000  7.66369118e-001  2.24813887e-002  3.52934622e-002
  1.40165263e+000  1.46148877e+001  3.33740596e-001  2.05704486e-001

Z= 74,  chisq=     0.009479
  9.24906855e-001  1.28663377e-001  2.75554557e+000  7.65826479e-001
  3.30440060e+000  1.34471170e+001  3.29973862e-001  1.98218895e-001
  1.09916444e+000  1.35087534e+001  2.06498883e-002  3.38918459e-002


Z= 75,  chisq=     0.004620
  1.96952105e+000  4.98830620e+001  1.21726619e+000  1.33243809e-001
  4.10391685e+000  1.84396916e+000  2.90791978e-002  2.84192813e-002
  2.30696669e-001  1.90968784e-001  6.08840299e-001  1.37090356e+000

Z= 76,  chisq=     0.003085
  2.06385867e+000  4.05671697e+001  1.29603406e+000  1.46559047e-001
  3.96920673e+000  1.82561596e+000  2.69835487e-002  2.84172045e-002
  2.31083999e-001  1.79765184e-001  6.30466774e-001  1.38911543e+000

Z= 77,  chisq=     0.003924
  2.21522726e+000  3.24464090e+001  1.37573155e+000  1.60920048e-001
  3.78244405e+000  1.78756553e+000  2.44643240e-002  2.82909938e-002
  2.36932016e-001  1.70692368e-001  6.48471412e-001  1.37928390e+000
```

```
Z= 78,  chisq=    0.003817
  9.84697940e-001  1.60910839e-001  2.73987079e+000  7.18971667e-001
  3.61696715e+000  1.29281016e+001  3.02885602e-001  1.70134854e-001
  2.78370726e-001  1.49862703e+000  1.52124129e-002  2.83510822e-002

Z= 79,  chisq=    0.003143
  9.61263398e-001  1.70932277e-001  3.69581030e+000  1.29335319e+001
  2.77567491e+000  6.89997070e-001  2.95414176e-001  1.63525510e-001
  3.11475743e-001  1.39200901e+000  1.43237267e-002  2.71265337e-002

Z= 80,  chisq=    0.002717
  1.29200491e+000  1.83432865e-001  2.75161478e+000  9.42368371e-001
  3.49387949e+000  1.46235654e+001  2.77304636e-001  1.55110144e-001
  4.30232810e-001  1.28871670e+000  1.48294351e-002  2.61903834e-002

Z= 81,  chisq=    0.003492
  3.75964730e+000  1.35041513e+001  3.21195904e+000  6.66330993e-001
  6.47767825e-001  9.22518234e-002  2.76123274e-001  1.50312897e-001
  3.18838810e-001  1.12565588e+000  1.31668419e-002  2.48879842e-002

Z= 82,  chisq=    0.001158
  1.00795975e+000  1.17268427e-001  3.09796153e+000  8.80453235e-001
  3.61296864e+000  1.47325812e+001  2.62401476e-001  1.43491014e-001
  4.05621995e-001  1.04103506e+000  1.31812509e-002  2.39575415e-002

Z= 83,  chisq=    0.026436
  1.59826875e+000  1.56897471e-001  4.38233925e+000  2.47094692e+000
  2.06074719e+000  5.72438972e+001  1.94426023e-001  1.32979109e-001
  8.22704978e-001  9.56532528e-001  2.33226953e-002  2.23038435e-002

Z= 84,  chisq=    0.008962
  1.71463223e+000  9.79262841e+001  2.14115960e+000  2.10193717e-001
  4.37512413e+000  3.66948812e+000  2.16216680e-002  1.98456144e-002
  1.97843837e-001  1.33758807e-001  6.52047920e-001  7.80432104e-001

Z= 85,  chisq=    0.033776
  1.48047794e+000  1.25943919e+002  2.09174630e+000  1.83803008e-001
  4.75246033e+000  4.19890596e+000  1.85643958e-002  1.81383503e-002
  2.05859375e-001  1.33035404e-001  7.13540948e-001  7.03031938e-001

Z= 86,  chisq=    0.050132
  6.30022295e-001  1.40909762e-001  3.80962881e+000  3.08515540e+001
  3.89756067e+000  6.51559763e-001  2.40755100e-001  1.08899672e-001
  2.62868577e+000  6.42383261e+000  3.14285931e-002  2.42346699e-002

Z= 87,  chisq=    0.056720
  5.23288135e+000  8.60599536e+000  2.48604205e+000  3.04543982e-001
  3.23431354e-001  3.87759096e-002  2.55403596e-001  1.28717724e-001
  5.53607228e-001  5.36977452e-001  5.75278889e-003  1.29417790e-002

Z= 88,  chisq=    0.081498
  1.44192685e+000  1.18740873e-001  3.55291725e+000  1.01739750e+000
  3.91259586e+000  6.31814783e+001  2.16173519e-001  9.55806441e-002
  3.94191605e+000  3.50602732e+001  4.60422605e-002  2.20850385e-002

Z= 89,  chisq=    0.077643
  1.45864127e+000  1.07760494e-001  4.18945405e+000  8.89090649e+001
  3.65866182e+000  1.05088931e+000  2.08479229e-001  9.09335557e-002
  3.16528117e+000  3.13297788e+001  5.23892556e-002  2.08807697e-002

Z= 90,  chisq=    0.048096
  1.19014064e+000  7.73468729e-002  2.55380607e+000  6.59693681e-001
  4.68110181e+000  1.28013896e+001  2.26121303e-001  1.08632194e-001
  3.58250545e-001  4.56765664e-001  7.82263950e-003  1.62623474e-002

Z= 91,  chisq=    0.070186
  4.68537504e+000  1.44503632e+001  2.98413708e+000  5.56438592e-001
```

```
  8.91988061e-001  6.69512914e-002  2.24825384e-001  1.03235396e-001
  3.04444846e-001  4.27255647e-001  9.48162708e-003  1.77730611e-002

Z= 92,  chisq=    0.072478
  4.63343606e+000  1.63377267e+001  3.18157056e+000  5.69517868e-001
  8.76455075e-001  6.88860012e-002  2.21685477e-001  9.84254550e-002
  2.72917100e-001  4.09470917e-001  1.11737298e-002  1.86215410e-002

Z= 93,  chisq=    0.074792
  4.56773888e+000  1.90992795e+001  3.40325179e+000  5.90099634e-001
  8.61841923e-001  7.03204851e-002  2.19728870e-001  9.36334280e-002
  2.38176903e-001  3.93554882e-001  1.38306499e-002  1.94437286e-002

Z= 94,  chisq=    0.071877
  5.45671123e+000  1.01892720e+001  1.11687906e-001  3.98131313e-002
  3.30260343e+000  3.14622212e-001  1.84568319e-001  1.04220860e-001
  4.93644263e-001  4.63080540e-001  3.57484743e+000  2.19369542e+001

Z= 95,  chisq=    0.062156
  5.38321999e+000  1.07289857e+001  1.23343236e-001  4.15137806e-002
  3.46469090e+000  3.39326208e-001  1.75437132e-001  9.98932346e-002
  3.39800073e+000  2.11601535e+001  4.69459519e-001  4.51996970e-001

Z= 96,  chisq=    0.050111
  5.38402377e+000  1.11211419e+001  3.49861264e+000  3.56750210e-001
  1.88039547e-001  5.39853583e-002  1.69143137e-001  9.60082633e-002
  3.19595016e+000  1.80694389e+001  4.64393059e-001  4.36318197e-001

Z= 97,  chisq=    0.044081
  3.66090688e+000  3.84420906e-001  2.03054678e-001  5.48547131e-002
  5.30697515e+000  1.17150262e+001  1.60934046e-001  9.21020329e-002
  3.04808401e+000  1.73525367e+001  4.43610295e-001  4.27132359e-001

Z= 98,  chisq=    0.041053
  3.94150390e+000  4.18246722e-001  5.16915345e+000  1.25201788e+001
  1.61941074e-001  4.81540117e-002  4.15299561e-001  4.24913856e-001
  2.91761325e+000  1.90899693e+001  1.51474927e-001  8.81568925e-002

Z= 99,  chisq=    0.036478
  4.09780623e+000  4.46021145e-001  5.10079393e+000  1.31768613e+001
  1.74617289e-001  5.02742829e-002  2.76774658e+000  1.84815393e+001
  1.44496639e-001  8.46232592e-002  4.02772109e-001  4.17640100e-001

Z=100,  chisq=    0.032651
  4.24934820e+000  4.75263933e-001  5.03556594e+000  1.38570834e+001
  1.88920613e-001  5.26975158e-002  3.94356058e-001  4.11193751e-001
  2.61213100e+000  1.78537905e+001  1.38001927e-001  8.12774434e-002

Z=101,  chisq=    0.029668
  2.00942931e-001  5.48366518e-002  4.40119869e+000  5.04248434e-001
  4.97250102e+000  1.45721366e+001  2.47530599e+000  1.72978308e+001
  3.86883197e-001  4.05043898e-001  1.31936095e-001  7.80821071e-002

Z=102,  chisq=    0.027320
  2.16052899e-001  5.83584058e-002  4.91106799e+000  1.53264212e+001
  4.54862870e+000  5.34434760e-001  2.36114249e+000  1.68164803e+001
  1.26277292e-001  7.50304633e-002  3.81364501e-001  3.99305852e-001

Z=103,  chisq=    0.024894
  4.86738014e+000  1.60320520e+001  3.19974401e-001  6.70871138e-002
  4.58872425e+000  5.77039373e-001  1.21482448e-001  7.22275899e-002
  2.31639872e+000  1.41279737e+001  3.79258137e-001  3.89973484e-001
```

Appendix D
Bilinear Interpolation

Abstract This appendix describes a mathematical approximation that can be used to interpolate in two dimensions such as in an image. This is a useful operation to compare images of different pixels sizes or spacings.

It is frequently necessary to combine or compare two different digital images with different sampling sizes (Δx or Δy). Interpolation in two dimensions is one method of reconciling the difference in sampling sizes. For example, most simulated electron microscope images will have a rectangular pixel with unequal spacings in x and y (to match the underlying periodicity of the specimen) but many image display devices (computer screens) will have square pixels with equal spacings in x and y. To properly display an image with rectangular pixels on a device with square pixels will require resampling the image.

The basic problem can be stated as: Given a set of image intensities sampled on a two-dimensional grid with spacing Δx_a and Δy_a generate another set of image intensities on a grid with a different spacing Δx_b and Δy_b. Interpolating the initial grid in two dimensions generates a function of two independent variables $f(x,y)$ that is continuous but may not have continuous derivatives. Calculating this function at each point in the new grid effectively samples the first image onto the second grid. This procedure works best if the first and second grids are nearly the sample spacing. If the spacing are dramatically different (i.e., more than about a factor of two different) then various artifacts can be produced.

The basic geometry of interpolation on a grid is shown in Fig. D.1. The initial image is only given at discrete points in (x,y). To find an interpolated value at an arbitrary point requires first locating the four grid points surrounding the point (x,y). In Fig. D.1 the new point is located between x_1 and x_2 in x and between y_1 and y_2 in y. The values of the initial image at the four grid point surrounding (x,y) are f_{11}, f_{12}, f_{22}, and f_{21}.

With four points there are four know conditions and the best interpolation available is a bilinear form:

$$f(x,y) = a + bx + cy + dxy. \tag{D.1}$$

E.J. Kirkland, *Advanced Computing in Electron Microscopy*,
DOI 10.1007/978-1-4419-6533-2_12, © Springer Science+Business Media, LLC 2010

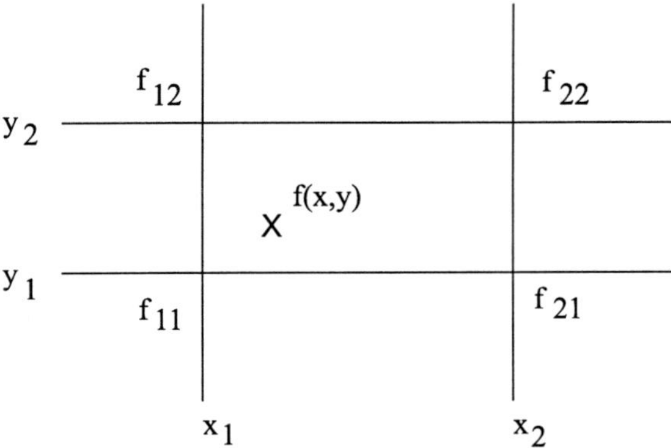

Fig. D.1 Interpolation in a two-dimensional rectangular grid. The function $f(x,y)$ can be found at an arbitrary point (x,y) given only the values sampled at discrete values of $x = x_1, x_2$ and $y = y_1, y_2$ surrounding point (x,y). The spacing in x does not need to be the same as that in y

The a, b, c, and d coefficient are determined by the surrounding image values as:

$$\begin{aligned} f_{11} &= a + bx_1 + cy_1 + dx_1y_1 \\ f_{12} &= a + bx_1 + cy_2 + dx_1y_2 \\ f_{22} &= a + bx_2 + cy_2 + dx_2y_2 \\ f_{21} &= a + bx_2 + cy_1 + dx_2y_1. \end{aligned} \tag{D.2}$$

Combining pairs of equations yields:

$$\begin{aligned} (f_{11} - f_{12}) &= (c + \mathrm{d}x_1)(y_1 - y_2) \\ (f_{21} - f_{22}) &= (c + \mathrm{d}x_2)(y_1 - y_2). \end{aligned} \tag{D.3}$$

The coefficients may be found one at a time. Subtracting (D.3) yields a value for d:

$$d = \frac{f_{11} - f_{12} - f_{21} + f_{22}}{(x_1 - x_2)(y_1 - y_2)}. \tag{D.4}$$

Using this value for d and one of (D.3) yields a value for c:

$$c = \left(\frac{f_{11} - f_{12}}{y_1 - y_2} \right) - dx_1. \tag{D.5}$$

Next combing (D.2) in a slightly different order gives:

$$f_{11} - f_{21} = b(x_1 - x_2) + dy_1(x_1 - x_2) \tag{D.6}$$

from which the a and b coefficients can be found.

$$b = \left(\frac{f_{11} - f_{21}}{x_1 - x_2}\right) - \mathrm{d}y_1 \tag{D.7}$$

$$a = f_{11} - bx_1 - cy_1 - \mathrm{d}x_1y_1. \tag{D.8}$$

Now given a value for all four coefficients (a, b, c, and d) a value for the function $f(x,y)$ at any point inside the four grid points may be obtained from (D.1). Repeating this process at each point of the second grid yields an interpolated value of the first image at each grid point of the second image.

Appendix E
3D Perspective View

Abstract This appendix describes a simple method to calculate a three dimensional perspective view on a two-dimensional screen of a collection of atom coordinates.

The specimen is clearly three dimensional (3D) but the electron microscope image is two dimensional (2D). The three dimensional structure of the specimen is projected into the final two-dimensional image making it difficult to determine if the original three dimensional structure of the specimen was properly described. The electron microscope image is in an x, y plane but the crystal structure is a set of (x_i, y_i, z_i) coordinates. The third dimension z can have a significant influence on the scattering within the specimen and influence the final image in rather nonintuitive ways. It is important to check that the full three dimensional structure of the specimen has been entered correctly.

A crystalline or amorphous specimen must be described as a detailed numerical list of atomic coordinates before an electron microscope image of the specimen can be simulated in the computer. Generating this set of numbers can be a rather tedious and difficult task. It is difficult to generate this list and even more difficult to determine if it is correct in the first place.

One particular type of diagnostic tool to determine if the specimen description is correct is to render a 3D perspective view of the entire specimen structure. Each atom can be drawn as a simple hard sphere and the entire structure can be rotated and viewed to inspect its three dimensional structure. A full rendering with shading and hidden surfaces can be difficult to calculate. There are a variety of sophisticated programs available for this procedure (for example RasMol [310] or jmol [189]). However, if some approximations are made there is a simple way to draw a reasonable approximation of the full 3D structure.

The specimen is assumed to be a collection of atoms in 3D. Each atom is drawn as a hard sphere at a particular set of coordinates (x_i, y_i, z_i) in 3D. This structure is viewed from a particular point in space as shown in Fig. E.1. The image seen by the viewer must be projected into a 2D image (the computer screen or a piece of paper).

E.J. Kirkland, *Advanced Computing in Electron Microscopy*,
DOI 10.1007/978-1-4419-6533-2_13, © Springer Science+Business Media, LLC 2010

If the viewer is close, then the structure will appear more distorted than if the viewer is far away. By varying the relative viewing position the three-dimensional nature of the structure can be investigated.

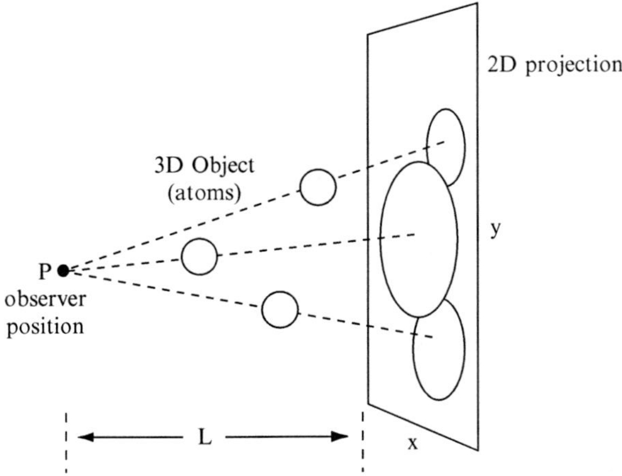

Fig. E.1 3D geometry of viewing a three dimensional collection of atoms (a specimen) as a two-dimensional projection

There are several approximations that make the 2D image simple to generate. First approximate a 3D sphere as a shaded circle. In 3D some atoms will be in front of other atoms and fully or partially hide those atoms in the back from the viewer. This is the so-called hidden surface problem. A simple approach to drawing hidden surfaces is to sort by depth and draw from the back forward. This is not particularly efficient but is simple to program and does a reasonable job of hiding the appropriate atoms. One particular situation that is not handled properly is the case where two adjacent atoms are at the same depth. One atom will be arbitrarily drawn on top of the other. This approach is simply enough to program on relatively simple and inexpensive personal computers (Kirkland [200]).

A 3D perspective view of the specimen is more useful if it can be rotated to see it from different angles. This should be done before drawing the 2D image for obvious reasons. Given a rotation angle ϕ and the tilt angle θ the initial set of atom coordinates (x_i, y_i, z_i) can be rotated about the point (x_0, y_0, z_0) (usually the center of the crystal) in two steps. First rotate by ϕ as:

$$\begin{aligned} x_i' &= (x_i - x_0)\cos\phi - (y_i - y_0)\sin\phi \\ y_i' &= (x_i - x_0)\sin\phi + (y_i - y_0)\cos\phi \end{aligned} \tag{E.1}$$

then tilt by θ as:

$$\begin{aligned} y_i'' &= y_i'\cos\theta + (z_i - z_0)\sin\theta \\ z_i' &= -y_i'\sin\theta + (z_i - z_0)\cos\theta. \end{aligned} \tag{E.2}$$

Rotation and tilt produce a new set of coordinates (x'_i, y''_i, z'_i). An angle of $(\phi, \theta) = (0,0)$ generates a view down the beam direction (the optic axis of the electron microscope). After rotation and tilt the z'_i coordinates are offset to yield $z > 0$ at the top or entrance surface and $z = 0$ at the bottom or exit surface. In Fig. E.1 the electrons are traveling from left to right with $z = 0$ on the right.

The position of each atom on the 2D viewing screen is not simply (x'_i, y''_i). The three dimensional geometry between the viewer, the specimen and 2D viewing screen must be taken into account (for example Newman and Sproull [263]). If L is the distance from the viewer to the viewing screen and the specimen is in between the viewer and the viewing screen (as in Fig. E.1) then the actual coordinates on the viewing screen $(x_i, y_i)_s$ are:

$$\begin{aligned} x_{si} &= \frac{Lx'_i}{L - z'_i} \\ y_{si} &= \frac{Ly''_i}{L - z'_i} \end{aligned} \tag{E.3}$$

by comparison of similar triangles as in Fig. E.2. The apparent size of each atom must also be scaled in a similar manner. If the actual 3D diameter of the atom is d_i then the diameter of the 2D circle is:

$$d_{si} = \frac{Ld_i}{L - z'_i} \tag{E.4}$$

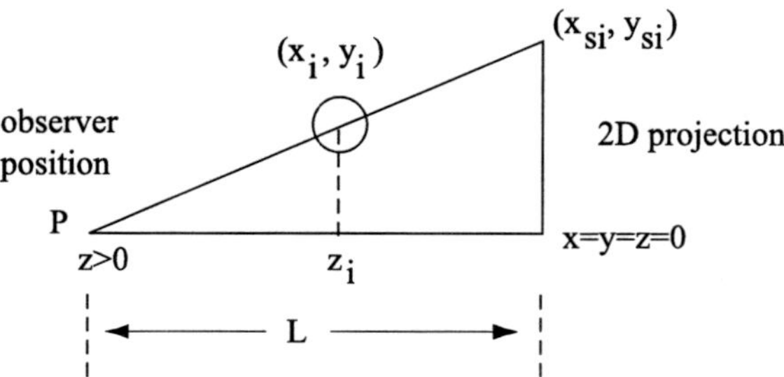

Fig. E.2 Similar triangles used to calculate the relative coordinates in a 3D perspective view

The final set of coordinates is sorted by depth z'_i and drawn from negative z'_i to positive z'_i. The sorting method of Shell is relatively easy to program and efficient enough for this purpose (Shell [318], Press et al. [288]). After sorting, each atom is drawn as a shaded circle. Each successive layer of atoms will overwrite the previous layer taking care of the hidden surface problem. Drawing a simple shaded circle is relatively easy and can mimic an actual 3D shaded sphere in a convincing manner. There are several possible ways to do this and two empirical schemes are listed later. If the output 2D perspective image is encoded as eight bits per pixel (integer valued) with 255 being white and 0 being black then the grey scale intensity g inside each circle can follow:

$$g = 255 - 150\frac{r^2}{r_{max}^2} \tag{E.5}$$

where r is the radius of the circle, and r_{max} is its maximum radius. This generates a white shaded circle on a black background. The entire background should be set to black before drawing any of the atoms. Alternately if an image is generated for printing on a postscript printer then the following postscript macros (Adobe Reference manual [170]) generates a black shaded circle on a white background:

```
%
% macro to make a unit circle at (0,0)
%
 /circle {newpath 0 0 1 0 360 arc
 closepath fill} def

%
%  macro to make a shaded sphere
%  call as-->  xscale yscale xpos ypos sp
%
/sp { gsave translate scale
   0.0 0.04 1 { sqrt 1 exch sub setgray circle
   0.98 0.98 scale  } for grestore } def
```

The first macro called “circle” draw a circle and fills it in with the current color or grey level. The second macros called “sp” draws a series of solid circle one on top of the other. Each successive circle is slightly smaller and slightly blacker to generate the shading. The “sp” macro is given a short name because it must be called many times and a shorter name will use less disk space and transfer quicker to the printer. Figure E.3 shows an example of a 3D view looking down the 110 direction of the silicon lattice.

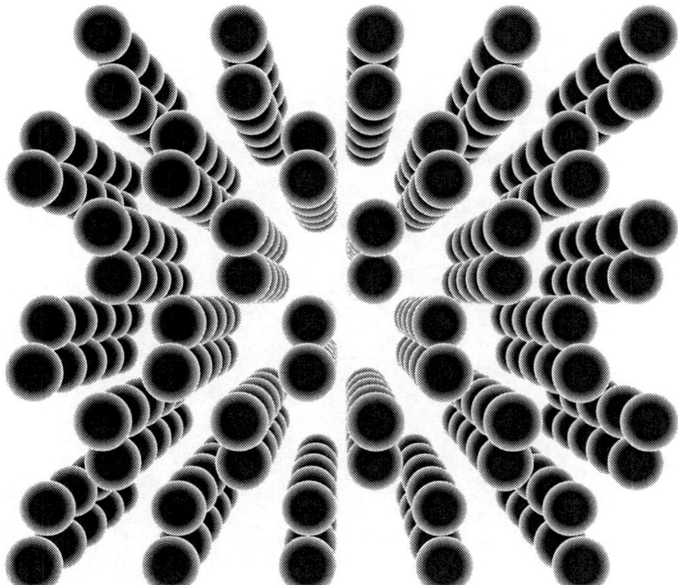

Fig. E.3 A 3D perspective view looking down the (110) direction of silicon (drawn in postscript format). Each silicon atom is drawn as a *shaded hard sphere (black)*

References

1. Abramowitz M, Stegun IA (1965) Handbook of mathematical functions. Dover, New York
2. Agrawal GP (1995) Nonlinear fiber optics, 2nd edn. Academic Press, San Diego
3. Akima H (1970) A new method of interpolation and smooth curve fitting based on local procedures. J Assoc Comput Mach 17:589–602
4. Akima H (1972) Algorithm 433: Interpolation and smooth curve fitting based on local procedures. Commun Assoc Comput Mach 15:914–918
5. Allen LJ, Faulkner HML, Leeb H (2000) Inversion of dynamical electron diffraction data including adsorption. Acta Cryst A56:119–126
6. Allen LJ, Findlay SD, Oxley MP, Rossouw CJ (2003) Lattice-resolution contrast from a focused coherent electron probe.part i. Ultramicroscopy 96:47–63
7. Allen LJ, Josefsson TW, Leeb H (1998) Obtaining the crystal potential by inversion from electron scattering intensities. Acta Cryst A54:388–398
8. Allen LJ, Leeb H, Spargo AEC (1999) Retrieval of the projected potential by inversion from the scattering matrix in electron-crystal scattering. Acta Cryst A55:105–111
9. Allen LJ, Rossouw CJ (1990) Absorptive potentials due to ionization and thermal diffuse scattering by fast electrons in crystals. Phys Rev B 42:11644–11654
10. Allpress JG, Hewat EA, Moodie AF, Sanders JV (1972) n-beam lattice images. I. experimental and computed images of $W_4Nb_{26}O_{77}$. Acta Cryst A28:528–536
11. Allpress JG, Sanders JV (1973) The direct observation of the structure of real crystals by lattice imaging. J Appl Cryst 6:165–190
12. Amali A, Rez P (1997) Theory of lattice resolution in high-angle annular dark-field images. Micro Microanal 3:28–46
13. Anderson E, Bai Z, Bischof C, Blackford S, Demmel J, Dongarra J, Du Croz J, Greenbaum A, Hammarling S, McKenney A, Sorensen D (1999) LAPACK Users' Guide, 3rd edn. Society for Industrial and Applied Mathematics, Philadelphia, PA
14. Anstis GR (2004) The s-state analysis applied to high-angle, annular dark-field image interpretation when can we use it? Micro Microanal 10:4–8
15. Anstis GR, Cai DQ, Cockayne DJH (2003) Limitations on the s-state approach to the interpretation of sub-angstrom resolution electron microscope images and microanalysis. Ultramicroscopy 94:309–327
16. Anstis GR, Cockayne DJH (1979) The calculation and interpretation of high-resolution electron microscope images of lattice defects. Acta Cryst A35:511–524
17. Ashcroft NW, Mermin ND (1976) Solid state physics. Holt, Rinehart and Winston, New York
18. Barakat R, Houston A (1965) Transfer function of an annular aperture in the presence of spherical aberration. J Optical Soc Am 55:538–541
19. Barry JC (1989) Semiquantitative image matching in HRTEM. In: Krakow W, O'Keefe M (eds) Computer Simulation of Electron Microscope Diffraction and Images, pp 57–78, Warrendale, Penn., The Minerals, Metals and Materials Society

20. Barry JC (1992) Image-matching as a means of atomic structure evaluation in high resolution transmission electron microscopy. In: Hawkes PW (ed) Signal and image processing in microscopy and microanalysis, scanning microscopy, supplement 6, pp 209–221, Chicago. Scanning Microscopy Intern.
21. Batterman BW, Cole H (1964) Dynamical diffraction of X-rays by perfect crystals. Rev Modern Phys 36:681–717
22. Beeching MJ, Spargo AEC (1993) A method for crystal potential retrieval in HRTEM. Ultramicroscopy 52:243–247
23. Beeching MJ, Spargo AEC (1998) Inversion of nonperiodic wavefields to determine localized defect structure. J. Microscopy 190:262–266
24. Bethe H (1928) Theorie der beugung von elektronen an kristallen. Annalen der Physik 87:55–129
25. Bethe H (1930) Zur theorie des durchgangs schneller korpuskularstrahlen durch materie. Annalen der Physik 5:325–400
26. Bethe HA (1997) Selected Works of Hans A. Bethe. World Scientific, Singapore
27. Bethe HA, Jackiw RW (1968) Intermediate quantum mechanics, 2nd edn. Benjamin, Reading, Mass.
28. Binnig G, Rohrer H (1987) Scanning tunneling microscopy - from birth to adolescence. Rev Modern Phys 59:615–625
29. Bird DM (1989) Theory of zone axis electron diffraction. J Elect Micros Tech 13:77–97
30. Black G, Linfoot EH (1957) Spherical aberration and the information content of optical images. Proc R Soc Lond A239:522–540
31. Bonevich JE, Marks LD (1988) Contrast transfer theory for non-linear imaging. Ultramicroscopy 26:313–320
32. Bonham RA, Fink M (1974) High energy electron scattering. Van Nostrand Reinhold, New York
33. Boothroyd CB (1998) Why don't high-resolution simulations and images match? J Micro 190:99–108
34. Born M, Wolf E (1980) Principles of optics, 6th edn. Pergamon Press, Oxford
35. Bowen DK, Hall CR (1975) Microscopy of materials. MacMillan Press, London
36. Bozzola JJ, Russell LD (1999) Electron Microscopy, Princ. and Tech. for Biologists, 2nd edn. Jones and Bartlett, Sudbury, Mass.
37. Bracewell RN (1986) The Fourier Transform and its applications, 2nd edn. McGraw-Hill, New York
38. Bragg WL, West J (1929) A technique for the X-ray examination of crystal structures with many parameters. Zeit fur Kristallographie 69:118–148
39. Brigham EO (1974) The fast fourier transform. Prentice-Hall, Englewood Cliffs, New Jersey
40. Budinger TF, Glaeser RM (1976) Measurement of focus and spherical aberration of an electron microscope objective lens. Ultramicroscopy 2:31–41
41. Burke VM, Grant IP (1967) The effect of relativity on atomic wave functions. Proc Phys Soc 90:297–576
42. Bursill LA, Wilson AR (1977) Electron-optical imaging of the hollandite structure at 3 å resolution. Acta Cryst A33:672–676
43. Buseck PR, Cowley JM, Eyring L (eds.) (1988) High-resolution transmission electron microscopy. Oxford Univniversity Press, New York
44. Buxton BF, Loveluck JE, Steeds JW (1978) Bloch waves and their corresponding atomic and molecular orbitals in high energy electron diffraction. Philosophical Magaz A 38:259–278
45. Carlino E, Grillo V, Palazzari P (2008) Accurate and fast multislice simulations of HAADF image contrast by parallel computing. In: Cullis AG, Midgley PA (eds.) Springer Proc. Phys., volume 120, pp 177–180
46. Castleman KR (1979) Digital image processing. Prentice Hall, Englewood Cliffs, New Jersey
47. Chen JH, Van Dyck D (1997) Accurate multislice theory for elastic electron scattering in transmission electron microscopy. Ultramicroscopy 70:29–44
48. Chen JH, Van Dyck D, Op de Beck M (1997) Multislice method for large beam tilt with applications to HOLZ effects in triclinic and monoclinic crystals. Acta Cryst. A53:576–589

49. Chen JH, Van Dyck D, Op de Beck M, Van Landuyt J (1997) Computational comparisons between the conventional multislice method and the third-order multislice method for calculating high-energy electron diffraction and imaging. Ultramicroscopy 69:219–240
50. Coene W, Van Dyck D (1984) The real space method for dynamical electron diffraction calculation in high resolution electron microscopy, II. critical analysis of the dependency on the input parameters. Ultramicroscopy 15:41–50
51. Coene W, Van Dyck D (1984) The real space method for dynamical electron diffraction calculations in high resolution electron microscopy III. a computational algorithm for the electron propagation with practical applications. Ultramicroscopy 15:287–300
52. Coene W, Janssen G, Op de Beck M, Van Dyck D (1992) Phase retrieval through focus variation for ultra-resolution in field-emission transmission electron microscopy. Phys Rev Lett 69:3743–3746
53. Cooley JW, Tukey JW (1965) An algorithm for the machine calculation of complex fourier series. Math Comput 19:297–301
54. Cordes MJ, Pidwerbetsky A, Lovelace RVE (1986) Refractive and diffractive scattering in the interstellar medium. Astrophys J 310:737–767
55. Aldus Corp. (1992) TIFF, Revision 6.0. Aldus Corp., Seattle, WA
56. Coulthard MA (1967) A relativistic hartree-fock atomic field calculation. Proc Phys Soc 91:44–49
57. Cowan RD (1981) The theory of atomic structure and spectra. University of California Press, Berkeley
58. Cowley JM (1969) Image contrast in a transmission scanning electron microscope. Appl Phys Lett 15:58–59
59. Cowley JM (1975) Diffraction physics, 2nd edn. North-Holland, Amsterdam
60. Cowley JM (1976) Scanning transmission electron microscopy of thin specimens. Ultramicroscopy 2:3–16
61. Cowley JM (1988) Electron microscopy of crystals with time-dependent perturbations. Acta Cryst A44:847–853
62. Cowley JM, Iijima S (1972) Electron microscope image contrast for thin crystals. Z. Naturforsch 27a:445–451
63. Cowley JM, Moodie AF (1957) The scattering of electrons by atoms and crystals. I. a new theoretical approach. Acta Cryst 10:609–619
64. Cowley JM, Murray RJ (1968) Diffuse scattering in electron diffraction patterns, II. short-range order scattering. Acta Cryst A24:329–336
65. Cowley JM, Pogany AP (1968) Diffuse scattering in electron diffraction patterns, I. general theory and computational methods. Acta Cryst A24:109–116
66. Cowley JM, Spence JCH (1979) Innovative imaging and microdiffraction in STEM. Ultramicroscopy 3:433–438
67. Cox HL, Bonham RA (1967) Elastic electron scattering amplitudes for neutral atoms calculated using the partial wave method at 10, 40, 70, and 100 kV for $Z = 1$ to $Z = 54$. J Chem Phys 47:2599–2608
68. Crewe AV, Langmore JP, Isaacson MS (1975) Resolution and contrast in the scanning transmission electron microscope. In: Siegel BM, Beaman DR (eds.) Physical aspects of electron microscopy and microbeam analysis, pp 47–62. Wiley, New York
69. Crewe AV, Wall J, Welter LM (1968) A high-resolution scanning transmission electron microscope. J Appl Phys 39:5861–5868
70. Croitoru MD, Van Dyck D, Van Aert S, Bals S, Verbeeck J (2006) An efficient way of including thermal diffuse scattering in simulation of scanning transmission electron microscope images. Ultramicroscopy 106:933–940
71. Cromer DT, Mann JB (1968) X-ray scattering factors computed from numerical Hartree-Fock wave functions. Acta Cryst A24:321–324
72. Cromer DT, Waber JT (1965) Scattering factors computed from relativistic Dirac-Slater wave functions. Acta Cryst 18:104–109

73. Cromer DT, Waber JT (1974) Atomic scattering factors for X-rays. In: Ibers JA, Hamilton WC (eds.) International Tables for X-Ray crystallography, Vol. IV, Section 2.2, pp 71–147. Kynoch Press, Birmingham
74. Dawson B (1961) Atomic scattering amplitudes for electrons for some of the lighter elements. Acta Cryst 14:1120–1124
75. Op de Beck M (1993) Comments on the use of the relativistic Schrodinger equation in high-energy electron diffraction. In: Bailey GW, Rieder CL (eds.) Proceedings of the 51th Annual Meeting of the Microscopy Society of America, pp 1212–1213. San Fransisco Press
76. Desclaux JP (1973) Relativistic dirac-fock expectation values for atoms with z=1 to z=120. Atomic Data Nuclear Data Tables 12:311–406
77. Desclaux JP (1975) A multiconfiguration relativistic dirac-fock program. Comput Phys Commun 9:31–45
78. Desseaux J, Renault A, Bourret A (1977) Multi-beam images of germanium oriented in (011). Phil Mag 35:357–372
79. Dinges C, Rose H (1995) Simulation of filtered and unfiltered TEM images and diffraction patterns. Phys. Stat. Sol. (A) 150:23–30
80. Downing KH, Grano DA (1982) Analysis of photographic emulsions for electron microscopy of two-dimensional crystalline specimens. Ultramicroscopy 7:381–404
81. Doyle PA, Cowley JM (1974) Scattering factors for the diffraction of electrons by crystalline solids. In: Ibers JA, Hamilton WC (eds.) International Tables for X-Ray Crystallography, Vol. IV, Section 2.4, pp 152–175. Kynoch Press, Birmingham
82. Doyle PA, Turner PS (1968) Relativistic Hartree-Fock x-ray and electron scattering factors. Acta Cryst A24:390–397
83. Dudgeon DE, Mersereau RM (1984) Multidimensional digital signal processing. Prentice Halls, New Jersey
84. Dwyer C (2009) Multislice simulation of scanning transmission electron microscope images. In: Brewer LN, McKernan S, Shields JP, Schmidt-jr FE, Woodward JH, Zaluzec NJ (eds.) Microscopy and microanalysis 2009, volume 15, suppl. 2, pp 754–755, Cambridge University Press, Cambridge, UK
85. Dwyer C, Etheridge J (2003) Scattering of Å-scale electron probes in silicon. Ultramicroscopy 96:343–360
86. Dyall, KG, Grant IP, Johnson CT, Parpia FA, Plummer EP (1989) Grasp: A general-purpose relativistic atomic structure program. Comput Phys Commun 55:425–456
87. Edington JW (1976) Practical electron microscopy in materials science. Van Nostrand Reinhold, New York
88. Eisberg R, Resnick R (1985) Quantum physics of atoms, molecules, solids, nuclei, and particles, 2nd edn. Wiley, New York
89. Eisenhandler CB, Siegel BM (1966) Imaging of single atoms with the electron microscope by phase contrast. J. Appl Phys 37:1613–1620
90. Eisenhandler CB, Siegel BM (1966) A zone-plate aperture for enhancing resolution in phase contrast electron microscopy. Appl Phys Lett 8:258–260
91. El-Kareh AB, El-Kareh JCJ (1970) Electron beams, lenses, and optics, Vol. 1,2. Academic Press, New York
92. Engel A (1974) The principle of reciprocity and its application to conventional and scanning dark field electron microscopy. Optik 41:117–126
93. Engel A (2009) Scanning transmission electron microscopy: Biological application. In: Hawkes PW (ed.) Adv. in Imaging and Electron Physics, vol. 159, pp 357–386. Academic Press, San Diego
94. Engel A, Colliex C (1993) Application of scanning transmission electron microscopy to the study of biological structure. Curr Opinion Biotechnol 4:403–411
95. Erikson HP (1973) The fourier transform of an electron micrograph - first order and second order theory of image formation. In: Barer R, Cosslett VE (eds.) Adv. in Optical and Electron Microscopy, Vol. 5, pp 163–199. Academic Press, London

96. Fan GY (1989) Multislice calculation of kikuchi patterns. In: Bailey GW (ed.) Proceedings of the 47th Annual Meeting of the Microscopy Society of America, pp 52–53. San Fransisco Press
97. Fan GY, Mercurio PJ, Young SJ, Ellisman MH (1993) Telemicroscopy. Ultramicroscopy 52:499–503
98. Feit MD, Fleck JA (1978) Light propagation in graded-index optical fibers. Appl Optics 17:3990–3998
99. Fejes PL (1977) Approximations for the calculation of high resolution electron microscope images of thin films. Acta Cryst A33:109–113
100. Fertig J, Rose H (1981) Resolution and contrast of crystalline objects in high-resolution scanning transmission electron microscopy. Optik 59:407–429
101. Ferwerda HA, Hoenders BJ, Slump CH (1986) The fully relativistic foundation of linear transfer theory in electron optics based on the Dirac equation. Optica Acta 33:159–183
102. Ferwerda HA, Hoenders BJ, Slump CH (1986) Fully relativistic treatment of electron-optical image formation based on the Dirac equation. Optica Acta 33:145–157
103. Ferwerda HA, Visser FPC (1973) Applications of Glauber's scattering theory to the scattering of electrons by heavy elements. In: Hawkes PW (ed.) Image Processing and Computer-Aided Design in Electron Optics, pp 212–219. Academic Press, London
104. Feynman RP (1951) An operator calculus having applications in quantum electrodynamics. Phys Rev 84:108–128
105. Fields PM, Cowley JM (1978) Computed electron microscope images of atomic defects in fcc metals. Acta Cryst A34:103–112
106. Froese Fischer C, Brage T, Jonsson P (1997) Computational atomic structure, an MCDF approach. Institute of Physics Publishing, Bristol and London
107. Fleck(Jr.) JA, Morris JR, Feit MD (1976) Time-dependent propagation of high energy laser beams through the atmosphere. Appl Phys 10:129–160
108. Forsyth JB, Wells M (1959) On the analytical approx. to the atomic scattering factor. Acta Cryst 12:412–415
109. Fox AG, O'Keefe MA, Tabbernor MA (1989) Relativistic Hartree-Fock X-ray and electron atomic scattering factors at high angles. Acta Cryst A45:786–793
110. Frank J (1972) A study on heavy/light atom discrimination in bright-field electron microscopy. Biophys J 12:484–511
111. Frank J (1973) The envelope of electron microscope transfer functions for partially coherent illumination. Optik 38:519–536
112. Freeman AJ (1959) Atomic scattering factors for spherical and aspherical charge distributions. Acta Cryst 12:261–270
113. Freeman AJ, Wood JH (1959) An atomic scattering factor for iron. Acta Cryst 12:271–273
114. French RG, Lovelace RVE (1983) Strong turbulence and atmospheric waves in stellar occultations. Icarus 56:122–146
115. Frigo M, Johnson SG (2003) FFTW. www.fftw.org
116. Frigo SP, Levine ZH, Zaluzec NJ (2002) Submicron imaging of buried integrated circuit structures using scanning confocal electron microscopy. Appl Phys Lett 81:53–60
117. Froese-Fischer C (1977) The Hartree-Fock method for atoms. Wiley, New York
118. Fujimoto F (1959) Dynamical theory of electron diffraction in Laue-case I. general theory. J. Phys Soc Jpn 14:1558–1568
119. Fujiwara K (1961) Relativistic dynamical theory of electron diffraction. J Phys Soc Jpn 16:2226–2238
120. Fultz B, Howe JM (2001) Transmission electron microscopy and diffractometry of materials. Springer-Verlag, Berlin
121. Glauber R, Schomaker V (1953) The theory of electron diffraction. Phys Rev 89:667–671
122. Goldstein JI, Newbury DE, Echlin P, Joy DC, Fiori C, Lifshin E (1981) Scanning electron microscopy and X-ray microanalysis. Plenum Press, New York
123. Gómez-Rodríguez A, Beltrán del Río LM, Herrera-Becerra R (2010) SimulaTEM: Multislice simulations for general objects. Ultramicroscopy 110:95–104

124. Gonzalez RC, Wintz P (1987) Digital image processing, 2nd edn. Addison-Wesley, Reading, Mass.
125. Gonzalez RC, Woods RE (2008) Digital image processing, 3rd edn. Prentice Hall, New York
126. Goodman JW (2005) Introduction to Fourier Optics, 3rd edn. Roberts and Co., New York
127. Goodman P, Moodie AF (1974) Numerical evaluation of N-beam wave functions in electron scattering by the multislice method. Acta Cryst A30:280–290
128. Gradshteyn IS, Ryzhik IM (1994) Table of integrals, series, and products, 5th edn. Academic Press, San Diego
129. De Graf M (2003) Introduction to conventional transmission electron microscopy. Cambridge University Press, Cambridge, UK
130. Grant IP (1961) Relativistic self-consistent fields. Proc. R Soc A 262:555–576
131. Grant IP (1970) Relativistic calculation of atomic structures. Adv Phys 19:747–811
132. Grant IP, McKenzie BJ, Norrington PH, Mayers DF, Pyper NC (1980) An atomic multiconfigurational Dirac-Fock package. Comput Phys Commun 21:207–231
133. Grillo V, Carlino E (2008) Influence of the static atomic displacement on atomic resolution Z-contrast imaging. Phys Rev B 77:054103
134. Grinton GR, Cowley JM (1971) Phase and amplitude contrast in electron microscopy of biological materials. Optik 34:221–233
135. Grivet P (1972) Electron optics, Parts 1 and 2, 2nd edn. Pergamon, Oxford
136. Haase J (1968) Berechnung der komplexen atomaren streufaktoren fur schnelle elektronen unter verwendung von Hartree-Fock-atompotentialen. Zeitschrift fur Naturforschung 23a:1000–1019
137. Haase J (1970) Zusammenstellung der koeffizienten fur die anpassung kompexer atomarer streufaktoren fur schnelle elektron durch polynome 1. mitteilung: Hartree-Fock fall. Zeitschrift fur Naturforschung 25a:936–945
138. Haase J (1970) Zusammenstellung der koeffizienten fur die anpassung kompexer atomarer streufaktoren fur schnelle elektron durch polynome 2. mitteilung: Thomas-Fermi-Dirac-fall. Zeitschrift fur Naturforschung 25a:1219–1235
139. Haider M, Uhlemann S, Zach J (2000) Upper limits for the residual aberrations of a high-resolution aberration-corrected STEM. Ultramicroscopy 81:163–175
140. Haken H, Wolf HC (1984) Atomic and quantum physics. Springer-Verlag, Berlin
141. Hall CE (1966) Introduction to electron microscopy, 2nd edn. McGraw-Hill, New York
142. Hall CR, Hirsch PB (1965) Effect of thermal diffuse scattering on propagation of high energy electrons through cyrstals. Proc R Soc Lond A286:158–177
143. Hall EL (1979) Computer image processing and recognition. Academic Press, New York
144. Hanai T, Yoshida H, Hibino M (1998) Characteristics and effectiveness of a foil lens for correction of spherical aberration in scanning transmission electron microscopy. J Elect Micros 47:185–192
145. Hanszen K.-J. (1971) The optical transfer theory of the electron microscope: fundamental principles and applications. In: Barer R, Cosslett VE (eds.) Advances in Optical and Electron Microscopy, Vol. 4, pp 1–84. Academic Press, New York
146. Hanszen K.-J., Trepte L (1971) Die kontrastubertragung im elektronenmikroskop bei partiell koharenter beleuchtung. Optik 33:166–198
147. Hartree DR (1957) The calculation of atomic structure. John Wiley and Sons, New York
148. Hawkes PW (1972) Electron optics and the electron microscope. Taylor and Francis, London
149. Hawkes PW (ed.) (1985) The beginnings of electron microscopy. Adv. in Electronics and Electron Physics, Suppl. 16. Academic Press, Orlando
150. Hawkes PW (ed.) (2008) Aberration-corrected electron microscopy. Adv. in Imaging and Electron Physics, Vol. 153. Academic Press, Orlando
151. Hawkes PW, Kasper E (1989) Principles of electron optics, volume 1. Academic Press, San Diego. Basic Geometrical Optics
152. Hawkes PW, Kasper E (1989) Principles of electron optics, volume 2. Academic Press, San Diego. Applied Geometrical Optics
153. Hawkes PW, Kasper E (1994) Principles of electron optics, volume 3. Academic Press, San Diego. Wave Optics

154. Heidenreich RD (1964) Fundamentals of transmission electron microscopy. Wiley, New York
155. Herman F, Skillman S (1963) Atomic structure calculations. Prentice-Hall, Englewood Cliffs, NJ
156. Herrmann K.-H. (1978) The present state of instrumentation in high-resolution electron microscopy. J Phys E: Sci Instrum 11:1076–1091
157. Herrmann K.-H. (1983) Instrumentational requirments for high resolution imaging. J Microscopy 131:67–78
158. Hillyard S, Silcox J (1995) Detector geometry, thermal diffuse scattering and strain effects in ADF-STEM imaging. Ultramicroscopy 58:6–17
159. Hirsch P, Howie A, Nicholson RB, Pashley DW, Whelan MJ (1977) Electron microscopy of thin crystals, 2nd edn. Krieger, Huntington, New York
160. Hoppe W (1970) Prin. of electron structure research at atomic resolution using conventional electron microscopes for the measurement of amplitudes and phases. Acta Cryst A26:414–426
161. Horiuchi S (1994) Fundamentals of high resolution transmission electron microscopy. North-Holland, Amsterdam
162. Howie A, Basinski ZS (1968) Approx. of the dynamical theory of diffraction contrast. Phil Mag 17:1039–1063
163. Howie A, Whelan MJ (1961) Diffraction contrast of electron microscope images of crystal lattice defects, II. the development of a dynamical theory. Proc R Soc Lond A263:217–237
164. Humphreys CJ (1979) The scattering of fast electrons by crystals. Rep Prog Phys 42:1825–1887
165. Hunt HC, Andrews BR (1977) Digital image restoration. Prentic-Hall, New York
166. Hutchison JL, Waddington WG (1988) Atomic images of silicon? Ultramicroscopy 25:93–96
167. Hÿtch MJ, Stobbs WM (1994) Quantitative comparison of high resolution TEM images with image simulation. Ultramicroscopy 53:191–203
168. Ibers JA (1958) Atomic scattering amplitudes for electrons. Acta Cryst 11:178–183
169. Ibers JA, Vainshtein BK (1962) Section 3.3.3. In: MacGillavry CH, Rieck GD, Lonsdale K (eds.) International tables for X-ray crystallography, Vol. III, pp 216–227. Kynoch Press, Birmingham
170. Adobe Systems Inc. (1990) Postscript language reference manual, 2nd edn. Addison-Wesley, Reading, Mass.
171. Intaraprasonk V, Xin HL, Muller DA (2008) Analytic derivation of optimal imaging conditions for incoherent imaging in aberration-corrected electron microscopes. Ultramicroscopy 108:1454–1466
172. Isaacson MS, Langmore J, Parker NW, Kopf D, Utlaut M (1976) The study of the adsorption and diffusion of heavy atoms on light element substrates by means of the atomic resolution STEM. Ultramicroscopy 1:359–376
173. Ishizuka K (1980) Contrast transfer of crystal images in TEM. Ultramicroscopy 5:55–65
174. Ishizuka K (1982) Multislice formula for inclined illumination. Acta Cryst A38:773–779
175. Ishizuka K (1994) Coma-free alignment of a high-resolution electron microscope with three-fold astigmatism. Ultramicroscopy 55:407–418
176. Ishizuka K (2002) A practical approach for STEM image simulation based on the FFT multislice method. Ultramicroscopy 90:71–83
177. Ishizuka K (2004) FFT multislice method - the silver anniversary. Microsc Microanal 10:34–40
178. Ishizuka K (2006) www.hremresearch.com.
179. Ishizuka K, Uyeda N (1977) A new theoretical and practical approach to the multislice method. Acta Cryst A33:740–749
180. Izui K, Furuno S, Otsu H (1977) Observations of crystal structure images of silicon. Jpn J Elect Micros 26:129–132
181. Jagannathan R (1990) Quantum theory of electron lenses based on the Dirac equation. Phys Rev A 42:6674–6689
182. Jagannathan R, Simon R, Sudarshan ECG, Mukunda N (1989) Quantum theory of magnetic electron lenses based on the Dirac equation. Phys Lett A 134:457–464

183. Jahne B (1995) Digital image processing, 3rd edn. Springer, Berlin
184. Jain AK (1989) Fundamentals of digital image processing. Prentice Hall, Englewood Cliffs, NJ
185. James RW, Brindley GW (1931) Some numerical calculations of atomic scattering factors. Phil Mag 12:81–112
186. Jap BK, Glaeser RM (1978) The scattering of high-energy electrons. I. Feynman path-integral formulation. Acta Cryst A34:94–102
187. Jesson DE, Pennycook SJ (1993) Incoherent imaging of thin specimens using coherent scattered electrons. Proc R Soc Lond A 441:261–281
188. Jia C-L, Lentzen M, Urban K (2004) High-resolution transmission electron microscopy using negative spherical aberration. Microsc Microanal 10:174–184
189. Jmol. (2008) Jmol: an open-source java viewer for chemical structure in 3d. www.jmol.org.
190. Joy DC (1995) Monte Carlo modeling for electron microscopy and analysis. Oxford University Press, New York
191. Kambe K, Lehmpfuhl G, Fujimoto F (1974) Interpretation of electron channeling by the dynamical theory of electron diffraction. Z Naturforsch 29a:1034–1044
192. Kernighan BW, Ritchie DM (1988) The C programming language, 2nd edn. Prentice Hall, Englewood Cliffs, New Jersey
193. Keyse RJ, Garratt-Reed AJ, Goodhew PJ, Lorimer GW (1998) Introduction to scanning transmission electron microscopy. Springer-Verlag, New York
194. Kilaas R (1987) Interactive simulation of high-resolution electron micrographs. In: Bailey GW (ed.) Proceedings of the 45th Annual Meeting of the Microscopy Society of America, pp 66–69. San Fransisco Press
195. Kilaas R (2006) www.totalresolution.com/index.html
196. Kilaas R, Gronsky R (1983) Real space image simulation in high resolution electron microscopy. Ultramicroscopy 11:289–298
197. Kilaas R, O'Keefe MA, Krishman KM (1987) On the inclusion of upper Laue layers in computational methods in high resolution transmission electron microscopy. Ultramicroscopy 21:47–62
198. King WE, Campbell GH (1994) Quantitative HREM using non-linear least-squares methods. Ultramicroscopy 56:46–53
199. Kirkland EJ (1984) Improved high resolution image processing of bright field electron micrographs I. theory. Ultramicroscopy 15:151–172
200. Kirkland EJ (1985) Viewing molecules with the macintosh. Byte Mag Feb.:251–259
201. Kirkland EJ (1988) Digital restoration of ADF-STEM images. In: Bailey GW (ed.) Proceedings of the 46th Annual Meeting of the Microscopy Society of America, pp 832–833. San Fransisco Press
202. Kirkland EJ (1988) High resolution image processing of electron micrographs. In: Hawkes PW, Ottensmeyer FP, Saxton WO, Rosenfeld A (eds.) Image and signal processing in electron microscopy, scanning microscopy, Supplement 2, pp 139–147, Chicago. Scanning Microscopy Intern.
203. Kirkland EJ (1990) An image and spectrum acquisition system for a VG HB501 STEM using a color graphics workstation. Ultramicroscopy 32:349–364
204. Kirkland EJ (1997) Z-contrast in a conventional TEM. In: Bailey GW, Dimlich RVW, Alexander KB, McCarthy JJ, Pretlow TP (eds.) Proceedings Microscopy and Microanalysis 1997, pp 1147–1148, Springer, New York
205. Kirkland EJ (1998) Advanced computing in electron microscopy. Plenum, New York
206. Kirkland EJ (2005) Some effects of electron channeling on electron energy loss spectroscopy. Ultramicroscopy 102:199–207
207. Kirkland EJ, Loane RF, Silcox J (1987) Simulation of annular dark field STEM images using a modified multislice method. Ultramicroscopy 23:77–96
208. Kirkland EJ, Loane RF, Xu P, Silcox J (1989) Multislice simulation of ADF-STEM and CBED images. In: Krakow W, O'Keefe M (eds.) Computer simulation of electron microscope diffraction and images, pp 13–31, Warrendale, Penn. The Minerals, Metals and Materials Society

209. Kirkland EJ, Siegel BM, Uyeda N, Fujiyoshi Y (1980) Digital reconstruction of bright field phase contrast images from high resolution electron micrographs. Ultramicroscopy 5:479–503
210. Kirkland EJ, Siegel BM, Uyeda N, Fujiyoshi Y (1985) Improved high resolution image processing of bright field electron micrographs II. experiment. Ultramicroscopy 17:87–104
211. Kittel C (1996) Introduction to solid state physics, 7th edn. Wiley, New York
212. Klemperer O, Barnett ME (1971) Electron optics, 3rd edn. Cambridge University Press, Cambridge, Great Britain
213. Klenov DO, Findlay SD, Allen LJ, Stemmer S (2007) Influence of orientation on the contrast of high-angle annular dark-field images of silicon. Phys Rev B 76:014111
214. Knoll M, Ruska E (1932) Das elektronenmikroskop. Z fur Physik 78:318–339
215. Koops H (1978) Aberration correction in electron microscopy. In: Sturgess JM (ed.) Proceedings of the ninth international congress on electron microscopy, vol. 3, pp 185–196. Imperial Press, Ontario, Canada
216. Koslof D, Kosloff R (1983) A fourier method solution for the time dependent schrodinger equation as a tool in molecular dynamics. J Comput Phys 52:35–53
217. Kosloff R (1988) Time-dependent quantum-mechanical methods for molecular dynamics. J Phys Chem 92:2087–2100
218. Krishna NG, Sirdeshmukh DB (1998) Compilation of temperature factors of hexagonal close packed elements. Acta Cryst A54:513–514
219. Krivanek OL (1976) A method for determining the coefficient of spherical aberration from a single electron micrograph. Optik 45:97–101
220. Krivanek OL (1994) Three fold astigmatism in high resolution transmission electron microscopy. Ultramicroscopy 55:419–433
221. Krivanek OL, Dellby N, Lupini AR (1999) Towards sub-Å electron beams. Ultramicroscopy 78:1–11
222. Langmore JP (1978) Electron microscopy of atoms. In: Hayat MA (ed.) Princ. and Tech. of Electron Microscopy (Biol. App.), Vol. 9, pp 1–63. Van Nostrand, New York
223. LeBeau JM, Findlay SD, Allen LJ, Stemmer S (2008) Quantitative atomic resolution scanning transmission electron microscopy. Phys Rev Lett 100:206101
224. Lenz FA (1971) Transfer of image information in the electron microscope. In: Valdre U (ed.) Electron microscopy in materials science, pp 541–569. Academic Press, New York
225. Lewis AL, Hammond RB, Villagrana RE (1975) The importance of second-order partial derivatives in the theory of high-energy-electron diffraction from imperfect crystals. Acta Cryst A31:221–227
226. Loane RF, Silcox J (1989) Hollow-cone illumination in simulated ADF STEM. In: Bailey GW (ed.) Proceedings of the 47th annual meeting of the microscopy society of America, pp 124–125. San Fransisco Press
227. Loane RF, Xu P, Silcox J (1991) Thermal vibrations in convergent-beam electron diffraction. Acta Cryst A47:267–278
228. Loane RF, Xu P, Silcox J (1992) Incoherent imaging of zone axis crystals with ADF STEM. Ultramicroscopy 40:121–138
229. Lucy LB (1974) An iterative technique for the rectification of observed distributions. Astron J 79:745–754
230. Lynch DF, Moodie AF (1972) Numerical evaluation of low energy electron diffraction intensities I. the perfect crystal with no upper layer lines and no absorption. Surface Sci 32:422–438
231. Lynch DF, O'Keefe MA (1972) n-beam lattice images II. methods of calculation. Acta Cryst A28:536–548
232. Maccagnano-Zacher SE, Mkhoyen KA, Kirkland EJ, Silcox J (2008) Effects of tilt on high-resolution ADF-STEM imaging. Ultramicroscopy 108:718–726
233. MacLagan DS, Bursill LA, Spargo AEC (1977) Experimental and calculated images of planar defects at high resolution. Phil Mag 35:757–780
234. Maréchal A (1947) Étude des effects combinés de la diffraction et des aberrations géométriques sur l'image d'un point lumineux. Rev Opt 26:257–277

235. Marks L, Kilass R (2006) www.numis.northwestern.edu/edm/documentation/edm.htm
236. Matsuhata H, Van Dyck D, Van Lanuyt J, Amelincjx S (1984) A practical approach to the periodic continuation method for the simulation of high resolution TEM images of crystal defects. Ultramicroscopy 13:343–348
237. McMulan D (1995) Scanning electron microscopy 1928-1965. Scanning 17:175–185
238. McWeeny R (1951) X-ray scattering by aggregates of bonded atoms, I. analytical approximations in single atom scattering. Acta Cryst 4:513–519
239. McWeeny R (1952) X-ray scattering by aggregates of bonded atoms, II. the effects of bonds with applications to H_2. Acta Cryst 5:463
240. Meek GA (1976) Practical electron microscopy for biologists, 2nd edn. Wiley, London
241. Megaw HD (1973) Crystal structures: a working approach. Saunders, Philadelphia
242. Misell DL, Stroke GW, Halioua M (1974) Coherent and incoherent imaging in the scanning transmission electron microscope. J Phys D: Appl Phys 7:L113–L117
243. Missel DL (1978) Image analysis, enhancement annd interpretation, volume 7 of *Prac. Methods in Electron Microscopy*. North Holland, Amsterdam
244. Mitsuishi K, Iakoubovskii K, Takeguchi M, Shimojo M, Hashimoto A, Furuya K (2008) Bloch wave-based calculations of imaging properties of high-resolution scanning confocal electron microscopy. Ultramicroscopy 108:981–988
245. Mitsuishi K, Takeguchi M, Yasuda M, Furuya K (2001) New scheme for calculation of annular dark-field STEM image including both elastically diffracted and TDS waves. J Electron Microsc 50:157–162
246. Mkhoyan KA, Maccagnano-Zacher SE, Kirkland EJ, Silcox J (2008) Effects of amorphous layers ADF-STEM imaging. Ultramicroscopy 108:791–803
247. Möbus G (1996) Retrieval of crystal defect structures from HRTEM images by simulated evolution I. basic technique. Ultramicroscopy 65:205–216
248. Möbus G, Dehm G (1996) Retrieval of crystal defect structures from HRTEM images by simulated evolution II. experimental image evaluation. Ultramicroscopy 65:217–228
249. Möbus G, Gemming T, Gumbsch P (1998) The influence of phonon scattering on HREM images. Acta Cryst A54:83–90
250. Möbus G, Rühle M (1994) Structure determination of metal-ceramic interfaces by numerical contrast evaluation of HRTEM micrographs. Ultramicroscopy 56:54–70
251. Möbus G, Schweinfest R, Gemming T, Wagner T, Rühle M (1998) Iterative structure retrieval techniques in HREM: a comparative study and a modular program package. J Microsc 190:109–130
252. Moliere G (1947) Theorie der streuung schneller gelandener teilchen I. einzelstreuung am abgeschirmten coulomb-field. Z fur Naturforsch 2A:133–145
253. Morrison MA, Estle TL, Lane NF (1976) Quantum states of atoms, molecules, and solids. Prentice-Hall, Englewood Cliffs, New Jersey
254. Mory C, Colliex C, Cowley JM (1987) Optimum defocus for STEM imaging and microanalysis. Ultramicroscopy 21:171–178
255. Mott NF (1930) The scattering of electrons by atoms. Proc R Soc A127:658–665
256. Mott NF, Massey HSW (1965) The theory of atomic collisions, 3rd edn. Claredon Press, Oxford
257. Muller DA, Edwards B, Kirkland EJ, Silcox J (1997) Detailed calculations of thermal diffuse scattering. In: Bailey GW, Dimlich RVW, Alexander KB, McCarthy JJ, Pretlow TP (eds.) Proceedings microscopy and microanalysis, pp 1153–1154. Springer, New York
258. Mulvey T (ed.) (1996) The growth of electron microscopy, volume 96 of Advances in imaging and electron physics. Academic Press, San Diego
259. Murray JD, vanRyper W (1996) Encyclopedia of graphics file format, 2nd edn. O'Reilly and Assoc., Sebastopol, CA
260. Nellist PD, Pennycook SJ (1999) Incoherent imaging using dynamically scattered coherent electrons. Ultramicroscopy 78:111–124
261. Nellist PD, Pennycook SJ (2000) The principles and interpretation of annular dark-field Z-contrast imaging. In: Hawkes PW (ed.) Advances in imaging and electron physics, vol. 113, pp 147–203. Academic Press, San Diego

262. Nellist PD, Cosgriff EC, Behan G, Kirkland AI (2008) Imaging modes for scanning confocal electron microscopy in a double aberration-corrected transmission electron microscope. Microsc Microanal 14:82–88
263. Newman WM, Sproull RF (1979) Princ. interactive computer graphics, 2nd edn. McGraw-Hill, New York
264. Niehrs H, Wagner EH (1955) Die amplituden der wellenfelder bei elektroneninterferenzen im Laue-fall. Z Physik 143:285–299
265. Oatley CW, McMullan D, Smith KCA (1985) The development of the scanning electron microscope. In: Marton L, Marton C (eds.) Advances in electronics and electron physics, Suppl. 16, pp 443–482. Academic Press, New York
266. Ohtomo A, Muller DA, Grazul JL, Hwang HY (2002) Artificial charge-modulation in atomic-scale perovskite titanate superlattices. Nature 419:378–380
267. O'Keefe MA (1979) Resolution-damping functions in non-linear imaging. In: Bailey GW (ed.) Proceedings of the 37st Annual Meeting of the Microscopy Society of America, pp 556–557, Cincinnati. San Fransisco Press
268. O'Keefe MA (1995) Advances in image simulation for high resolution TEM. In: Bailey GW, Ellisman MH, Hennigar RA, Zaluzec NJ (eds.) Proc. Micros. Microanal. 1995, pp 38–39. Jones and Begell, New York
269. O'Keefe MA, Buseck PR (1979) Computation of high resolution TEM images of materials. Trans Am Crystallog Assoc 15:27–46
270. O'Keefe MA, Buseck PR, Iijima S (1978) Computed crystal structure images for high resolution electron microscopy. Nature 274:322–324
271. O'Keefe MA, Kilaas R (1988) Advances in high-resolution image simulation. In: Hawkes PW, Ottensmeyer FP, Saxton WO, Rosenfeld A (eds.) Image and signal processing in electron microscopy, scanning microscopy, Supplement 2, pp 225–244. Scanning Microscopy Intern, Chicago
272. O'Keefe MA, Sanders JV (1975) n-beam lattice images. VI. degradation of image resolution by a combination of incident-beam divergence and spherical aberration. Acta Cryst A31:307–310
273. O'Keefe MA, Taylor J, Owen D, Westmacott KH, Johnston W, Dahmen U (1996) Remote on-line control of a high-voltage in situ transmission electron microscope with a rational user interface. In: Bailey GW, Corbett JM, Dimlich RVW, Michael JR, Zaluzec NJ (eds.) Proceedings of the 54th Annual Meeting of the Microscopy Society of America, pp 384–385. San Fransisco Press
274. Oldfiled LC (1973) Computer design of high frequency electron-optical systems. In: Hawkes PW (ed.) Image processing and computer-aided design in electron optics, pp 370–399. Academic Press, London
275. O'Neil EL (1963) Introduction to statistical optics. Addison-Wesley (Dover reprinted 1992), Reading, Mass.
276. Orloff J (ed.) (2009) Handbook of charged particle optics, 2nd edn. CRC Press, Taylor and Francis, Boca Raton
277. Otten MT, Coene WMJ (1993) High resolution imaging on a field emission TEM. Ultramicroscopy 48:77–91
278. Ourmazd A, Baumann FH, Bode M, Kim Y (1990) Quantitative chemical lattice matching: theory and practice. Ultramicroscopy 34:237–255
279. Pan M, Rez P, Cowley JM (1989) A new method of computing the phase-grating function in multislice calculation. In: Bailey GW (ed.) Proceedings of the 47th annual meeting of the microscopy society of America, pp 478–479. San Fransisco Press
280. Peng LM, Cowley JM (1988) Errors arising from numerical use of the Mott formula in electron image simulation. Acta Cryst A44:1–5
281. Peng LM, Ren G, Dudarev SL, Whelan MJ (1996) Robust parameterization of elastic and absorptive electron atomic scattering factors. Acta Cryst A52:257–276
282. Pennycook SJ, Jesson DE (1990) High-resolution incoherent imaging of crystals. Phys Rev Lett 64:938–941

283. Pennycook SJ, Jesson DE (1991) High resolution Z-contrast imaging of crystals. Ultramicroscopy 37:14–38
284. Pennycook SJ, Jesson DE (1992) Atomic resolution Z-contrast imaging of interfaces. Acta Metall Mater 40:S149–S159
285. Pidwerbetsky A, Lovelace RVE (1989) Chaotic wave propagation in a random medium. Phys Lett A 140:411–415
286. Pogany AP, Turner PS (1968) Reciprocity in electron diffraction and microscopy. Acta Cryst A24:103–109
287. Pratt WK (1978) Digital image processing. Wiley, New York
288. Press WH, Teukolsky SA, Vetterling WT, Flannery BP (1992) Numerical recipes in C, 2nd edn. Cambridge University Press, Cambridge
289. Press WH, Teukolsky SA, Vetterling WT, Flannery BP (2007) Numerical recipes, 3rd edn. Cambridge University Press, Cambridge
290. Pulvermacher H (1981) Der transmissions-kreuz-koeffizient fur die elektronenmikroskopische abbildung bei partiell koharenter beleuchtung und elektischer instabilitat. Optik 60:45–60
291. Quinn MJ (2004) Parallel Programming in C with MPI and openMP. McGraw Hill, New York
292. Lord Rayleigh (1896) On the theory of optical images, with special reference to the microscope. Phil Mag, XLII-fifth series:167–195
293. Reid JS (1983) Debye-waller factors of zinc-blende-structure materials - a lattice dynamical comparison. Acta Cryst A39:1–13
294. Reimer L (1985) Scanning electron microscopy, volume 45 of Spring Series in Optical Sciences. Springer-Verlag, New York
295. Reimer L (1993) Transmission electron microscopy, volume 36 of Spring Series in Optical Sciences, 3rd edn. Springer-Verlag, New York
296. Reimer L (1995) Energy-filtering transmission electron microscopy, volume 71 of Spring Series in Optical Sciences. Springer-Verlag, New York
297. Reimer L, Gilde H (1973) Scattering theory and image formation in the electron microscope. In: Hawkes PW (ed.) Image processing and computer-aided design in electron optics, pp 138–167. Academic Press, London
298. Rez D, Rez P (1993) Dirac-Fock calculations of scattering factors and mean inner potentials for neutral and ionized atoms. In: Bailey GW, Rieder CL (eds.) Proceedings of the 51st annual meeting of the microscopy society of America, pp 1202–1203. San Fransisco Press
299. Rez D, Rez P, Grant I (1994) Dirac-Fock calculations of X-ray scattering factors and contributions to the mean inner potential for electron scattering. Acta Cryst A50:481–497
300. Rez D, Rez P, Grant I (1997) Dirac-Fock calculations of X-ray scattering factors and contributions to the mean inner potential for electron scattering. erratum. Acta Cryst A53:522
301. Rez P, Humphreys CJ, Whelan MJ (1977) The distribution of intensity in electron diffraction patterns due to phonon scattering. Phil Mag 35:81–96
302. Richardson WH (1972) Bayesian-based iterative method of image restoration. J Opt Soc Am 62:55–59
303. Robertson MD, Bennett JC, Burns MMJ, Currie D (2006) The simulation of annular dark field images of InAs/InP quantum dots. In: Kotula P, Marko M, Scott J-H, Gauvin R, Beniac D, Lucas G, McKernan S, Shields J (eds.) Microscopy and microanalysis 2006, volume 12, suppl. 2, pp 714–715. Cambridge University Press, Cambridge, UK
304. Rose H (2008) History of direct aberration correction. In: Aberration-corrected electron microscopy, volume 153 of Advances in imaging and electron physics, pp 3–39. Academic Press, Amsterdam
305. Rosenfeld A, Kak AC (1976) Digital picture processing. Academic Press, New York
306. Rother A, Scheerschmidt K (2009) Relativistic effects in elastic scattering of electrons in TEM. Ultramicroscopy 109:154–160
307. Ruska E (1987) The development of the electron microscope and of electron microscopy. Rev. Modern Phys 59:627–638

308. Sawada H, Sannomiya T, Hosokawa F, Nakamichi T, Kaneyama T, Tomita T, Kondo Y, Tanaka T, Oshima Y, Tanishiro Y, Takayanagi K (2008) Measurement method of aberration from ronchigram by autocorrelation function. Ultramicroscopy 108:1467–1475
309. Saxton WO (1978) Computer techniques for image processing in electron microscopy. Advances in electronics and electron physics, Supplement 10. Academic Press, New York
310. Sayle R (1999) RasMol and OpenRasMol. rasmol.org
311. Scherzer O (1949) The theoretical resolution limit of the electron microscope. J Appl Phys 20:20–29
312. Schiff LI (1968) Quantum mechanics, 3rd edn. McGraw-Hill, New York
313. Schowalter M, Rosenauer A, Titantah JT, Lamoen D (2008) Computation and parametrization of the temperature dependence of debye-waller factors for group IV, III-V and II-VI semiconductors. Acta Cryst A65:1–13
314. Sears VF, Shelley SA (1991) Debye-Waller factor for elemental crystals. Acta Cryst A47:441–446
315. Self PG, O'Keefe MA, Buseck PR, Spargo AEC (1983) Practical computation of amplitudes and phases in electron diffraction. Ultramicroscopy 11:35–52
316. Septier A (1966) The struggle to overcome spherical aberration in electron optics. In: volume 1 of *Adv. in Optical and Electron Microscopy*, pp 204–274. Academic Press, London
317. Septier A (ed.) (1980) Applied charged particle optics, volume 13A,B of Advanced in Electronics and Electron Physics. Academic Press, New York
318. Shell DL (1959) A high speed sorting procedure. Comm. ACM 2:30–32
319. Sheppard CJR, Wilson T (1978) Image formation in scanning microscopes with partially coherent source and detector. Optica Acta 25:315–325
320. Sheppard CJR (2004) Orthogonal aberration functions for high-aperture optical systems. J Opt Soc Am A 21:832–838
321. Skarnulis AJ (1982) A computer system for on-line image capture and analysis. J Microsc 127:39–46
322. Smith DJ (1989) Instrumentation and operation for high-resolution electron microscopy. In: Mulvey T, Sheppard CJR (eds.) Advances in Optical and Electron Microscopy, volume 11, pages 1–55. Academic Press, London
323. Smith DJ (1997) The realization of atomic resolution with the electron microscope. Rep Prog Phys 60:1513–1580
324. Smith DJ (2008) Development of aberration-corrected electron microscopy. Microsc. Microanal 14:2–15
325. Smith DJ (2008) Progress and perspectives for atomic-resolution electron microscopy. Ultramicroscopy 108:159–166
326. Smith GH, Burge RE (1962) The analytical representation of atomic scattering amplitudes for electrons. Acta Cryst 15:182–186
327. Smith KCA (1982) On-line digital computer techniques in electron microscopy: general intro. J Microsc 127:3–16
328. Song L, Hobaugh MR, Shustak C, Cheley S, Bagley H, Gouaux JE (1996) Structure of staphylococcal α-hemolysin, a heptameric transmembrane pore. Science 274:1859–1866
329. Spence JCH (1998) Direct inversion of dynamical electron diffraction patterns to structure factors. Acta Cryst A54:7–18
330. Spence JCH (2003) Experimental high-resolution electron microscopy, 2nd edn. Oxford University Press, New York
331. Spence JCH, Calef B, Zuo JM (1999) Dynamic inversion by the method of generalized projections. Acta Cryst A55:112–118
332. Spence JCH, Cowley JM (1978) Lattice imaging in STEM. Optik 50:129–142
333. Spence JCH, O'Keefe MA, Kolar H (1977) High resolution image interpretation in crystalline germanium. Optik 49:307–323
334. Spence JCH, Zuo JM (1992) Electron microdiffraction. Plenum Press, New York
335. Stadelmann PA (1987) EMS - a software package for electron diffraction analysis and HREM image simulation in materials science. Ultramicroscopy 21:131–146

336. Stadelmann PA (2004) JEMS - EMS java version. cimewww.epfl.ch/people/stadelmann/jemsWebSite/jems.html.
337. Sturkey L (1962) The calculation of electron diffraction intensities. Proc Phys Soc 80: 321–354
338. Su Z, Coppens P (1997) Relativistic X-ray elastic scattering factors for neutral atoms Z=1-54 from multiconfiguration Dirac-Fock wavefunctions in the 0-12 å$^{-1}$ $\sin\theta/\lambda$ range, and six-Gaussian analytical expressions in the 0-6å$^{-1}$ range. Acta Cryst A53:749–762
339. Szasz L (1992) The electronic structure of atoms. Wiley, New York
340. Takeguchil M, Hashimotol A, Shimojo M, Mitsuishi K, Furuya K (2008) Development of a stage-scanning system for high-resolution confocal stemn. J Elect Micr 57:123–127
341. Thomson MGR (1973) Resolution and constrast in the conventional and the scanning high resolution transmission electron microscope. Optik 39:15–38
342. Thon F (1971) Phase contrast electron microscopy. In: Valdre U (ed.) Electron microscopy in materials science, pages 571–625. Academic Press, New York
343. Thust A (2009) High-resolution transmission electron microscopy on an absolute contrast scale. Phys Rev Lett 102:220801
344. Thust A, Urban K (1992) Quantitative high-speed matching of high-resolution electron microscopy images. Ultramicroscopy 45:23–42
345. Tillmann K, Thust A, Urban K (2004) Spherical aberration correction in tandem with exit-plane wave function reconstruction: Interlocking tools for atomic scale imaging of lattice defects in GaAs. Microsc. Microanal 10:185–198
346. Tournaire M (1962) Recent developments of the matrical and semi-reciprocal formulation in the field of dynamical theory. J. Phys Soc Jpn, Suppl. B II, 17:98–100
347. Treacy MMJ, Gibson JM (1993) Coherence and multiple scattering in Z-contrast images. Ultramicroscopy 52:31–53
348. Urban K (2008) Studying atomic structures by aberration-corrected transmission electron microscopy. Science 321:506–510
349. Urban K, Houben L, Jia C-L, Mi M. Lentzen S-B, Thust A, Tillman K (2008) Atomic-resolution aberration-corrected transmission electron microscopy. In: Aberration-corrected electron microscopy, volume 153 of Advances in Imaging and Electron Physics, pp 439–480. Academic Press, Amsterdam
350. Vainshtein BK (1994) Modern crystallography I, fundamentals of crystals, 2nd edn. Springer-Verlag, Berlin
351. Vainshtein BK, Fridkin VM, Indenbom VL (1982) Modern crystallography II, structure of crystals. Springer-Verlag, Berlin
352. van Dyck D (1975) The path integral formalism as a new description for the diffraction of high-energy electrons in crystals. Phys Stat Sol B72:321–336
353. van Dyck D (1978) On the optimisation of methods for the computation of many-beam structure images. In: Sturgess JM, Kalnins VI, Ottensmeyer FP, Simon GT (eds.) Electron Microscopy 1978, Vol. 1, Ninth Intern. Congress on Electron Microscopy (Toronto), pp 196–197. The Imperial Press, Ontario
354. van Dyck D (1979) Improved methods for the high speed calculation of electron microscopic structure images. Phys Stat Sol A52:283–292
355. van Dyck D (1980) Fast computational procedures for the simulation of structures in complex or disordered crystal: A new approach. J Microsc 119:141–152
356. van Dyck D (1983) High-speed computation techniques for the simulation of high resolution electron micrographs. J. Microsc 132:31–42
357. van Dyck D (1985) Image calculations in high-resolution electron microscopy: Problems, progress, and prospects. In: Hawkes PW (ed.) Advances in electronics and electron physics, Vol. 65, pp 295–355. Academic Press, Orlando
358. van Dyck D, Coene W (1984) The real space method for dynamical electron diffraction calculation in high resolution electron microscopy, I. principles of the method. Ultramicroscopy 15:29–40
359. Vand V, Eiland PF, Pepinsky R (1957) Analytical representation of atomic scattering factors. Acta Cryst 10:303–306

360. vonArdenne M (1938) Das elektronen-rastermikroskop. Z fur Physik 109:553–572
361. Wade RH, Frank J (1977) Electron microscope transfer functions for partially coherent axial illumination and chromatic defocus spread. Optik 49:81–92
362. Walker JS (1996) Fast Fourier Transforms, 2nd edn. CRC Press, Boca Raton
363. Wall J (2006) Simulation of coherent scattering in STEM. In: Kotula P, Marko M, Scott J-H, Gauvin R, Beniac D, Lucas G, McKernan S, Shields J (eds.) Microscopy and microanalysis 2006, volume 12, suppl. 2, pages 1350–1351. Cambridge University Press, Cambridge, UK
364. Wang J, Smith VH, Bunge CF, Jauregui R (1996) Relativistic X-ray elastic scattering factors for He-Ar from Dirac-Hartree-Fock wave functions. Acta Cryst A52:649–658
365. Wang ZL (1992) Dynamics of thermal diffuse scattering in high-energy electron diffraction and imaging: Theory and experiments. Phil Mag B65:559–587
366. Wang ZL (1995) Dynamical theories of dark-field imaging using diffusely scattered electrons in STEM and TEM. Acta Cryst A51:569–585
367. Wang ZL (1996) Electron statistical dynamical diffuse scattering in crystals containing short range order point defects. Acta Cryst A52:717–729
368. Wang ZL (1998) An optical potential approach to incoherent multiple thermal diffuse scattering in quantitative HRTEM. Ultramicroscopy 74:7–26
369. Wang ZL, Cowley JM (1989) Simulating high-angle annular dark-field STEM images including inelastic thermal diffuse scattering. Ultramicroscopy 31:437–454
370. Wang ZL, Cowley JM (1990) Dynamical theory of high-angle annular dark-field STEM lattice images for a Ge/Si interface. Ultramicroscopy 32:275–289
371. Warren BE (1969,1990) X-ray diffraction. Dover, New York
372. Watanabe K (1993) n-beam dynamical calculations. In: Hawkes PW (ed.) Advances in electronics and electron physics, Vol. 86, pages 173–224. Academic Press, San Diego
373. Watanabe K, Kikuchi Y, Hiratsuka K, Yamaguchi H (1988) A new approach for n-beam lattice image calculation. Phys Status Solidi A109:119–126
374. Watanabe K, Kikuchi Y, Hiratsuka K, Yamaguchi H (1990) A new approach for n-beam dynamical calculations. Acta Cryst A46:94–98
375. Watanabe K, Yamazaki T, Hashimoto I, Shiojiri M (2001) Atomic-resolution annular dark-field STEM image calculations. Phys Rev B 64:115432
376. Weickenmeier A, Kohl H (1991) Computation of absorptive form factors for high-energy electron diffraction. Acta Cryst A47:590–597
377. Weiss GH, Maradudin AA (1962) The Baker-Hausdorff formula and a problem in crystal physics. J Math Phys 3:771–777
378. Wiese WL, Martin GA (1989) Atomic spectroscopy. In: Anderson HL (ed.) A Physicists Desk Reference, the second edition of Physics Vade Mecum, pages 92–102. American Institute of Physics, New York
379. Wilcox RM (1967) Exponential operators and parameter differentiation in quantum physics. J Math Phys 8:962–982
380. Williams DB, Carter CB (1996) Transmission electron microscopy, a textbook for materials science. Plenum Press, New York
381. Wilson AR, Spargo AEC (1982) Calculation of the scattering from defects using periodic continuation methods. Phil Mag A46:435–449
382. Wilson AR, Spargo AEC, Smith DJ (1982) The characterisation of instrumental parameters in high resolution electron microscopy. Optik 61:63–78
383. Wilson T, Sheppard C (1984) Theory and practice of scanning optical microscopy. Academic Press, London
384. Wong K, Kirkland E, Xu P, Loane R, Silcox J (1992) Measurement of spherical aberration in STEM. Ultramicroscopy 40:139–150
385. Wyckoff RWG (1963,1964) Crystal structures, Vol. 1 and 2, 2nd edn. Wiley, New York
386. Zaluzec NJ (1996) Tele-presence microscopy/labspace: An interactive collaboratory for use in education and research. In: Bailey GW, Corbett JM, Dimlich RVW, Michael JR, Zaluzec NJ (eds.) Proceedings of the 54th Annual Meeting of the Microscopy Society of America, pp 382–383. San Fransisco Press

387. Zaluzec NJ (2003) The scanning confocal electron microscope. Microscopy Today Nov./Dec.:8–12
388. Zeitler E, Olsen H (1964) Screening effects in elastic scattering. Phys Rev 136:A1546–A1552
389. Zeitler E, Olsen H (1967) Complex scattering amplitudes in elastic electron scattering. Phys Rev 162:1439–1447
390. Zeitler E, Thomson MGR (1970) Scanning transmission electron microscopy. Optik 31: 258–366
391. Zemlin F, Weiss K, Schiske P, Kunath W, Herrman K-H (1978) Coma free alignment of high resolution electron microscopes with the aid of optical diffractograms. Ultramicroscopy 3:49–60
392. Zhang H, Marks LD, Wang YY, Zhang H, Dravid VP, Han P, Payne DA (1995) Structure of planar defects in $(Sr_{0.9}Ca_{0.3})_{1.1}Cuo_2$ infinite-layer superconductors by quantitative high-resolution electron microscopy. Ultramicroscopy 57:103–111
393. Zuo JM (2009) Web electron microscopy applications software (WebEMAPS). emaps.mrl.uiuc.edu/
394. Zworykin VK, Morton GA, Ramberg EG, Hillier J, Vance AW (1945) Electron optics and the electron microscope. Wiley, New York

Index

CPSIA information can be obtained at www.ICGtesting.com
Printed in the USA
LVOW080133151211

259506LV00003B/32/P

9 781441 965325